Principles of
emergency planning
and management

Principles of emergency planning and management

David Alexander

University of Massachusetts, Amherst

TERRA

First published in 2002 by Terra Publishing

Terra Publishing
PO Box 315, Harpenden, Hertfordshire AL5 2ZD, England
Telephone: +44 (0)1582 762413
Fax: +44 (0)870 055 8105
Website: www.terrapublishing.net
E-mail: publishing@rjpc.demon.co.uk

ISBNs: 1-903544-10-6 paperback

10 09 08 07 06 05 04 03 02
11 10 9 8 7 6 5 4 3 2 1

British Library Cataloguing-in-Publication Data
A CIP record for this book is available from the British Library

Library of Congress Cataloging-in-Publication Data are available

Typeset in Palatino and Helvetica
Printed and bound by Biddles Limited, Guildford and King's Lynn, England

Ai miei medici con affetto, stima e gratitudine.
Prof. Giampiero Bellesi, Prof. Sergio Boncinelli, Dr Paolo Fontanari,
Prof. Paola Lorenzi, Prof. Massimo Marsili, Dr Sergio Pittino

Contents

Preface

There is a curious paradox about disasters. On the one hand they are extraordinary events that require special organization and resources to tackle the
damage, casualties and disruption that they cause, and on the other hand they
are sufficiently frequent and similar to each other to be normal, not abnormal,
events. Although emergency powers and special measures are needed when
disaster strikes, the requirements and exigencies are predictable enough to be
planned for. Indeed, disaster planning is both eminently possible and an obligation of the civil authorities responsible for the safety of workers, patients,
inmates and members of the public.

The field of civil protection (known as emergency preparedness in the USA)
is relatively new and rapidly evolving. It was born out of the civil defence
organizations that were set up at the beginning of the Cold War with the perhaps futile aim of protecting the population and vital institutions against the
effects of nuclear attack (its earlier origins can be found in the air-raid precautions brigades of the Second World War, and in pre-war measures to protect
civilians against armed aggression). In the 1970s the emphasis began to shift to
disasters not caused by warfare, and this gradually led to the creation of an
entirely new field concerned with the provision of aid, relief, mitigation and
preparedness measures to combat natural disasters (e.g. floods, hurricanes and
earthquakes) and technological ones (e.g. radiation emissions, transportation
accidents and toxic spills). Society has become progressively more complex
and, apparently, more willing to put its wealth at risk. Disasters have thus
become inexorably more frequent and more costly. At the same time, increases
in connectivity and communications power have given disasters an increasing
sense of immediacy to people who are not directly involved in any particular
event but can follow it through television images, newspaper articles or Internet distribution.

Increasing public interest in disasters is gradually being turned into demand
and support for official efforts to mitigate them. To begin with, continuing high
levels of casualties and steeply increasing costs of disasters have made
catastrophe prevention socially and economically imperative. Future risks are

heavy and pose serious questions about how damage will be paid for in times of increasing fiscal stringency. At the same time, virtually all cost–benefit studies carried out on sensible programmes of disaster preparedness and mitigation have shown that the benefits outweigh the costs, often by two or three times, in terms of damage prevented and lives saved. Further impetus has been given by conferences, publications and rising academic interest in the field of disaster prevention and management. At the world level or through national committees, many initiatives have been launched under the aegis of the UN's International Decade for Natural Disaster Reduction (1990–2000) and its permanent successor, the International Strategy for Disaster Reduction (ISDR).

Emergency and disaster planning (herein treated as synonymous) took off in the 1990s and began to spread at all levels of government, industry and society. Demand for training courses and trained personnel has accelerated to the point that, at the time of writing, it far exceeds supply. As a result, civil protection is practised in many places by people who are not adequately trained. Emergency plans written under such circumstances can be haphazard, partial or unworkable. In a certain proportion of cases, lack of expertise has led to a total lack of planning. But emergency management cannot be improvised, even though much can be learned from the impact of each disaster that occurs. Advance plans are needed wherever there is a chance that catastrophes will occur, which means most places in the world. The plans must be robust, durable and flexible, capable of being tested and modified, and functional in difficult circumstances.

My aim in writing this book is to provide a general introduction to comprehensive disaster plans, with some reference to more specific sorts of plan, such as those needed for factories and hospitals. The methodology described is not intended to be specific to any level of government or any particular set of legislation. Instead, it provides a blueprint that can be adapted to jurisdictions of various sizes, at various levels, and which are subject to various laws. Taken too literally, the prescriptions outlined in this book might seem to be a counsel of perfection, which would require huge resources and vast amounts of time to realize, and be hopelessly unwieldy. Hence, the measures described need to be taken selectively, by adopting the suggestions that are appropriate to each given situation and leaving the others until conditions are more propitious.

The book begins by introducing a rationale for emergency planning, which provides a context of motivation and justification for what is to follow. The second and third chapters describe methods used in developing, implementing and applying a plan: in effect it assembles the tools of the trade and explains how they are used. Chapter 4 begins with a review of the structure of a generic emergency plan and moves on to explain how one is activated, tested, revised and updated. Disaster plans are living documents that need constant attention to ensure that they remain functional as conditions change, knowledge about hazards and risks increases, and the balance of available resources alters. Chapter 5 is devoted to a review of what is needed during emergencies and how they are managed during both the crisis period and the subsequent initial phase of recovery. This book is not intended to be a treatise on disasters *per se*, as there are plenty of other works that fulfil that role. However, what actually happens in disaster is obviously the substance of what needs to be planned for and managed. Hence, it must be treated systematically as a series of events, developments and evolving needs. The sixth chapter describes particular sorts of disaster planning and management: for medical facilities, industrial plants, libraries, art treasures and architectural heritage, to manage the press, and to guide post-disaster reconstruction. Chapter 7 deals with the longer term of reconstruction planning, and Chapter 8 offers some reflections on training methods for emergency planners and managers. The threads are drawn together in Chapter 9, which offers a brief conclusion and some observations about what the future holds in store for emergency planners and managers.

Although I have tried to be comprehensive in my treatment of the subject matter, it is impossible to cover all eventualities in this field. Hence, individual situations will inevitably suggest ground that has not been covered herein. For the rest, I have endeavoured to make this a practical guide, not an academic treatise. Thus, aspiring disaster planners and managers are urged to acquire a grounding in theories of emergencies and hazards. Disasters are complex events and to understand them one needs a comprehensive multidisciplinary training. Theory needs to be learned, because it is, in the words of the eminent sociologist of disasters, Thomas E. Drabek, the road map that enables us to navigate our way through complicated emergencies. The rewards for learning this material are reaped in terms of greater security, more lives saved and more damage avoided.

David Alexander
catastrophe@tiscalinet.it
San Casciano in Val di Pesa
February 2002

Acknowledgements

The author and publisher wish to acknowledge the following sources, from which the illustrations listed are derived:

Figures 3.4 and 3.5   Source: "Managing the response to disasters using microcomputers", S. Belardo, K. R. Karwan, W. A. Wallace, *Interfaces* **14**(2), 29–39 (1984).

Figure 3.6   Adapted from figure 8.2 in Coburn & Spence (1992: see Bibliography, p. 327).

Figure 3.8   Source: "Societal benefit versus technological risk", C. Starr, *Science* **165**, 1234 (1969).

Figure 3.9   Source: "Handling hazards", B. Fischhoff, C. Hohenemser, R. E. Kasperson, R. W. Kates, *Environment* **20**(7), 34 (1978).

Figure 3.12   Source: adapted from Cosgrove (1997: see Bibliography, p. 321).

Figure 4.3   Source: courtesy of Italian Department of Civil Protection.

Figure 4.7   Source: p.67 in *Flood hazard in the United States: a research assessment*, G. F. White. Report PB-262-023. Institute of Behavioural Sciences, University of Colorado, Boulder (1975).

Figure 5.1   After *Ignorance and uncertainty: emerging paradigms*, M. J. Smithson (New York: Springer, 1989); and "Ignorance and disasters", M. J. Smithson, *International Journal of Mass Emergencies and Disasters* **8**(3), 207–235 (1989).

Figure 5.3   Source: Partly after Foster (1980: see Bibliography, p. 322).

Figure 6.1   Modified after p. 13 in PAHO (1981: see Bibliography, p. 326).

Figure 6.5   Partly based on diagrams in Croce Rossa Francese/Associazione. Italiana Medicina delle Catastrofi 1994. *Manuale di protezione civile*. Piemme, Casale Monferrato, Italy.

Figure 6.6   Source: Manni, C. 1982. Italian earthquake. Source: *Mass casualties: a lessons learned approach*, R. A. Cowley, S. Edelstein, M. Silverstein (eds), 195–201. Washington DC: US Department of Transportation.

Figure 6.7   Courtesy of Mettag, Inc., P. O. Box 910, Starke, Florida 32091, USA.

Figure 6.13   Partly based on *Valutazione probabilistica di rischio*, S. Messina, N. Piccinini, G. Zappellini, pp. 90–91. Rome: Associazione degli Analisti di Affidibilità e Sicurezza (1987).

Figure 7.3   Source: *Paying the price: the status and role of insurance against natural disasters in the United States*, H. Kunreuther & R. J. Roth Sr (eds) 1998. Washington DC: National Academy Press.

CHAPTER 1

Aims, purpose and scope of emergency planning

We are all part of the civil-protection process. Emergencies can be planned for by experts, but they will be experienced by the relief community and the general public alike. Therefore, we should all prepare for the next disaster as remarkably few of us will be able entirely to avoid it.
(author's maxim)

1.1 Introduction

This chapter begins with a few words on the nature and definition of concepts related to emergencies and disasters. It then offers some definitions and primary observations on the question of how to plan for and manage emergencies. Next, brief overviews are developed of the process of planning for disasters in both the **short term** and the **long term**.

1.1.1 The nature of disaster

For the purposes of this work, an **emergency** is defined as an exceptional event that exceeds the capacity of normal resources and organization to cope with it. Physical extremes are involved and the outcome is at least potentially, and often actually, dangerous, damaging or lethal. Four levels of emergency can be distinguished. The lowest level involves those cases, perhaps best exemplified by the crash of a single passenger car or by an individual who suffers a heart attack in the street, which are the subject of routine **dispatch** of an ambulance or a **fire appliance**. This sort of event is not really within the scope of this book, which deals with the other three levels. The second level is that of **incidents** that

1

can be dealt with by a single municipality, or jurisdiction of similar size, without significant need for resources from outside. The third level is that of a major incident or disaster, which must be dealt with using regional or interjurisdictional resources. In comparison with lower levels of emergency, it requires higher degrees of coordination. The final level is that of the national (or international) disaster, an event of such **magnitude** and seriousness that it can be managed only with the full participation of the national government, and perhaps also international aid.

Although the concept of an emergency is relatively straightforward, disaster has long eluded rigorous definition.[*] In this work it will be treated as synonymous with **catastrophe** and calamity. Some authors prefer to define a catastrophe as an exceptionally large disaster, but as no quantitative measures or operational thresholds have ever been found to distinguish either term, I regard this as unwise. Generally, disasters involve substantial destruction. They may also be mass-casualty events. But no-one has ever succeeded in finding universal minimum values to define "substantial" or "mass". This is partly because relatively small monetary losses can lead to major suffering and hardship or, conversely, large losses can be fairly sustainable, according to the ratio of losses to reserves. It thus follows that people who are poor, disadvantaged or marginalized from the socio-economic mainstream tend to suffer the most in disasters, leading to a problem of *equity* in emergency relief.

It is important to note that disasters always involve the interaction of physical extremes (perhaps tempered by human wantonness or carelessness) with the human system. There is not always a proportionate relationship between the size of the physical forces unleashed and the magnitude of the human suffering and losses that result. Chains of adverse circumstances or coincidences can turn small physical events into large disasters (Fig. 1.1). For instance, if a minor earthquake causes an unstable bridge to collapse, the consequences will be different if the bridge is unoccupied or if vehicles are on it.

Disasters involve both direct and indirect losses. The former include damage to buildings and their contents; the latter include loss of employment, revenue or sales. Hence, the **impact** of a disaster can persist for years, as indirect losses continue to be created. This obviously complicates the issue of planning and management, which must be concerned with several timescales.

A basic distinction can be made between **natural disasters** and **technological disasters** (see Table 1.1), although they may require similar responses. The former include sudden catastrophes, such as earthquakes and tornadoes, and so-called **creeping disasters**, such as slow landslides, which may take years to

---

* For an extensive treatment of this problem, see *What is a disaster? Perspectives on the question*, E. L. Quarantelli (ed.) (London: Routledge, 1998).

Physical impact   **Human consequences**

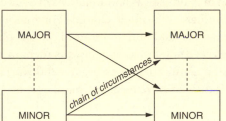

Figure 1.1   Relationship between physical impact and human consequences of disaster. A minor physical event can lead to a major disaster in terms of casualties and losses if circumstances combine in unfavourable ways.

develop. In between these two extremes there are events of various durations; for example, there are floods that rise over several days, or volcanic eruptions that go on intermittently for months. Technological disasters include explosions, toxic spills, emissions of radio-isotopes, and transportation accidents. Riots, guerrilla attacks and terrorist incidents are examples of other forms of **anthropogenic hazard** that give rise to **social disasters** (Table 1.1). These will receive limited coverage in this book, which is mainly about coping with natural and technological incidents and disasters. Terrorism, riots and crowd incidents are germane because emergency personnel often have to provide back-up services and care for victims. However, such events are problematic in

Table 1.1   Classes of natural, technological and social hazard.

Class of hazard	Examples
Natural (geophysical)	
Geological	Earthquake, volcanic eruption, landslide (including rockfall, debris avalanche, mudflow), episode of accelerated erosion, subsidence
Meteorological	Hurricane, tornado, icestorm, blizzard, lightning, intense rainstorm, hailstorm, fog, drought, snow avalanche
Oceanographic	Tsunami (geological origins), sea storm (meteorological origins)
Hydrological	Flood, flashflood
Biological	Wildfire (forest or range fire), crop blight, insect infestation, epizootic, disease outbreaks (meningitis, cholera, etc.)
Technological	
Hazardous materials and processes	Carcinogens, mutagens, heavy metals, other toxins
Dangerous processes	Structural failure, radiation emissions, refining and transporting hazardous materials
Devices and machines	Explosives, unexploded ordnance, vehicles, trains, aircraft
Installations and plant	Bridges, dams, mines, refineries, power plants, oil and gas terminals and storage plants, power lines, pipelines, high-rise buildings
Social	
Terrorist incidents	Bombings, shootings, hostage taking, hijacking
Crowd incidents	Riots, demonstrations, crowd crushes and stampedes

Source: Partly based on table 4.3 on p.101 in *Regions of risk: a geographical introduction to disasters*, K. Hewitt (Harlow, England: Addison-Wesley-Longman, 1997).

that, more than any other kind of eventuality, they risk compromising the impartiality of emergency workers, who may be seen as handmaidens of a group of oppressors, such as policemen or soldiers, who are in charge of operations.

In reality, the distinction between natural and human-induced catastrophe is less clear-cut than it may seem: it has been suggested that the cause of natural catastrophe lies as much, or more, in the failings of human organization (i.e. in human **vulnerability** to disaster) as it does in extreme geophysical events.

1.1.2 Preparing for disaster

This book concentrates on planning and preparation for short-term emergencies: events that develop rapidly or abruptly with initial impacts that last a relatively short period of time, from seconds to days. These transient phenomena will require some rapid form of adaptation on the part of the socio-economic system in order to absorb, fend off or amortize their impacts. If normal institutions and levels of organization, staffing and resources cannot adequately cope with their impacts, then the social system may be thrown into **crisis**, in which case workaday patterns of activity and "normal" patterns of response must be replaced with something more appropriate that improves society's ability to cope. As a general rule, this should be done as much as possible by strengthening existing organizations and procedures, by supplementing rather than supplanting them. The reasons for this are explained in full later, but they relate to a need to preserve continuity in the way that life is lived and problems are solved in disaster areas. There is also a need to utilize resources as efficiently as possible, which can best be done by tried and tested means.

Current wisdom inclines towards the view that disasters are *not* exceptional events. They tend to be repetitive and to concentrate in particular places. With regard to natural catastrophe, seismic and volcanic belts, hurricane-generating areas, unstable slopes and tornado zones are well known. Moreover, the **frequency** of events and therefore their statistical **recurrence intervals** are often fairly well established, at least for the smaller and more frequent occurrences, even if the short-term ability to **forecast** natural hazards is variable (Table 1.2). Many **technological hazards** also follow more or less predictable patterns, although these may become apparent only when research reveals them. Finally, intelligence gathering, strategic studies and policy analyses can help us to understand the pattern of emergencies resulting from conflict and insurgence. Thus, there is little excuse for being caught unprepared.

The main scope of emergency planning is to reduce the **risk** to life and limb posed by actual and potential disasters. Secondary motives involve reducing damage, ensuring public safety during the aftermath of disaster, and caring for

Table 1.2 Short-term predictability of sudden-impact natural disasters.

Potentially high	Potentially medium	Potentially low	Moderate or variable
Fog	Coastal erosion	Earthquake	Avalanche
Hail	Drought	Lightning	Crop disease
Hurricane	Flood		Insect infestation
Landslide	Frost hazard to		Snowstorm
Tsunami	agriculture		Windstorm
Volcanic eruption	Icestorm		
Wildfire	Intense rainstorm		
	Subsidence		
	Tornado		

survivors and the disadvantaged. Inefficiencies in planning translate very easily into loss of life, injuries or damage that could have been avoided. Thus, emergency planning is at least a moral, and perhaps also a legal, responsibility for all those who are involved with the safety of the public or employees. When a known significant risk exists, failure to plan can be taken as culpable negligence. Moreover, planning cannot successfully be improvised during emergencies: this represents one of the worst forms of inefficiency and most likely sources of error and confusion. Fortunately, however, 50 years of intensive research and accumulated experience have furnished an ample basis for planning, as is outlined in this book.

Given that disasters tend to be repetitive events, they form a cycle that can be divided into phases of mitigation, preparedness, response and recovery, including reconstruction (Fig. 1.2). The first two stages occur before catastrophe strikes and the last two afterwards. The actions taken, and therefore the planning procedures that predetermine them, differ for each of the periods, as different needs are tackled. **Mitigation** comprises all actions designed to reduce the impact of future disasters. These usually divide into structural measures (the engineering solutions to problems of safety) and non-structural measures, which include land-use planning, insurance, legislation and **evacuation** planning. The term **preparedness** refers to actions taken to reduce the impact of disasters when they are forecast or imminent. They include security measures, such as the evacuation of vulnerable populations and sandbagging of river **levees** as floodwaters begin to rise (thus the *planning* of evacuation is a mitigation measure, whereas its *execution* is a form of preparedness). **Response** refers to emergency actions taken during both the impact of a disaster and the short-term aftermath. The principal emphasis is on saving and safeguarding human lives. Victims are rescued and the immediate needs of survivors are attended to. **Recovery** is the process of repairing damage, restoring services and reconstructing facilities after disaster has struck. After major catastrophes it may take as long as 25 years, although much less time is needed in lighter impacts or disasters that strike smaller areas.

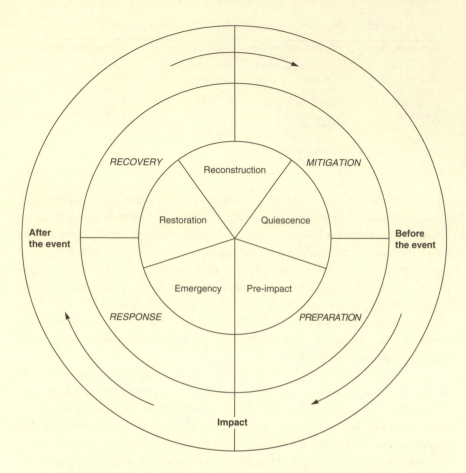

Figure 1.2 The disaster cycle.

It is worth noting at the outset that the type of emergency planning required may vary considerably with the kind of society to which it will be applied and the level of economic development (crudely speaking, the availability of resources) that backs it up. The industrialized world has tended to take one route through the emergency-planning process, the developing countries have taken another. However, it should be noted that neither category constitutes a bloc of nations: there is as much variation within each as there is between the two groups, not least because, in reality, the Third World consists of richer and poorer nations, and even the richest industrialized countries have more- and less-developed regions. However, the tendency in the developed world is to seek a form of emergency planning adaptable to a wide range of exigencies – the so-called **all-hazards approach**, which is increasingly based on advanced technologies, especially in the field of communications. Thus, information technology will doubtless become a key tool in the management of future disasters. However, it remains to be seen to what extent the technologically based approach to disaster management is transferable to the world's poorer nations, although some of these have nevertheless begun to invest in the necessary communications infrastructure. Meanwhile, some countries (especially in Africa, southeastern Europe and the Caucasus region) have become embroiled in **complex emergencies**. These involve forms of political and military instability, coupled with socio-economic breakdown, that are extremely hard to plan for. Or rather, planning requires a different approach, one marked by extreme flexibility and a primary objective of restoring peace and stability. Under such circumstances, aid and relief planners must not be afraid to face up to hard moral dilemmas and complex socio-economic problems. For it is widely recognized that disaster prevention and mitigation go hand in hand with economic development, which should aim to reduce human vulnerability to catastrophe. This has led to a lively debate about what form development should take and what it should mean in terms of environmental sustainability.

In this book I concentrate on emergency planning of the kind that is needed in countries with highly developed economies. Planning for complex emergencies in less-developed nations is no less important a topic, but has been covered well by other authors.[*] Nevertheless, it is something of a false distinction, for all emergencies are to some extent complex, and it should not be thought that experience in developing nations is of no value in determining how things are to be done in the industrialized world. On the contrary, "development" is a very relative term: with respect to emergency planning there are many places and regions that, although economically rich and sophisticated, are still at an early stage of organizing resources to combat the impacts of extreme events.

Before developing an approach to emergency planning, it is as well to recognize that two fundamentally different timescales form its context: the long term of mitigation, **hazard** reduction and methodical preparation for the next disaster, and the short term of reactions to abrupt crisis. Now let us examine what is involved in each context.

1.2 Long-term planning

The timing of the impacts of many extreme natural events cannot be predicted exactly, neither can the timing and location of transportation accidents, toxic spills and terrorist incidents. Nevertheless, all such events form patterns

---

[*] See the journal *Disasters* (Oxford: Blackwell) for papers on this theme.

characterized by their repetitiveness. In general, the longer the timespan considered, the more regular the pattern appears to be. This is not to say that extreme events are fundamentally regular; indeed, time-series of natural disasters are highly irregular. However, distinct intervals of time are likely between events, and seasonal **natural hazards** offer the most regularity.

The interval is the period in which a concerted effort must be made to clear up the damage caused by the last event and prepare for the next one. Mitigation studies show that resources invested in appropriate protection measures almost invariably pay for themselves by reducing damage and casualties the next time that disaster strikes. The reductions are usually disproportionate, and, in purely economic terms, the cost–benefit ratio of hazard mitigation may vary from about 1:1.3 to 1:9, according to circumstance. Experience shows that *ad hoc* and haphazard efforts at mitigation are an inefficient way to protect lives and property. It is better to make a concerted plan, in which hazards (the like-lihood of physical impact) and risks (the likelihood of adverse consequences resulting from such impact) are carefully evaluated, compared with one another, and reduced in such a way as to derive maximum possible benefit for a given investment of time, money, manpower and other resources.

Three types of the long-term planning are needed, dealing respectively with reconstruction, mitigation and general preparedness. Although each of these phases or categories imposes distinctive needs and constraints upon the plan-ner, it is as well that plans be interlinked in order to take account of the overlap between each sector. The long-term aftermath of disasters may last up to 25 years, depending on the seriousness and extent of damage. It begins with a phase known as the **window of opportunity**, in which public opinion is sen-sitized to the problem of hazards and disasters to the extent that there is sub-stantial demand for measures designed to increase safety and reduce future risks. Thus, it is hardly surprising that much of the legislation that governs the field was designed and passed after specific events that led to calls for improved regulation and norms. For example, in the USA the basic federal law governing disaster relief, the Stafford Act of 1993, was passed in the wake of the country's most expensive disaster to date, Hurricane Andrew (1992). As the disaster recedes in time, it becomes less salient among the problems of every-day life and, as public interest wanes, so does the level of political support for expenditure on mitigation. Reconstruction planning must take account of this change by being able to capitalize on the opportunities offered by the window of opportunity: mitigation planning must take into account the fact that public and political interest in reducing the impact of disasters waxes and wanes.

Reconstruction planning should try to impart an orderly progression to the processes of repair, rebuilding and post-disaster development. It should en-sure that these do not re-create past vulnerability but ameliorate it by building in new forms of protection. Mitigation planning is generally divided into two categories: structural and non-structural. In the evolution of approaches to disaster abatement, the former has usually preceded the latter, although there is no inherent reason why this need be so. **Non-structural mitigation** has often been brought in when engineering solutions have become very costly and have not resulted in a substantial reduction in losses (indeed, as structural measures often encourage further development of vulnerable areas but seldom or never guarantee absolute protection, losses may rise inexorably). As a result, insurance, evacuation planning and **land-use control** are utilized. Current wisdom suggests that hazards are best mitigated by using a combination of structural and non-structural measures. This is likely to be complicated and, to carry it out efficiently, integrated planning is required.

Short-term emergency planning often has to cope with situations that are fairly well defined and impose specific demands on available resources. Its longer-term counterpart, preparedness planning, is less clear cut and usually involves a certain amount of **prediction** of what may eventually be needed. Preparedness planning encompasses the marshalling of resources and the process of prediction, forecasting and **warning** against short-term disaster events. It also involves education and training initiatives, and planning to evacuate vulnerable populations from threatened areas. It often takes place against a background of attempts to increase public and political awareness of potential disaster and to garner support for increased funding of mitigation efforts. Like other forms of long-term planning, flexibility is needed in order to cope with changes in the salience of hazards and, above all, to capitalize on periods in which there is a high level of support for mitigation.

All forms of long-term planning can be considered as a form of preliminary to the inevitable moment when a short-term plan will be put into operation.

1.3 Planning for the short term

Emergencies tend to be transient events, but, given adequate attention to the problem, there is no reason to regard them as unpredictable. Although each catastrophe involves a unique mixture of impacts and consequences, it is also likely to have much in common with similar events in different areas and past events in the same area. Given the repetitive nature of disasters, there is no excuse for failing to plan for them and, to be more positive, there is every reason for being able to plan successfully. Short-term planning cannot be improvised efficiently during a crisis, as it is unlikely that there will be enough time to gather information and analyze it. This shows that information is a key resource in both planning for and managing emergencies. Without adequate attention to information flows, it is highly likely that this commodity will run short in key areas, especially concerning the pattern of damage and casualties and the immediate disposition of resources needed to cope with such problems. Short-term plans must guarantee the rapid deployment of the available resources in the most efficient manner with respect to immediate needs. Danger should be reduced, safety improved and, where necessary, victims should be rescued as promptly as possible. A great deal of foresight is required.

Generally, what we know about hazards and disasters, and therefore what we can do about them, is based on accumulated theory and practical experience. The latter is vital; its absence can be counteracted only partly by substitute methods: whereas no-one would wish a disaster upon people, the experience of living through one is often an incomparable aid to future planning.

Over time, short-term emergency planning has become more widespread, more systematic and more technologically based. It is officially encouraged by the agencies of the United Nations and other international bodies. Models and examples are easily accessible on the Internet. Academic expertise has been summarized in an increasing number of books; national and regional legislation increasingly requires it, and software to aid it is now available commercially. There are signs that information technology will stimulate a revolution in both the methodology and the execution of emergency plans, although it remains to be seen whether innovations will be adopted universally.

Emergency planning is a reaction to both a need for improved safety and advances in the scientific knowledge of hazards. Monitoring of the geophysical processes that are liable to produce extreme events, accumulation of case studies of past disasters, analysis of risk (including how it is perceived by the risk takers), and improved understanding of the social basis of hazard and risk are all developments that have stimulated planning. Rising costs of damage (Fig. 1.3) have made it uneconomic to ignore risk and have fostered the search for ways of reducing future losses through mitigation. As knowledge improves, so risks and hazards need to be re-evaluated and plans written to take account of the new findings. This underscores the fact that both long- and short-term emergency planning are continuing processes rather than final objectives. Continuous effort is required, for disasters can be abated only by constant vigilance and frequent improvements in method.

The short-term planning process encompasses a range of different sorts of event. Examples of these are:

- Seasonal or other repetitive events, such as hurricanes or snowmelt floods. The size and consequences of the event may vary each time it occurs, but the general pattern of impacts is predictable and must be planned for.
- Very large, infrequent events, such as major volcanic eruptions. These pose

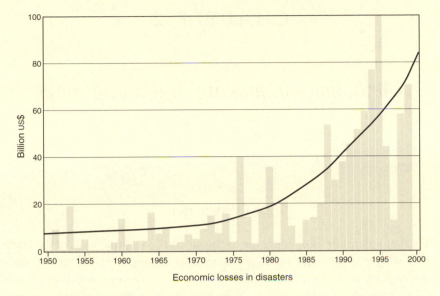

Figure 1.3 The consistent increase in twentieth century disaster losses.

a dilemma for planning: on the one hand it does not make economic sense to invest resources in preparing for an event that may not occur for generations; on the other hand one cannot afford to be negligent about an event that has at least some probability of occurring within the near future.

• Events that are unforeseen or unlikely to repeat themselves. These include dam bursts, some forms of toxic spills, explosions or catastrophic pollution episodes, and some types of transportation accident. Despite the predominance of uniqueness in these events, at least insofar as they are viewed at the local scale, there is generally a reasonable basis for hypothesizing their occurrence (in approximate terms) and for drawing on experience from similar events elsewhere. Planning is thus still possible in the case of such events.

By way of conclusion, it is emphasized that emergencies are not exactly unforeseen events. Neither are they abnormal. Rather, they tend to conform to broadly predictable patterns of occurrence and evolution. Given this, there is a moral and a practical imperative to plan in order to reduce their effects. If the proponents of emergency planning are faced with scepticism or outright opposition, a feasibility study may be warranted. This will probably demonstrate that planning is eminently justified on cost–benefit grounds, even if the moral arguments are not accepted. If, on the other hand, planning goes ahead on the basis of legislation that calls for it, it is as well to remember that laws often lag behind experience and scientific findings. Highly successful emergency plans will go beyond minimum legal requirements and will anticipate the future upgrading of laws and norms, thus creating an additional margin of safety.

11

CHAPTER 2

Methodology: making and using maps

2.1 Introduction

This chapter and the next cover the technical and conceptual tools of the trade as used by emergency planners and managers. When using the tools to construct a plan, the first task is to assess and map situations of risk as they appear in the geographical area under analysis. Risk involves the exposure to hazards of vulnerable populations, building stocks and human activities. Therefore, each of these concepts – hazard, vulnerability, risk and exposure (see Glossary) – must be assessed systematically for each of the elements that may sooner or later be impacted by disaster. Under the modern approach to emergency planning, this is done comprehensively for all hazards that are significant in the area and all the principal approaches that will be used to manage them. This requires that the planner be familiar with methods of characterizing, estimating and depicting the factors quantitatively, spatially and conceptually. It also requires the ability to build logical projections of what will happen in future events, using **scenarios**, Delphic questionnaires[*] or computer simulations. Then, as emergencies are tackled by matching needs with available resources, the latter must be configured and deployed efficiently (Fig. 2.1).

Emergency planning needs to be based on a firm foundation of research. This will illuminate the physical and social processes involved in impact and response. It will also indicate the most appropriate planning strategies to adopt. At the most basic level, planning needs to be backed by a comprehensive and detailed study of the resources to be protected. General plans will thus require study of demographic and geographical factors – the pattern and characteristics of the local population and its geographical distribution.

This will mean considering the geography and typology of buildings, infrastructure, economic activities and so on. It will also involve establishing the *static* pattern of geographical layouts of these phenomena and the *dynamic*

---

[*] See §2.4.2; also **Delphi technique** in the Glossary.

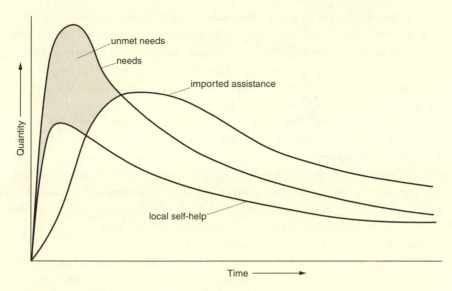

Figure 2.1 Emergency planning must meet the urgent needs by integrating locally generated aid with relief from outside the disaster area.

picture of change and development – population growth, changes in the demographic pyramid of the age and sex characteristics of the population, daily and seasonal population movement, trends in urban and industrial decay and renewal, and many other factors.

2.2 The emergency planner's choice of methodology

The emergency planner has much in common with the urban planner, a person who is trained as a geographer or architect. But as the methodologies used may eventually differ between the two fields, it is worth giving some consideration as to what sort of professional training best befits the emergency planner and what sort of background the aspiring emergency planner should have.

The methods used in planning for emergencies vary from the technical to the intuitive, and from the academic to the practical. The spectrum is exceedingly broad and hence versatility is the key to success in the role. This means that the planner should have as broad a training as possible. If the appropriate equipment is available, it is helpful, perhaps essential, to be versed in the use of personal computers for data and information management. Familiarity with the use of **geographic information systems** is also a definite asset, as these will play an increasingly important role in emergency planning and management

13

in the future. In this context, one of the most important skills in this field is the basic ability to visualize and depict spatial information – in other words, to read and draw maps. Understanding of Earth and environmental science factors will be useful for interpreting the physical aspects of hazard planning, as will a training in engineering with respect to industrial hazards or the risk of building-failure. Techniques of demographic, statistical and economic analysis can also be used profitably in emergency planning. However, the non-technical aspects of drawing up a plan require knowledge of human behaviour and perception, and possibly of medical, sanitary or paramedical operations if the plan refers to these aspects. Legal knowledge can be highly beneficial where the emergency plan responds to complex legislation and embodies important legal requirements associated with protecting the public. Finally, all plans require fluent writing and good communication with the users.

The reality of the situation is that relatively few people who are called upon to devise and operate emergency plans have the full training that is required, although many will acquire some or all of it on the job. The type and level of qualifications of emergency planners tend to vary markedly from place to place and with the kind of institution for which they work. In some instances, geologists tend to be preferred on the grounds that they have specialized knowledge of certain natural hazards: in others, engineers tend to be hired because of their technical competence. Be this as it may, the prime requisite is versatility and the ability to think without allowing oneself to be constrained by disciplinary boundaries. Geographers are often appropriate candidates here because their view tends to be lateral, or integrative, rather than longitudinal, or specialized. However, no background should exclude a candidate from the job *a priori*. The best characteristics of a successful emergency planner will include not merely a high level of academic training in an appropriate discipline but also the ability to learn rapidly from experience and adapt to the demands of the field. When the job of emergency planner is being assigned, candidates should be assessed on the basis of their appreciation of, and sensitivity to, these prerequisites. Most of the rest can be learned when it is needed.

The following sections will outline the methodological ingredients of the emergency-planning process. These essentially consist of the ability to acquire information, manage and analyze it in certain ways, and present it in certain forms, especially in maps, whose value lies in the fact that they express territorial and logistical aspects of emergency operations.

2.3 Cartographic methods

The basic dimensions of emergencies are time, space, magnitude and intensity. *Time* is the backbone of events, the linear measure of their unfolding (although it is not a process in itself, merely an abstract concept), whereas *geographical space* is the medium in which the events take place, the paper on which they are written. *Magnitude* refers to the physical forces at work, and *intensity* to the effects and reactions that these forces cause (although the two terms are often used in complex and interchangeable ways).

In order to plan for emergencies, one must have some idea of how they are likely to unfold in time and space, and about the evolution and distribution of magnitudes and intensities. Cartography, the depiction of spatial relations, is thus fundamental to the understanding, prediction, prevention and management off disasters. Indeed, an understanding of where things are is fundamental to all stages of the disaster cycle (see Fig. 1.2).

The object of this section is to elucidate the various kinds of cartographic work and to explain how they interact with the processes of emergencies and their management. It is also important to take note of the flexibility of modern cartographic methods. For example, traditionally one has been able to map a spatial situation only as it appears at a single moment in time. Concepts and rates of change could be depicted only as symbols on the map or by comparing maps that represented the same places at different moments in time. But with the advent of digital processing and refresher graphics it is now possible to map situations that evolve by iteration. If there are iterations, and the rate of screen refreshment is rapid enough, the process of change will appear smooth.

Cartographic methods are useful at all stages of the disaster cycle. In *mitigation* they help indicate where the physical phenomena to be prevented or protected against are located, and where to place structural defences. In both mitigation and *preparedness*, it can help determine the zones of risk and vulnerability. It can also locate resources that are to be used when disaster strikes. In *response*, cartography is fundamental. To begin with, it is essential to start operations with a good idea of where the boundaries of the incident or disaster area are, and where the greatest damage and most casualties are located. A new development is that of **emergency mapping**. In this, the spatial pattern of unstable situations is depicted as they develop, or afterwards for the purposes of hindsight analysis or **debriefing**. Maps can also depict how participants in the emergency perceive the situation. If this is expressed in formal terms and with sufficient rigour, it can add a useful new dimension to the process of managing the disaster. Lastly, recovery from catastrophe also makes use of cartography, for maps are essential to the reconstruction process. Thus, geologists map hazards and sites that are suitable for new construction, and urban planners map the pattern of reconstruction in areas that have suffered damage.

In deciding what sort of maps to draw, some fundamental choices must be made. Increasingly, one tends to opt for digital methods on the personal computer, as this facilitates making copies and modifications. It also offers greater flexibility in choice of cartographic methods than does drawing maps by hand. Other choices relate to the use of point, line and areal symbols, and shading, and to the scale of the map. With regard to the last of these, there are no hard and fast rules, but the following may serve as a rule of thumb and guide:

- 1:200000: regional maps with highly generalized local details
- 1:100000: sub-regional maps with crudely generalized local details
- 1:50000: sub-regional maps with a greater level of local details
- 1:25000: local overview maps with some generalization of local details
- 1:10000: a basic scale for depicting local features over small areas (say 10 km)
- 1:5000: localized maps with a high degree of detail, suitable for depicting, for example, a village (5×3 km)
- 1:2000: highly detailed localized maps for depicting complex, highly concentrated features.

Scales of 1:50000 or smaller are essentially for synoptic maps, whereas 1:25000 and larger are for detailed local work. However, computer mapping has to some extent blurred the distinction, as maps derived from very large files can be depicted on screen, and subsequently printed, at a variety of scales related more to line width and pixel size than to anything else.

Another choice that must be made relates to the type of scale used to measure the phenomena shown in the map. With increasing complexity, data can be classified by scale as nominal, ordinal, interval or ratio. Nominal-scale data are binomial in character, which means a simple distinction between presence and absence, yes or no, on or off. This is appropriate where the phenomena to be mapped are known only in simple terms (e.g. presence/absence of unstable slopes). Ordinal-scale data involve ranked categories, such as "extremely", "very", "moderately", "slightly" and so on. This is appropriate when one wants to map simple phenomena according to their intensity (e.g. landslides that are highly active, moderately active or inactive). Obviously, more detail is included than is the case with nominal data. Ordinal scales are often used where the seriousness, or strength, of the phenomena to be mapped can be assessed only according to the judgement of the observer. On occasion, however, it is helpful to divide numerical data into categories and map these in order to generalize the phenomenon in question, especially if its distribution is highly complex. Interval data are those that increase in fully quantitative gradations that have no absolute zero. In practice, temperature is one of the only common examples: although we talk about absolute zero (−273K) in physical terms, in other respects there is no point on the scale at which temperature does not exist. The bar scale used to measure atmospheric pressure is analogous to this, in that zero pressure is in practice never achieved at or close to Earth's surface. Much more important is the ratio scale, in which the quantitative gradations are tied to a zero value (length is the primary example of this). Most detailed maps of physical variables and many of the socio-economic ones use this scale.

A further choice, which is restricted to digital computer methods, is that between raster and vector methods of depicting mapped phenomena. The former divides the mapped area into a mosaic of equal-sized cells known as picture elements (pixels). The colour or texture of groups of these represents the phenomena shown on the map (it is as well to note that the single pixel has no meaning on its own, but only in relation to its neighbours). In contrast, vector data establish coordinates (x_i, y_i, where x is width and y depth) as single pairs for dots and double pairs for the beginning and end of a line. This is a much more efficient way of filing the contents of a digital map, as it does not have to include every point on the map; raster maps tend to be either very crude or demand extremely large files.

In synthesis, a map is a low-level model of reality. The purpose of models is to simplify complex situations so that they can be understood adequately. Hence, the art of mapping is to ensure that the essence of the phenomenon or problem to be depicted is conveyed adequately to the map user, without being obscured by extraneous data. In most instances this will involve a hierarchy of mapping processes. Basic elements of physical and human geography will be shown on basemaps. Good examples of these will exclude features that do not contribute to the process of locating what needs to be known. Roads, rivers, urban settlements and some elements of topography (perhaps by using land contours) are usually essential ingredients of maps. **Thematic maps** represent a higher order of the spatial modelling process. They show the distribution of particular phenomena, such as landslides, earthquake **epicentres**, floodable areas or geological formations. Whereas basemaps tend to remain immutable, or largely so, thematic maps can be made with refresher graphics, or other forms of timestep iteration or symbols, in order to show changes in the phenomena to be depicted. Further up the hierarchy are **planning maps** and other forms of descriptive cartography that show not what is actually there but what ought to be or will be. Here the mapping process enters the realm of hypothesis, and flexibility is imparted by freedom from the bounds of actual spatial relations. An extreme form of non-literal cartography is represented by mental maps, or cognitive cartography, in which geographical space is depicted as it is perceived. As perception varies in accuracy with degree of proximity to and experience of phenomena, so the verisimilitude of mental maps varies in comparison with "ground truth", the real spatial pattern of phenomena. In practice, for our purposes, this is most relevant to emergency mapping, although all forms of cartography are likely to involve a certain lack of absolute objectivity.

2.3.1 Cartography, GIS and remote sensing

The digital revolution has enabled new and more powerful techniques of cartography to be developed. The foremost of these involves geographic information systems (GIS). These first establish a set of digital coordinates for the area to be mapped: $x_1, x_i, \ldots x_n$ and $y_1, y_i, \ldots y_n$. All further data $\{z_1, z_i, \ldots z_n\}$ are referenced to each position on the map $\{x_i, y_i\}$. A single variable is mapped as a series of variates $\{x_i, y_i, z_i\}$ in geographical space. Multiple datasets can be mapped as multiple or compound variates on a single map or as a set of overlays to the digital basemap, or to the most fundamental thematic map. Thereafter, data can be manipulated in a wide variety of ways to produce mapped distributions that are more or less literal according to what they display.

The initial investment in GIS technology can be daunting for the emergency planner. It requires a computer, workstation, software and printing or plotting equipment that are above the level found in the average suburban home or even home office. If there is more than one user, unit costs can be reduced by linking computers into networks that use the same software and peripherals. Alternatively, there are cut-price and reduced-scope versions of GIS software that are suitable for basic usage at a simple level. A further alternative, where connections permit, is to use GIS software over the Internet. The length of time that one can remain connected, and the volume of data that one can transmit, may be limited by cost and speed factors. One may also be daunted by the investment of time and effort required to master the commands and processes needed to run each GIS package.

However, once these hurdles have been overcome, much can be achieved. GIS has been applied to an enormously wide range of problems, from the esoteric to the homely. In most places its use in emergency planning and management is very limited, but it is gaining rapidly in popularity.

The emergency planner who wonders whether it is appropriate to use GIS needs to ask some fundamental questions. To begin with:
- Is the problem of hazards and their mitigation in the local area sufficiently complex to warrant a sophisticated digital approach?
- Will GIS eventually help save time and effort in making decisions or locating vital information?
- How many in the department that buys a system will have to learn to use it?
- Are the resources available to purchase the hardware and software and to train the users?

- Do other departments have GIS, such that many of the spatial data are already encoded and resources can be shared?
- Assuming that the decision is taken to invest in a system, is it necessary to choose the components with a view to integrating them with other systems in use locally, such as a municipal-services GIS?
- What can GIS be expected to produce for the emergency planner?

The answer to the last of these questions may depend on the planner's creativity and the availability and quality of input data.

The first stage in creating a GIS for **emergency management** is to compile a basemap. This may necessitate the construction of a **digital elevation matrix** (**DEM**, also known as a **digital terrain matrix**, **DTM**). In this, contours or elevations are converted to spot heights $\{z_i\}$ on a regular grid of two-dimensional spatial coordinates $\{x_i, y_i\}$. Contours can then be reconstructed digitally by interpolation at whatever scale they are required. This is most likely to be useful where hilly terrain is involved, or where there are mountain hazards such as landslides or avalanches. It is especially important where the ruggedness or steepness of terrain might seriously affect accessibility during the emergency phase or might have an adverse influence on **search-and-rescue (SAR)** operations. In other cases, such as relatively flat-lying cityscapes, it is sufficient to digitize the road network and other salient features, such as watercourses. The advantage of using a DEM is that GIS subroutines can transform it into a three-dimensional view, in the form of a block diagram either composed of lines or made up of shading that represents relief. Features such as roads, streams, villages or landslides can be "draped" over the block diagram so that they can be seen in relief. This is especially valuable where a perspective view helps visualize the geographical problem. The angle of view, or azimuth, can be changed and the block diagram rotated to change the perspective. This also enables the diagram to be fed into a "virtual reality" **simulation**, which can be valuable for training, although its use in emergency management is still in its infancy.

The principal value of GIS systems is their ability to relate datasets to each other in the form of spatially referenced overlays (Fig. 2.2). The emergency manager who has a full set of data in a GIS can call up overlays at will and superimpose them on a screen. For example, the location of hospitals can be added to the pattern of roads in order to calculate the quickest or simplest ways to transport victims from the site of a **mass-casualty incident** to places where their injuries can be treated. The main limitations on this procedure are the quality and extensiveness of the data, how up-to-date they are, speed of processing, storage capacity of the computer system, and the creativity, experience and judgement of the user. A good system and a reliable set of data, manipulated by an experienced user, mean that considerable flexibility can be achieved in using and interpreting the information contained in the GIS, such that it is a

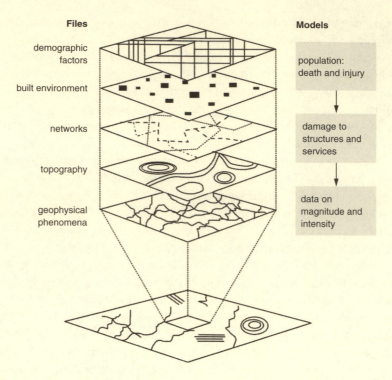

Figure 2.2 Use of geographic information system overlays in order to build predictive models of hazard.

valuable aid to decision making. A further advantage of GIS is that it can be updated rapidly. In emergencies, information can be fed into the system as it is received from the field. For instance, if an earthquake has caused some bridges and overpasses to collapse, information on their location can be added to the GIS as it is received at the **emergency-operations centre**, and routes for emergency vehicles can be replanned in order to avoid the blockages.

In synthesis, GIS can be useful to the emergency planner in a wide variety of situations in which data must be spatially referenced. Complex vulnerability mapping can be achieved by relating diverse information on physical hazards to that on human socio-economic risks. Concrete factual information, such as the distribution of buildings or bridges, can be mapped, but so can more conceptual information, such as the forecast distribution of future landslides or earthquake epicentres. Even information that is purely notional can be added in, such as the results from surveys of how residents perceive floods. Moreover, if it has been properly set up and supplied with data before disaster strikes, GIS can be an invaluable aid during an emergency. In this respect it can be incorporated into more general emergency-management software, most of which is based upon spreadsheets and relational **databases**. When data on the emergency have been added to the GIS (e.g. to show information on the degree of functionality of each hospital in a region affected by an earthquake), the results can be printed out as maps, or can be sent as faxes by telephone or as computer files over the Internet. Finally, GIS can be used in planning both reconstruction and mitigation efforts. Reconstruction plans, for example, may require information on terrain and surviving features of the human landscape. They may need overlays that detail the pattern of geophysical hazards, the location of safe or stable land for rebuilding, or the distribution of environmental impacts caused by the planned reconstruction.

Remote sensing is a related tool. It is defined quite simply as the acquisition of data from afar without physical contact with the subject that is being observed. In environmental terms, this usually means aerial or satellite imagery. This begs the question of what can be sensed from the air or from space that is useful to the emergency planner. The answer is dependent on the scale and resolution level (i.e. the level of detail) of the imagery and on whether phenomena can be sensed and interpreted from aloft. In general, resolution levels dictate that satellite imagery will give a broad, synoptic picture of regions, whereas **aerial photography** from a much lower level will provide some of the details that are lost in the relatively low resolution of satellite images.

A basic choice is that between using images obtained in the past and stored in archives or commissioning new imagery, perhaps by paying for an aerial photograph sortie. Archives can furnish images of, for example, flooded areas or the distribution of landslides following a period of intense rainfall. New images may be required where no imagery currently exists or none is adequate to needs. Digital files of satellite images may cost thousands of dollars to purchase, and so would an aerial photography overflight, whereas prints derived from archived images are generally cheap, although not as detailed or amenable to analysis as their digital counterparts.

It is important to recognize both the potential and the limitations of remote sensing. Regional flooding can be mapped from satellite images that are geocorrected (i.e. rectified from distortion caused by topography and Earth's curvature). Thematic mapper (TM) and advanced very high resolution radiometry (AVHRR) sensors can furnish the data, although the former cannot pierce the banks of clouds that often cover a newly flooded area. Radar imagery can pierce soil cover and reveal details of geological structure, but it cannot directly illuminate the dynamics of structural processes. Generally, Earth resources satellite images, such as those obtained from Landsats I–V, are not capable of sensing landslides, although they can depict elements of terrain that are useful to both geologists and seismologists, and land-surface texture may thus become a partial surrogate for slope instability.

Resolution level, which determines the detail that can be extracted from an image, is dependent on the dimensions of each picture element, or pixel, in the mosaic that composes the image. Computer displays of digital images enable one to zoom in on detail, but nothing can be determined from the individual pixel, or even from very small agglomerations of pixels. Thus, the main value of Earth resources satellites is in giving rapid synoptic pictures of land use or land cover. This is determined by finding the signature of particular surface features (i.e. the spectral frequency in the stream of remotely sensed data to which each feature uniquely responds), and by confirming it with fieldwork, which is the search for **ground truth**. Meteorological satellites give a synoptic picture of conditions in and above Earth's atmospheric boundary layer.

Both meteorological satellites and the new generation of volcanological observing modules can provide important information on extreme atmospheric disturbances, such as hurricanes, and plumes of volcanic ash and gas; repetitive sensing can establish a pattern. With short-lived phenomena such as eruption plumes, this is not a problem, but long-term changes can be sensed only from the 1970s onwards: Earth resources satellites provide relatively crude images that date from 1973, with improvements in quality taking place as each new satellite module was launched thereafter.

Although the resolution level of satellite images is increasing as new and more precise sensing equipment is sent into space, it cannot yet match that of aerial images. Overlapping stereographic pairs of vertical grey-tone aerial photographs are obtained, usually in black and white, by a camera, mounted on an aircraft fuselage, that points at the ground and automatically takes a series of photographs at measured intervals. A skilled interpreter can use these images to identify areas subject to intense erosion, slope instability phenomena, water-saturated ground, and other useful phenomena. After earthquakes or other damaging phenomena, collapsed buildings can be seen on low-level aerial photographs with print scales of 1:20000 or larger. However, damage cannot be identified unless it can be photographed from above and from several thousand metres distant.

Both satellite and aerial surveys can make use of the thermal infrared band of electromagnetic radiation that is sensitive to variations in surface moisture. Provided that cloud cover does not impede the sensing process, this can be very useful in monitoring floods, drought, or soil-moisture concentrations associated with accelerated erosion or landsliding.

In synthesis, besides questions of the cost and quality of images, and what they depict, there is a trade-off between resolution, area covered and volume of data. The user who wants highly detailed information on a wide variety of land conditions will have to deal with large amounts of data, which may require substantial computing power or excessive time.

Digital remote-sensing data are arrays of numbers corresponding to varying intensities of particular frequencies of the electromagnetic spectrum, as represented in each pixel. They must be filtered to separate the signal from noise, corrected for distortion, and assigned a value that corresponds to some visible part of the spectrum (i.e. a colour); then they can be plotted. At this point, as the matrix of plottable values in the x and y directions, the image can be combined with a GIS. Specifically, it can be "draped" over a digital elevation matrix (see p. 19) to give a three-dimensional perspective view of the area covered. This, again, is useful in visualization and virtual-reality exercises.

Manuals and treatises on GIS and remote sensing abound, although the link between these two methodologies has not yet been fully explored, let alone fully exploited. As with GIS, the emergency planner should consider remote sensing in terms of what it can contribute to the planning process. This may include spatial data on past hazards and disasters, sometimes even on past patterns of damage. But remote sensing will mainly be useful if it is able to contribute information on the geographical distribution of future hazards and on current vulnerability to them. As image interpretation is a specialized skill, and often a highly technical process, the intending user needs first to ascertain what imagery is available or can be commissioned; secondly, what it is likely to show; and, thirdly, whether the technical help is available to interpret it. In this respect, aerial photographs tend to be the easiest to use, as information can often be transferred from them directly to existing topographic basemaps with minimal loss of accuracy. In contrast, highly accurate mapping and satellite-image interpretations tend to require the help of an expert and the acquisition of appropriate equipment.

2.3.2 Hazard maps

Most forms of natural and technological hazard cannot be predicted exactly in terms of short-term impacts, but their general distributions are usually either well known or at least possible to determine. Hazard maps express the geographical pattern of forms of danger and they are thus essential raw material for emergency planning. Indeed, if they do not already exist, they must be created before planning can successfully be accomplished.

Hazard is not the same thing as risk (see §3.2). The former refers to a kind of danger irrespective of whether anyone is actually experiencing it; the latter expresses the correspondence between danger and exposure to it. Thus, hazard maps should not be taken to be portraits of risk distributions. They are loosely allied to these, but in ways that depend heavily on patterns of human settlement and activity – that is, human risk taking.

Before developing an emergency plan, it is important to assess which hazards are significant in the area that the plan will cover. These fall into the following broad categories:

- atmospheric – severe storms, etc.
- biological – pest infestations, epidemics, epizootics, etc.
- geological – earthquakes, landslides, etc.
- hydrological – floods, etc.
- social – riots, crowd problems, terrorist incidents, etc.
- technological – transportation accidents, industrial hazards, catastrophic pollution, etc.

Warfare is generally excluded from civil emergency planning, as it tends to require a different and more Draconian approach. In any case, not all forms of hazard will be amenable to emergency planning, especially those that create certain forms of long drawn-out diffuse risks. Having identified those hazards that are amenable to mitigation by planning, one needs to decide whether to tackle them singly or in a group. The latter approach has steadily gained favour, as the all-hazards planning is safer, more efficient and subject to useful economies of scale in comparison with trying to confront hazards singly. However, in some areas, usually small ones, only one form of hazard will be significant.

Let us assume for the sake of argument that hazard information is comprehensively available in a given area, but that it has not been adequately mapped (alternatively that such information can be collected reliably and quickly). The main question is then one of how to map the information in such a way that it will be useful to the emergency planner. We must therefore ask what characteristics of hazardous physical, technological or social processes best describe them in these terms? Natural processes tend to follow distinct rules of magnitude and frequency. High-magnitude events are rare, low-magnitude ones are more common (Fig. 2.3). It is necessary to define a threshold above which events have a magnitude (an energy expenditure, a geographical extension, or however else it is expressed) that is capable of causing a disaster. For example, the modified Mercalli, MCS and MSK earthquake-intensity scales each express disaster potential as intensities VIII–XII. Even though seismic intensity is not strictly correlated with physical variables such as duration of strong motion, magnitude (which is partially a function of the size of the largest seismic wave) and maximum acceleration of the ground, a given size of earthquake is capable of generating a given intensity of damage for each area with which it is associated. More scientifically, maximum acceleration (e.g. 20 per cent of $g = 9.81 \, \text{m/s}^2$) will be correlated with a minimum threshold of damage potential.

The principal characteristics of natural hazards that relate to their disaster potential are the quantity of energy liberated at Earth's surface, its concentration and distribution, and the average recurrence interval, or **return period**, of events of a given size. Thus, with sufficient data one can map the pattern of

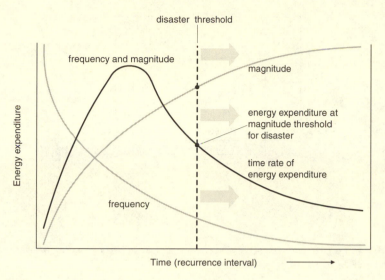

Figure 2.3 Frequency and magnitude relationships for extreme natural events. The disaster threshold is arbitrarily imposed on the basis of what society regards as catastrophic.

recurrence times for earthquakes that are capable of generating damage to, for nstance, intensity IX, although, given the role of the quality of buildings in permitting damage to occur, the result is as much a risk map as it is a hazard map. Alternatively, one can map the pattern and intensity of factors that give rise to the hazard. Thus, slope steepness, rock type and groundwater concentration can be mapped in relation to landslide hazard.

In the case of recurrent floods, decades of data have enabled predictions to be made about the areas with likelihood of being inundated once in a given period. In the USA and elsewhere much hazard mapping and mitigation is based on the 100-year flood event (the event of a size that has a probability of occurrence of 1 per cent per year), which defines the **floodplain** in legal terms. For more frequent floods, the floodway is defined by the 20-year event (Fig. 2.4). Flood-rating maps are produced to define flood-insurance districts with respect to the 100-year floodplain. However, the mere standardization of flood recurrences does not mean that the 100-year event is the threshold for disaster.

In fact, particular problems arise when the hazard is either a very long-term one or is in the process of dynamically changing its character. Thus, the threat of a volcanic eruption can often be ascertained by dating rocks such as welded ash deposits and lava flows situated on the flanks of an active volcano, but it may reveal only a handful of major eruptions over a period of, say, 10 000 years. This leads to the low-probability high-consequence hazard, whose exact nature is very hard to ascertain. Moreover, it may not make sense to plan *at all* for an event with an extremely small chance of occurring over an interval of 100 years.

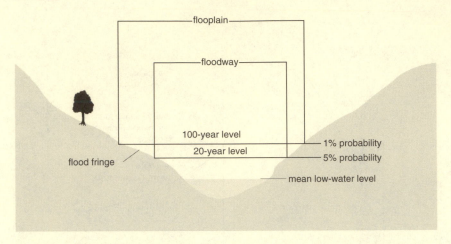

Figure 2.4 Legal or operational definition of a floodplain and floodway. Probabilities refer to annual likelihood of flood inundation.

The other problematic kind of event is that whose probability of occurrence is in the process of changing. For example, large tracts of northern-hemisphere coast are currently subsiding in response to post-glacial isostatic adjustment (the land-level changes that take place when a great weight of ice is lost from the continents and a corresponding weight of water is added to the ocean basins). At the same time, the greenhouse effect may possibly be making coastal storms more intense or more frequent. The result is a worsening pattern of coastal flooding and storm damage, an increasing hazard. In such cases the only solution is either to re-create the hazard maps at intervals or to draw up predictive maps of future hazard levels.

Maps of individual hazards may thus take various forms. At their simplest, these include locational descriptions of where the hazards are likely to manifest themselves, and perhaps approximately when they are likely to occur. That also means maps of past impacts or manifestations. It may include maps of the likelihood of given physical consequences, such as landslides, floods or lava flows, and this may be expressed in percentage terms, per unit of time. Alternatively, it may be appropriate to select the characteristics of the physical hazard that are most significant to its disaster potential and map that. Examples include seismic acceleration of the ground, density of volcanic vents, presence of unstable rock types and steep slopes, and low-lying floodable land.

However, emergency planning needs to be based on the concept of the "hazardousness of place", the collection of hazards that threatens each location in the area. At its simplest this means that one must recognize the presence of compound hazards, or chains of events. Earthquakes may cause dams to rupture or tsunamis (seismic sea waves) to overwhelm coasts. Hence, one needs to look at the distribution of the primary hazard (an earthquake) and that of any **secondary hazards** (such as tsunamis or landslides) that it may cause. But when the situation is more complex, it may be necessary to consider the distribution of hazards that are not linked to one another. On occasion, this may result in a map of general hazardousness (i.e. destructive potential) derived from the superimposition of various hazard maps that cover the same area.

Nevertheless, different hazards often require varied emergency strategies, and hence the raw material of planning will consist of a series of different maps. A necessary prelude to writing the plan may be either to commission studies of hazards in order to get them mapped or simply to search out existing maps. These may be held by geological surveys, meteorological forecasting centres, urban and regional planning offices, university departments of Earth sciences, or inspectorates of industrial hazard, among others. The acquired collection of maps must then be interpreted in terms of the emergency planning needs of the area they cover.

Let us conclude this section with an example of a simple methodology for mapping the hazardousness of place using minimal facilities and equipment. It can be made more sophisticated by applying GIS methodology (Fig. 2.5).

The area to be mapped must first be determined. Let us assume it is about 10–12 km^2, which is an appropriate size to deal with. Mapping will take place at the 1:10 000 scale, which will give a fair amount of local detail, although probably not enough to differentiate the risk to individual buildings. Next, base-maps of the area are divided into 1 ha quadrats (squares), of which there will be 10 per km and 100 per km^2, 12 000 in all. Individual hazards are next assessed from maps or aerial photographs, and occasionally from high-resolution satellite images. Intensities of mappable phenomena (such as flood depths for the event with an average recurrence interval of once in 100 years) are assessed quadrat by quadrat as an index number. This may reflect the average value of the phenomenon under study for each 1 ha quadrat, or the proportion (in tenths) of the quadrat that is covered by the phenomenon (e.g. a potential lava flow). Using the following formula, values of the phenomenon in each of the 12 000 quadrats are standardized so as to vary between zero and one:

$$x_i = (x_i - x_{min}) / x_{max}$$

This is done for each significant hazard in order to produce a series of matrices 100–120 cells in size.

The compiler must then make a judgement about the relative importance of each hazard and assign a weight to the values in each matrix. For instance, flood hazard × 1.0, storm hazard × 1.8, landslide hazard × 3.1, industrial explosion hazard × 0.5. The weighted matrices are added to each other cell by cell and the

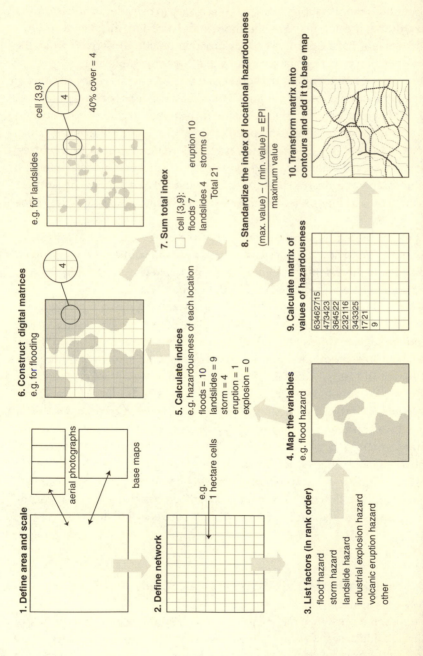

Figure 2.5 A simple methodology for systematically assessing and mapping hazards in a particular area. This procedure can be made more efficient by digitizing the input data and using geographic information system overlays in place of the manual techniques shown here.

1. Define area and scale

aerial photographs

base maps

2. Define network

e.g.
1 hectare cells

3. List factors (in rank order)
flood hazard
storm hazard
landslide hazard
industrial explosion hazard
volcanic eruption hazard
other

4. Map the variables
e.g. flood hazard

5. Calculate indices
e.g. hazardousness of each location
floods = 10
landslides = 9
storm = 4
eruption = 1
explosion = 0

6. Construct digital matrices
e.g. for flooding

4

e.g. for landslides

cell (3,9)

4

40% cover = 4

7. Sum total index

cell (3,9):
floods 7   eruption 10
landslides 4   storms 0
Total 21

8. Standardize the index of locational hazardousness

$$\frac{(\text{max. value}) - (\text{min. value})}{\text{maximum value}} = EPI$$

9. Calculate matrix of values of hazardousness

6	3	4	6	2	7	1	5
4	7	3	4	2	3		
3	6	4	5	2	2		
2	3	2	1	1	6		
3	4	3	3	2	5		
1	7	2	1				
9							

10. Transform matrix into contours and add it to base map

result is standardized using the formula given above, so that it varies between zero and one. Values for each cell of the matrix can be plotted on a grid overlay to the basemap and turned into contours or shaded isopleths, using manual interpretation or a computer program, such as Surfer. Choices must be made as to the spacing of contours, as these need to be simple but not greatly lacking in detail. The resulting map expresses the general spatial variation of hazard in the study area. It is perhaps crude and somewhat arbitrary, but at least it is fairly simple to produce. Compilation can be speeded up by using computer routines, and the map can be improved by refining the input data and starting assumptions, and by varying the scale of analysis. The results of the analysis can sometimes be quite startling in terms of what they reveal about the spatial variation of threats to people's security.

2.3.3 Vulnerability maps

There is considerable overlap between the concepts of hazard, vulnerability and risk. Hazard, the expression of danger, was covered in the previous section: risk, the likelihood of impacts, is dealt with in the next one. In this section, we will deal with vulnerability, that is, the susceptibility of people and things to losses attributable to a given level of danger, a given probability that a hazard will manifest itself at a particular time, in a particular place, in a particular way and with a particular magnitude (Fig. 2.6).

The assessment and mapping of vulnerability can be conceived of in a wide variety of ways that vary from the simple to the highly complex, in relation to the seriousness of the problem and the availability of resources with which to tackle it. The first task is to put together a scenario (see §3.1.1) which expresses the nature, location and magnitude of the event that causes the vulnerability. Thus, the hazards are a trigger of vulnerability: different magnitudes of a

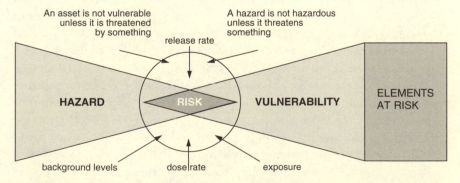

Figure 2.6 Relationship between hazard, vulnerability, risk and associated concepts.

particular hazard will expose vulnerability to different degrees. For instance, we might ask how many two-storey brick-built houses in a given town are likely to collapse in an **earthquake** of **magnitude** 7, maximum acceleration 0.35 g, duration of strong motion 25 s, and hypocentral depth 15 km (i.e. depth in Earth's crust to the point at which faulting begins), which becomes the **reference event** for the vulnerability analysis.

It is appropriate to assess vulnerability in both aggregate form, as statistics and tables, and as a spatial distribution, using maps. One may wish to assess the vulnerability to single hazards, or attempt a comprehensive evaluation, in relation to the "hazardousness of place", which considers all kinds of susceptibility to damage, losses and casualties. At the very least it may be necessary to consider the possible effects of secondary hazards, as when floods cause pollution, landslides or mudflows.

Perhaps more than at any other stage in the mapping exercises that form an input to the emergency-planning process, one needs to consider carefully the objectives and implications of vulnerability assessment. Once again one may either use existing data or generate one's own if basemaps and resources of time and manpower are available. Let us now concentrate on the forms that vulnerability may take and how to map them.

For the built environment, structures may demonstrate some vulnerability to damage or collapse (i.e. limited or total structural failure). This susceptibility can be assessed by eye (operator's judgement) or using more sophisticated engineering tests or formulae. It is not the purpose of this book to explain the latter, and the reader should refer to appropriate texts on structural and anti-seismic engineering. Visual methods usually rely on some knowledge of past building failures and a comparative typology of construction methods, building-design and so on. Here are some factors that help describe a building's liability to collapse when it is subjected to forces that are larger and more dynamic than those for which it was designed (e.g. as a result of an earthquake):

- *Construction material*   Generally, buildings with wood, steel or reinforced concrete frames perform better under seismic loading and other forms of dynamic stress than non-frame ones do (adobe, rubble masonry, dressed stone masonry or load-bearing brick). However, buildings may be composed of mixed construction types (e.g. stone walls and steel-joist floors), in many cases because their size, appearance and function have changed since they were first constructed. This complicates their response to dynamic forces, usually by worsening it.
- *Dimensions and shape*   Tall buildings tend to suffer inertial sway and possibly excessive vertical hammer effects during earthquakes, and substantial damage to window glass during hurricanes (often as a result of the distortion of window casements, or racking). Dynamic loading tends to have least

effect on compact buildings with simple well integrated structures and uninteresting plan and profile shapes. It has greater effect on buildings of irregular shape that present many faces to the forces that confront them and which shift their loads in irregular ways and have potentially fatal weak spots, such as isolated structural columns or open-plan ground floors (the "soft storey").

- *State of maintenance*   In general, well maintained buildings perform better in disaster than do poorly maintained ones, which may suffer from weaknesses caused by the degradation of structural materials.
- *Patterns of usage*   The danger of structural collapse is obviously also a danger of casualties. This problem is described later in this section.

In synthesis, buildings are vulnerable when they may have to withstand forces that they have not been designed to resist or when they may be affected in ways that put occupants or their activities at risk. A building that sustains a strong physical impact without significant damage will have an adequate combination of strength (to resist part of the forces) and ductility (to absorb part of them). On the basis of some knowledge of the past performance of buildings, one may wish to map their geographical distribution in respect of resistance to disaster. For earthquakes this may mean mapping the number of storeys (tall buildings may be more at risk of collapse than others), construction materials (masonry buildings may be more at risk than reinforced concrete ones), and their approximate ages (as a surrogate for estimates of state of maintenance or time-induced degradation of their fabric). It may be necessary to construct an index, or to overlay maps of each variable, in order to come up with a single distribution of the vulnerability of buildings in the area.

A somewhat more elusive aspect of vulnerability is represented by the threat to people and their activities. The greatest risk is obviously that of death or serious injury when disaster strikes. If this is governed primarily by the susceptibility of buildings to collapse, then one needs to look at patterns of building occupancy and how they vary over time (diurnally, between working days and holidays, seasonally, and so on). Similarly, the vulnerability of a bridge to flood or earthquake impact may be judged to be much higher if it is thronging with passengers at the time of impact than if it is deserted. Thus, the assessment of human vulnerability depends on predicting the aggregat behaviour and decision-making patterns of individuals. Where patterns of activity are well established, this does not present insuperable problems. For instance, if churches or mosques are especially vulnerable to structural collapse in disaster, then maximum vulnerability may be reached when they are full at Sunday or Friday prayer time. In this context one also has to deal with less obvious patterns of behaviour. For example, history reveals that the response of survivors to a devastating earthquake has often been to rush into the nearest

Box 2.1 Urban building collapse

It can be argued that we learn as much or more from things that go wrong as we do from exemplary practice. The following is an example of a poorly planned and poorly managed relief effort after an unpredicted sudden-impact disaster. It illustrates how serious consequences can be when emergency relief is badly designed, but it also shows how easy it is to derive lessons from the hindsight review of such a case.

The event
The incident occurs in the middle of a city of 160 000 inhabitants. It involves two interlinked apartment buildings, each six storeys high, which were constructed with reinforced concrete frames between 1968 and 1971. The building comprises 26 apartments, two of which are empty. It is estimated that there are 70–75 occupants in the part that collapses.

At 02.30 h on a November night in 1999 the custodian of one of the apartment blocks hears cracking noises coming from the structure. He calls the city's fire brigade and tries to rouse the sleeping occupants of both buildings. Not all of them leave promptly, as some spend precious minutes dressing themselves and collecting valuables. Half an hour later one of the two buildings collapses, raising a thick cloud of cement dust.

At 03.10 h the first fire appliances and ambulances arrive. No field-command post is set up and the early operations are therefore chaotic. Firemen recover the first lifeless body five minutes later. Volunteer ambulance services are not alerted during the early phase of the emergency, as they lack a cooperation agreement with state services. Instead, the city hospital sends many of its own ambulances to the site, leaving its normal services seriously depleted of vehicles and much of the region without regular ambulance coverage. By mid-day regular transportation of hospital patients has all but ceased, creating chaos in the wards, where there are infirm patients who need to be discharged but have no transportation. Only at this time does the hospital call up reserve forces.

Nevertheless, by 07.00 h word of the tragedy has got around and large numbers of rescuers are present at the site. Making an apparently arbitrary decision, the city's prefect (i.e. chief representative of the state apparatus of government) bans all volunteer rescuers from the site, allowing only firemen, soldiers and local Red Cross workers. After several hours of negotiations, groups of trained civil-protection volunteers are allowed back on site, where they carry on searching for survivors under the rubble.

By 12.30 h 7 bodies and 9 injured people have been recovered. Suddenly, fire breaks out in what had been the underground garages of the collapsed building. Trapped people are carbonized by the flames: some may have been alive at the time the fire reached them.

Rescuers continue to search the rubble for survivors and bodies using acoustic sensors. At 03.00 h on the following day, 24 h after the collapse, an explosion occurs amid the rubble: the fire, incompletely extinguished, appears to have ignited bottles of cooking gas.

At 07.00 h on the second day, 27 bodies and 10 injured people have been recovered. A team of specialists arrives with infrared sensing equipment. However, no more survivors are located. The emergency phase concludes 70 h after the collapse, with a final total of 62 deaths, 9 seriously injured people (one other having died of injuries sustained), and 5 people unaccounted for.

The living and dead victims are all transferred to the city's general hospital. At first, no arrangements are made to accommodate the bodies or grieving relatives. Eventually, these are accommodated in a nearby trade exhibitions hall. They are grouped by age and sex, divided by screens, and subjected to electrocardiogram monitoring for 20 minutes to verify that death has taken place. The police establish an office in an adjacent room in order to deal with the paperwork for identification and death certification.

Box 2.1 (continued) Urban building collapse

Days later, a judicial inquiry ascertains that the building collapsed because it lacked adequate drains, such that constant infiltration of water (some from a car-washing business on the ground floor) eroded the foundations away. The medical examiner states that half the victims died rapidly or instantaneously as a result of crush injuries, about one third were asphyxiated and one sixth died slowly as a result of injuries coupled with entrapment. Some of this last group may have been burned to death.

Evaluation

The copious but poorly organized response to this disaster leaves one with the nagging suspicion that a more efficient relief effort could have saved the lives of some of the victims who died under the rubble.

The problem began with deficiencies of general planning. It appears that the authorities had either no appropriate contingency plan or a patently inadequate one. It is evident that the hospital lacked plans for all aspects of the emergency. The result could easily have been a second disaster if anything else had happened that would have required large-scale ambulance coverage when so many emergency vehicles were tied up, unnecessarily as it happens, at the site of the building collapse.

The prefect's decision to restrict access to the site was not entirely irrational. However, it would not have been necessary if there had been more prior planning and coordination. In reality, many of the forces present were poorly equipped, and work at the site was undeniably dangerous: besides fire and explosions, the other half of the building threatened to collapse. Yet it was vitally necessary that search and rescue be carried out as intensively and quickly as possible, using all available means.

To avoid any future repetition of the problems described above, the following provisions would be needed:

- full scenarios for this and all similar events that might occur in the area, including assessments of the consequences of providing emergency assistance (e.g. the impact on regular services)
- a comprehensive disaster plan that deploys and coordinates forces rationally in the field and is based upon formal arrangements for mutual support between agencies, including those from far afield, that would converge on the site of the disaster
- plans would need substantial consideration of what sorts of command structures and procedures are appropriate in the area and with the available forces
- hospital emergency plans and their integration with the main city-wide plan
- more investment in emergency services, and greater incentives to volunteers and their organizations
- creation locally of fully trained urban heavy-rescue teams and purchase of specialized equipment; dog teams were lacking in this disaster and would have been helpful
- field exercises and other measures designed to improve timing and to familiarize participants with field-command procedures.

These provisions break down into planning, investment, coordination and training. One additional factor was noted in this disaster: inhabitants of the collapsed building did not react in optimal ways. Indeed, many could have been saved if they had not waited too long to leave the building. Although it is easy to criticize actions with the benefit of hindsight, it is also clear that, as the area where the building collapse occurred is also subject to earthquakes, more effort needed to be made to sensitize inhabitants to risks and prepare them to take self-protective actions.

Source of information: Valerio Esposito, 2000. "Crolla un palazzo a Foggia: quando una macro-emergenza diviene caos", *N&A Mensile Italiano Del Soccorso* Year 9, **99**, 14–19; and contemporary news reports.

place of worship in order to pray for salvation, but it may then collapse upon them in the next large **aftershock**.

The problem of behaviour is probably at its most acute when attempting to assess vulnerability to terrorist attacks; although such incidents have a certain degree of predictability, they are governed by uncertainty, irrationality, clandestine decision making, furtiveness and a desire to outwit the authorities. In any case, mapping vulnerability in terms of risks to public safety may require the cartographer to present several maps that reflect the average concentration of people at different times and in different places.

Another aspect of vulnerability as conceived in behavioural terms concerns the ability to save people once disaster has struck. Highly efficient rescue and medical services may be able to reduce death tolls significantly (in terms of at least numbers of people saved if not percentages), whereas inefficient services may lead to avoidable loss of life. For this reason, emergency planning places much emphasis on the maintenance of **lifelines** in disaster – the continued provision of vital services and communications. Moreover, it is usually essential to assess the vulnerability of hospitals, clinics and medical centres, and their staffs, given that they will have to save human lives when these are threatened.

There are, of course, other forms of vulnerability that may be worth mapping. Ecological vulnerability includes the susceptibility of trees and other vegetation to storm damage and the impact of disasters on populations of fauna. Economic vulnerability refers to the fragility of commercial or industrial enterprises, and patterns of employment and wealth generation, when disaster threatens.

Although there are reliable ways of measuring vulnerability, it depends enough on chance and complexity that it tends to be an elusive concept. It is therefore often considered in terms of surrogates. One of these is poverty, which frequently bears a strong relationship with susceptibility to loss and damage in disasters. Poor people tend to occupy housing that is least resistant to disaster. They have few economic resources that would enable them to avoid disaster impacts or recover quickly from them. In some regions the poor are marginalized to the areas of greatest physical hazard of disaster. Thus, one may find that income or some other measure of social status is used as a surrogate for vulnerability to disaster. This may be acceptable, but it incurs a certain degree of risk. In the first place, if poor people have a strong sense of community, they may evolve coping mechanisms that help them to resist the impact of disaster. In the second place, rich people indulge in more activities than poor people (e.g. they travel farther and undertake journeys more often), which may put them more at risk.

There is a danger here of remaining inconclusive. Unfortunately, local risks and hazards are sufficiently variable that there is no standard strategy for the vulnerability assessment. However, it is usually appropriate to proceed as follows:

- Assess and map the area's hazards and develop an idea of how they make the human environment susceptible to damage, loss and casualties.
- Map selected characteristics of the vulnerability of the built environment: state and quality of **vernacular** housing, public buildings, shops, offices, factories, bridges, etc.
- Make some assumptions about patterns of human activity and map these; for example, the density of occupancy of a town centre during a working day and at night and weekends.
- Assess and map vulnerability to secondary hazards, such as seismic landslides or dam failures resulting from floods.
- Assess other aspects of vulnerability, such as weaknesses in the capacity to respond to emergencies, and map these.
- Compare or overlay the maps and develop comprehensive ideas of vulnerability in terms of single and of overall "hazardousness of place".

2.3.4 Macrozonation and microzonation

Zonation is the process of dividing up an area into domains of roughly homogeneous risk, hazard or impact. This can be done at a fairly small coarse scale (**macrozonation**), or at a highly detailed large scale (**microzonation**).

The archetypal methodology is that of seismic macrozonation of earthquake impacts. This uses site surveys and questionnaires to develop a regional map of intensities in which tracts of land and human settlements are divided up on the basis of lines of equal **earthquake intensity** (**isoseismals**). Comparison of all historical data on the pattern of isoseismals can furnish a map of recurrence intervals for earthquakes of a given intensity. For the disaster planner, macrozonation is generally a given factor: if maps and data exist, they can be used to assess the hazardousness of the region that must be planned for; if not, they are seldom created under the aegis of an emergency plan.

Microzonation involves the detailed field investigation of hazard, vulnerability and risk. It is an expensive and time-consuming procedure, but one that is highly effective in delineating the spatial variations in these factors. In synthesis, microzonation involves the application of hazard and risk scenarios across the study area. First, these must be developed and explicated. For instance, buildings located on alluvial fans may be subject to a risk of high acceleration of the ground, plus consolidation–compaction of soft sediments or **liquefaction** failure during earthquakes, or to mudflow damage during floods. The first stage of a microzonation exercise is to make a fundamental choice between different strategies. This involves deciding whether to tackle single or multiple hazards and risks, and whether to do so for single or multiple purposes. A large difference of expenditure exists between microzonation of single and multiple hazards. The former is appropriate in areas entirely dominated by a single hazard, such as on a floodable alluvial plain that is not threatened by any other extreme natural or anthropogenic phenomena. However, it is more usual to microzone all significant hazards that are present in the area. This enhances safety, achieves economies of scale and increases efficiency, although it can be a much more complex process. Likewise, a safe environment is better ensured by using multiple-purpose microzonation, although again at considerable cost. Single-purpose microzonation is the most appropriate in specific analyses, such as in order to assess the likelihood that vernacular housing will be damaged in earthquakes.

There are various methods of microzonation. A typical analysis would begin with a visit to the local area's "hot spots", the localities that appear to demonstrate the greatest risks or the greatest degree of overlap between strong hazards and serious vulnerability. The investigator then draws up a series of risk scenarios, which represent archetypal local dangers. There is often no need to do this in a complex manner. For a simple microzonation exercise it will be enough to use pictorial designs of the hazards, with some level of annotation to help explain them. Concurrently, the investigator would draw up a list of factors that influence hazard and vulnerability, and rank them by importance.

The microzonation mapping exercise then involves one of two approaches. The first is to map each phenomenon or factor that influences local risk levels and to overlay the maps in order to get an idea of the concentration of risk levels. The alternative is to map combinations of factors in terms of their susceptibility to the scenarios drawn up beforehand. In both cases a central principle of mapping is invoked: the investigator must delineate areas on the map in such a way as to minimize differences within and maximize differences between one area and another. This usually involves **choropleth maps** with a series of ranked categories, such as high, medium and low (i.e. on an ordinal scale).

By way of example, consider an area that is subject to landslides and **flash-floods**. Risk categories can be designated for each locality on the basis of:
• the likelihood of slope instability or sudden inundation
• the probable dimensions of the physical hazard
• the presence of buildings, roads or bridges located in the path of hazards and possibly
• an assessment of the ability to mitigate the risks before disaster strikes or provide relief during it in such a way as to reduce the impact.

As microzonation can be a complex process, it is much facilitated by using a geographic information system to compile, overlay and display the data.

Generally speaking, microzonation has been most widely used as a method of mapping earthquake risks. It may incorporate a wide variety of physical information, including the seismic response of different surface geological formations, consolidation–compaction potential, liquefaction potential, topographic amplification of seismic waves, and their tendency to cause slope instability. The vulnerability of the built environment is generally assessed from a comparison of the dimensions of buildings, and their construction forms and materials, with data on the known performance of similar structures. The analysis may include some estimate of response capability after major damage has occurred, such as the probable quality of search-and-rescue operations and medical care.

If microzonation maps have been constructed to high standards, they can be invaluable bases for planning, because they help predict the impact of future disasters and they show where to concentrate planning efforts, in both topical and geographical terms.

2.3.5 Emergency maps and impact maps

Emergency mapping is an interactive form of cartography designed to assist with the real-time management of disasters. It is not generally used before disaster strikes, except perhaps for scenario building, but it must be designed beforehand so that it can be used during the emergency. Several elements commonly appear in emergency maps:

- lifelines, including roads that will be used by emergency vehicles, airports and air corridors along which airborne relief will travel, and the main conduits of essential utilities, such as water, gas and electricity
- the location of key facilities, such as hospitals, fire, police and ambulance stations, and emergency-operations centres
- points of special vulnerability, such as buildings that may collapse onto roads and block them with rubble, bridges that may fall down, floodable underpasses, and river embankments that may burst under the weight of high water discharge
- the principal telecommunications nodes and their paths.

In an emergency it may be necessary to relate these elements to the distribution of hazards and of vulnerable populations. Although emergency maps can be drawn and redrawn manually (either as sketch maps or by annotating copies of a basemap), it is preferable to use refresher graphics. Indeed, most programmes for emergency management include a mapping component. In the most sophisticated examples of this, the elements (roads, hospitals, etc.) can be called up as overlays. When a bridge has collapsed, it can be located on the

Box 2.2 Urban landslide disaster

At 03.30 h on 26 July, 1986 a landslide occurred at Senise in Basilicata Region, southern Italy, killing eight people who were asleep in two houses that collapsed and who were therefore unaware of what was happening. In technical terms, the landslide involved rapid, spontaneous sliding–toppling motions in compacted sands about 15 m deep. Conditions were perfectly dry and so the triggering event was simply a change from delicately poised static equilibrium to sudden movement. Nine buildings collapsed and, in addition to the 8 victims, 3 people were injured and 158 had to be evacuated from their homes. The site was subsequently cordoned off, placed under police guard and subjected to an inquiry by the district magistrate's office.

The two buildings that collapsed upon their occupants were single family homes of considerable age, probably dating from before the twentieth century. They were located at the top of the slope in the vicinity of the landslide headscarp. Farther down slope, seven apartment buildings were in the process of being completed. No doubt the excavations for the foundations of these buildings disturbed the rather frail equilibrium of the slope.

In the mid-1980s the municipality of Senise (population 11 000) did not have an urban plan as such, merely an urbanization program that indicated in which areas new development was permissible. It was based on a geological survey conducted in 1973 by a geologist from a nearby university. This document included a highly generalized 1:25 000-scale map of areas where surface instability could be expected to occur. On this, the slope that failed in 1986 was not shown as a zone of particular risk.

Aerial photograph analysis conducted after the 1986 landslide revealed that there is a high incidence of palaeo-landslides in urbanized parts of the municipality of Senise. The risk that these will re-mobilize is considerable, but this does not happen without warning. In 1984, slow, progressive slope instability began to tear apart a clinic and a school. A geotechnical report completed a year later revealed that the damage was entirely predictable. In the case of the 1986 landslide, movement had begun years before and accelerated significantly by the time of the disaster.

The case of Senise underlines the importance of coupling risk-mitigation measures to urban and regional planning instruments. The town lacked the rigorous planning instruments that Italian law required it to have, but in any case it is doubtful whether the law would have required full-scale hazard assessments and avoidance measures to have been incorporated into the plan. In short, in 1986 Senise was living with a false sense of security.

As is so often the case, it took a disaster to illustrate the dangers of ignoring information on hazard and risk. All the elements were there for a reliable landslide-hazard scenario to be constructed, but there were no incentives to build it. Central government ended up by disbursing 65 times the value of the damage in grants and loans for environmental protection in the area around Senise. This illustrates the importance of the post-disaster "window of opportunity" for attracting funds to an area and it also shows the gross inefficiency of tackling hazards after the event.

The Senise landslide is a reminder that the precursor of any scenario modelling for making an emergency plan (or an urban plan) should be a realistic assessment of the opportunities offered, and limitations posed, by official culture. Local populations have a fundamental right to live in a safe environment, but this is no guarantee that either the people or the politicians will be aware of the risks and determined to abate them. The disaster planner must be able to identify situations (and Senise was one of them) in which disaster is "waiting to happen", to analyze them rigorously, and to present convincing information on them to political authorities and citizens' groups.

Source of information: D. Alexander, 1991. Applied geomorphology and the impact of natural hazards on the built environment. *Natural Hazards* **4**(1), 57–80.

map, flagged, and an alternative route planned. This technique is especially important because, during a sudden-impact disaster, information tends to arrive at the emergency-operations centre gradually over time. Early reports may turn out to be unfounded or inaccurate and hence later versions of the map will show corrections. In addition, it takes time to establish the boundaries of the impact area and the critical points at which damage is concentrated. Detailed computer maps are helpful here because the scale can be altered by zooming in or out to show more detail or a wider area.

It is clear that impact maps cannot be constructed until after a disaster has occurred, but it is still important to plan for them before it happens. A typical impact map will show aspects such as floodable areas, places invaded and damaged by landslides, houses collapsed in earthquakes or major explosions, or areas burned by fire. After earthquakes, it is usually necessary to conduct rapid but reasonably accurate surveys that place housing into three categories:

- intact or lightly damaged and therefore inhabitable
- repairable but in need of evacuation until repairs can be carried out
- dangerous, in need of permanent evacuation and demolition or rebuilding.

On maps that are detailed enough to show individual houses (e.g. at the 1:5000 or 1:2000 scales) these categories are often represented by green, yellow and red shading, respectively.

The emergency planner needs to be able to foresee the need for impact maps. This means at the very least ensuring an adequate supply of basemaps that have the appropriate features clearly and unambiguously marked on them and are printed at a scale appropriate to the impact-mapping exercise. Such maps must be robust and manageable in the field, easily reproduced (e.g. as blue-prints or photocopies) and easy for the trained operator to interpret. In synthesis, impact mapping is widely used, not merely as a guide to where to direct post-disaster recovery operations but also as part of an archive on damage and losses in past disasters that will help predict their future pattern. For example, after a tsunami has damaged a coast, studies of areas inundated by the waves ("run-up studies") can be conducted, including the compilation of maps of damage, depths of flooding, scour effects by running water, and the extent of inundation. Hawaii and Japan both have archives of run-up maps that go back for decades and collectively give a clear picture of exactly where along the coasts tsunamis of various magnitudes are likely to invade.

One other aspect of emergency and impact mapping deserves to be considered here. A few researchers have studied the perception of geographical space by relief workers in disasters. In one instance, a fire chief arrived at the scene of a major train crash and toxic spill, and drew a sketch map of the site of the incident, the area likely to be affected by an explosion or the diffusion of toxic chemicals, and the system of approach roads. The researchers were interested to compare, with hindsight, the perceptual map sketched out by the fire chief with the actual spatial pattern of the items that he had mapped. There were some surprising differences, although based on factors that are well known to students of **mental mapping**, as the process is known.

We tend to have a somewhat distorted view of geographical space: in general terms we expand areas we are familiar with and fill them with details, and we compress unfamiliar areas and deprive them of details. Often this leads us to expand nearby areas and compress distant ones when we draw perceptual maps. It is a useful training exercise to give participants a fairly detailed scenario and then get them to map it without basemaps and with only a minimum of predetermined spatial referents. The resulting mental map can be very revealing about an individual's perception of space and place. It will probably reflect the person's training (e.g. geologists may show rock outcrops or geological hazards such as landslides) and his or her familiarity with the area and its details, which are not likely to be identical for any two people.

In synthesis, perceptual maps have little direct use in emergency planning but potentially great value in emergency management. However, it should be borne in mind that perception is an important component of the design and execution of any map.

Reconstruction mapping is the last significant form of cartography for emergency planning and it is dealt with in §7.3.

CHAPTER 3

Methodology: analytical techniques

3.1 Modelling

Models simplify reality in order to render it comprehensible. Well constructed, well thought-out models are invaluable aids to understanding the process of disaster. Moreover, they are flexible, in that changes in basic assumptions (or inputs) can lead to changed outcomes, a process of trial and error that allows one to investigate situations safely and efficiently. The three main kinds of model are conceptual, physical and numerical (which generally means digital). We will deal here with conceptual and numerical models, as physical hardware models have little direct application in emergency planning and management.

Successful modelling simplifies reality down to its essentials without losing credibility. Thus, it is a trade-off between including detail in order to make the model more realistic, and excluding it to preserve functional simplicity. The simplest model contains no details and merely connects inputs with outputs: a **black-box model**. But most models are of the **grey box** variety, because they include a certain amount of carefully selected detail. In addition, models can be subjected to **sensitivity analysis**, in which input variables are subject to controlled variations, in order to assess how output values respond to the changes. If the model is unstable, small variations in the inputs will cause inordinately large variations in outputs, so that sensitivity analysis can also be used to verify the model's stability under a series of carefully chosen conditions.

Perhaps the greatest pitfall of modelling is the "garbage in, garbage out" principle, in which bad input data generate unreliable outputs. This is often a more significant source of error than that offered by the model's internal processes. In summary, all forms of modelling involve some essential artificiality, but this needs to be constrained to certain reasonable limits by careful use of data and their transformation in the modelling process. And with respect to data, one should note the difference between variables (streams of variable data points, or variates), parameters (values that remain fixed for the duration of the **exercise**, or for a part of it) and constants (values that never vary).

41

3.1.1 Scenario methods and games simulation

In the context of disasters, a scenario is a hypothetical progression of circumstances and events designed to illustrate the consequences of some decision, action or impact. It is usually designed to provide a logical answer to the question "What would happen if . . . ?", although scenarios have also been used for backcasting rather than forecasting (i.e. as a means of hindsight review of events, not to forecast them).

A scenario is a model of conditions and circumstances, and it is usually designed to illustrate the connection between the two: how conditions influence circumstances and how circumstances alter conditions. Scenario methods can be used to show the consequences of disaster impacts, of mitigation decisions, or of post-disaster search-and-rescue strategies. Vulnerability can be studied by constructing scenarios of how it interacts with hazards to produce risks and disasters. Scenarios of losses in disaster can be modified on the basis of scenarios of risk-reduction measures and how they perform during impact. Table 3.1 lists some of the uses of scenario methods in emergency preparedness.

It is obviously essential to construct scenarios as plausible and reliable

Table 3.1 Some uses of scenario methodology in civil protection and disaster prevention.

Mitigation phase
- Scenario building as a training method for emergency planners.
- Assessment of the vulnerability of structures and communities by exploring their susceptibility to damage and destruction.
- Reconstruction of past disaster impacts and responses in order to learn lessons for future preparedness.

Preparedness phase
- Scenario building and exploration as a training method for emergency-response personnel.
- Study of the probable future effect of hazards by building conceptual models of impacts and responses.
- Scenarios for the probable performance of monitoring and alarm equipment.
- Warning and evacuation scenarios employed to design warning and evacuation systems.

Emergency response phase
- Study of the efficiency of emergency responses and disaster plans by exploring the progress of future post-disaster emergency operations.
- Logistical scenarios to estimate the efficacy and viability of certain relief operations.
- Post-disaster scenario reconstruction for debriefing emergency personnel.

Recovery phase
- Use of scenarios to estimate the magnitude and location of damage in future disasters for mitigation purposes and to calculate repair needs.
- Economic-scenario modelling of post-disaster employment and unemployment patterns.

Reconstruction phase
- Use of scenarios to model the social and economic conditions that either inhibit or facilitate reconstruction.
- Financial and fiscal planning of reconstruction using economic scenarios.

Source: Based on "Scenario methodology for teaching principles of emergency management", D. E. Alexander, *Disaster Prevention and Management* **9**(2), 89–97, 2000.

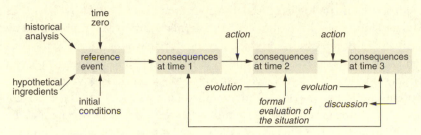

Figure 3.1 Temporal evolution of a scenario of an emergency situation.

chains of events. This can often best be done by using a systems approach (Fig. 3.1). The inputs are specified as a set of conditions and forces that provoke change. Further conditions are specified within the system; these constrain the mechanisms of change and together with them constitute the system's **forcing function**. The output from the model is the scenario as constructed and the changes that it embodies.

Scenario methods are invaluable to emergency planners because they can be used to clarify both the conditions that necessitate planning (i.e. hazards and their probable effects) and the impact of preparedness in terms of hazard and risk reduction. Indeed, studies of hazard, vulnerability and exposure to risk can be integrated into the development of scenarios for disaster losses and casualties. These can then form the basis of loss-mitigation measures and emergency planning (Fig. 3.2).

The first step in the construction of a scenario is to specify the inputs and initial constraints, or **boundary conditions**. For example, one may wish to

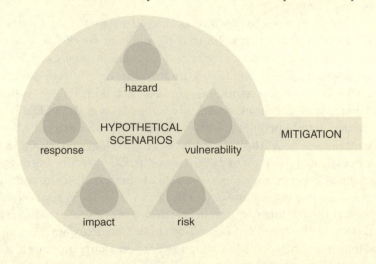

Figure 3.2 Use of scenario methodology in the mitigation of hazards.

develop a scenario for losses caused by a magnitude 7 earthquake, which forms the hazard input that drives the scenario. Initial conditions are represented by the vulnerability of the building stock, aggregate patterns of activity at the time when the earthquake strikes, and ability to respond to it in a timely way with adequate emergency forces. Development of the scenario is governed by the temporal unfolding of events. Vulnerability data may tell us that a certain proportion of houses may collapse and a certain number of people may be killed. One can then examine the relief efforts, or assess what would happen if preparedness levels were improved.

The scenario proceeds through a series of stages, which are usually defined by events (e.g. from emergency isolation to search-and-rescue operations and then to the end of the emergency phase and the start of the recovery process). At each stage it is possible to take stock and "freeze time" in order to obtain a cross section through the situation and examine progress made, and problems faced, by the major actors (e.g. disaster managers, rescue brigades, medical doctors, and so on). Once the full progression has been worked out, and a final outcome has been obtained, one can start to vary the scenario by asking such questions as "What if we double the number of relief workers?" or "What if the earthquake causes landslides that block the main roads needed for access by emergency services?" The answers are then obtained in terms of modifications to the outcome of the scenario.

It is highly recommended that the emergency planner construct written scenarios, if necessary with help from experts in particular fields, for all major risk conditions that fall under his or her jurisdiction. These will illuminate the situations that need to be planned for and will set the parameters of the emergency-planning process.

For most *a priori* scenario construction there are no means of verifying the results, unless the exercise be followed by a similar set of real events (the scenario akin to a self-fulfilling prophecy). One therefore seeks **verisimilitude** in the plausibility of facts, conditions, mechanisms and outcomes within the scenario. While constructing this, one must ask "Is this likely?" or "Is this possible?"; implausible scenarios, or unlikely turns of events, should be avoided. So, of course, should prejudice, fantasy, irrationality and excessive detail. A good written scenario will take the reader clearly and logically through a series of events, explain the pattern of decisions and actions, describe outcomes and justify them with logical reasoning. It will stick to the broad picture and will not indulge in long digressions into detail. Its basis will be factual data on hazard and risk, and also the turn of events in past disasters of a similar nature. Hence, scenarios should be based on meticulous study of source data. The outcome should thus be entirely plausible, although not necessarily immune to debate.

The scenario for a future repeat of the 1923 Great Kanto earthquake in Japan

44

gives the following figures for losses: 40 000–60 000 deaths, up to US$1.2 trillion in property losses, lifeline damage valued at up to $900 billion and business interruption losses of up to $1 trillion. However, much depends on the assumptions used in developing the scenario: the figures have been criticized as being too high. Nevertheless, they do suggest how much is at stake, as overall losses could conceivably equal half of Japan's annual GNP, and world financial markets could go into recession as a result of the earthquake. Most scenarios are likely to be much less grandiose than this, so let us consider a simplified, hypothetical example. For the sake of brevity, the background detail is omitted.

At 16.00 h on a Friday afternoon in May, an earthquake of magnitude 6.7 occurs. The shaking lasts for 33 seconds and generates surface horizontal acceleration of up to 0.31 g. Damage is serious, although not universal, in a town of 50 000 inhabitants. To begin with, a 200 m section of a viaduct carrying a motorway has collapsed onto a parking lot, crushing vehicles and possibly some of their occupants. Secondly, brick and concrete-panel cladding has fallen off the façade of a department store and is blocking the town's main street with rubble. Again, there may be victims underneath. Thirdly, a steel-frame warehouse in the town's industrial area appears to have collapsed (it was in use at the time of the tremors) and an apartment complex occupied by retired people has suffered partial collapse. Fires have broken out at three locations around town where gas mains have ruptured and escaping gas has ignited. Water mains have ruptured and water is gushing out at high velocity across the main street. Fire hydrants are beginning to run dry. The absence of **mutual aid** suggests that neighbouring communities have also been badly affected by the earthquake. Lastly, there is only one general hospital in the immediate area, and its capacity to take many injured victims is limited by perilous structural damage and the absence of key personnel.

The damage described in this scenario is clearly based upon vulnerability surveys that have been carried out with a given magnitude of earthquake in mind (and other given seismic coordinates and parameters). The description refers to a point in time about half an hour after the tremors. It is not complete because at this time much information will be lacking; for example, the number of dead and injured will not be known. Further development of the scenario depends on a series of assumptions about how the relief effort will proceed, and what the situation is, with respect to homeless people, living casualties and trapped victims. The result is a graphic expression of the sudden need for resources, something that the emergency plan will have to sort out. In doing so, it will in effect develop the scenario towards resolving the problems that it has initially set up.

One possible accessory of the scenario method is **games simulation**, although it is seldom used as such. In this, the scenario is developed by a

Box 3.1 A future eruption of Mount Vesuvius

Although Mount Vesuvius in southern Italy has not erupted since 1944, it remains one of the world's most dangerous volcanoes. At least 600 000 people live on its flanks, and the city of Naples (population 1 million) is only 8 km away from its summit. So great are the risks associated with a possible future eruption, and so large must be the needed response, that emergency planning has been coordinated at the national level.

Volcanic hazards

The eruption of AD 79 lasted two days and destroyed the towns of Pompeii, Herculaneum, Oplontis and Stabiae, with the loss of 18 000 lives. It initiated a long period of intermittent eruptive activity, which ceased in 1139. Another violent eruption, although less so than that of AD 79, began on 15 December 1631 and continued for some weeks. Pyroclastic flows (fluxes of superheated gas, ash and rock debris) swept down the flanks of the volcano and killed 4000 people in the town of Portici at its south-western margin. Smaller eruptions occurred intermittently until 1944, when the last eruptive cycle ceased. It is generally thought that the longer the interval before the next eruption, the larger and more violent this will be. Although micro-earthquakes and emissions of gas and steam occur on Vesuvius, at the time of writing the volcano is quiescent. But in scientific terms it remains one of the most intensively studied and monitored volcanoes.

The scenario for emergency planning is based on the largest expected event, which is deemed equivalent to the 1631 eruption in terms of energy released, volume of magma emitted and pattern of eruption. Volcanologists term this a medium-intensity event (sub-Plinian or Vesuvian according to classifications of its explosive power). Pyroclastic flows would probably occur on the southern and perhaps eastern flanks of the volcano.

Zonation

In 1994, vulnerability studies were carried out on the circum-Vesuvian area and in 1995 the main emergency plan was published. This divides the area into a red zone of about 200 km^2, where the worst effects of an eruption would be concentrated. In 1631, about one fifth of this area was affected by pyroclastic flows. The yellow zone 1100 km^2 would suffer accumulations of fallen ash on roofs to weights of more than 300 kg per m^2, which is the average threshold for collapse. In 1631, about one tenth of this area suffered major ash accumulations, but in a future case the location of impact would depend on the height of the column of erupting ash and the direction and strength of prevailing winds. Finally, the blue zone covers 100 km^2 and is the area where both major ashfalls and lahars (highly destructive volcanic mudflows) may occur.

Detailed volcanological modelling has established a full sequence of events for a probable major eruption, although it is unable to predict the precise locations or timing of impacts. Nevertheless, it is believed that earthquakes, gas emissions, heat fluxes, swelling of the ground surface and other physical phenomena would precede the eruption, probably by some days, and thus enable short-term preparations to be made. These will be essential, as 600 000 people (comprising 172 000 families) live in the red zone.

For the vulnerability analysis, the 18 municipalities of the circum-Vesuvian area were divided into 120 microzones, which were each assigned vulnerability ratings on the basis of their concentrations of high-risk buildings. For example, the urban centre of Somma Vesuviana, a municipality on the southeast side of the volcano, contained 1738 buildings, mostly of modest size (only 5 per cent exceed 15 m in height). About one third of the buildings were classed as high-vulnerability structures (i.e. liable to collapse in a major eruption). These were concentrated in particular places at densities of up to 80 per cent.

Box 3.1 (continued) A future eruption of Mount Vesuvius

The emergency plan

The principal problem of planning for a major eruption of Vesuvius is how to safeguard the populations of the 18 circum-Vesuvian municipalities. These towns include Portici (population 67 500), which has an urban density of more than 17 500 people per km^2, a figure equalled no where else in Europe (even central Milan has a figure of only 3900). The problem is rendered particularly difficult by the narrowness of streets and the chronic congestion everywhere. At Somma Vesuviana there is an average of two cars per household, but a notable lack of space to drive and park them.

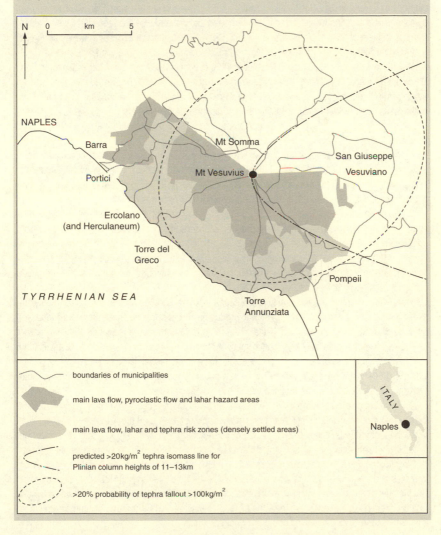

N

0   km   5

NAPLES

Barra

Mt Somma

San Giuseppe

Vesuviano

Mt Vesuvius

Portici

Ercolano
(and Herculaneum)

Torre del
Greco

Pompeii

T Y R R H E N I A N   S E A

Torre
Annunziata

boundaries of municipalities

main lava flow, pyroclastic flow and lahar hazard areas

main lava flow, lahar and tephra risk zones (densely settled areas)

predicted >20kg/m^2 tephra isomass line for
Plinian column heights of 11–13km

>20% probability of tephra fallout >100kg/m^2

ITALY

Naples

Box 3.1 (continued) A future eruption of Mount Vesuvius

As the region of Campania, in which Vesuvius is situated, would be unable to cope with a flux of 600 000 refugees, they will be evacuated by private car, train, bus and ship to 18 of the other regions of Italy in quantities of 20 000–25 000 per region. People from Somma Vesuviana, for example, would go to the nearby region of Abruzzo. Inhabitants of the red zone would be evacuated automatically after a short-term prediction of an eruption. People from the other two zones would be evacuated according to needs posed by the developing risk situation. In each case, residents have been asked to specify their mode of evacuation. At Somma Vesuviana, 8971 families have indicated that they will leave by car and 22 154 people expect to be taken away by bus. This would involve more than 3000 departures a day for a period of one week, and so it is to be hoped that the lead time between identification of the threat and onset of the eruption will be sufficient. Departures are timed to occur in waves and to be managed through special traffic-flow patterns on roads, motorways, railways and sea lanes.

A field exercise was conducted in November 1999 to evacuate between 500 and 600 people from Somma Vesuviana to the town of Avezzano in Abruzzo Region farther north. It was successfully accomplished, and it inaugurated a process of twinning between Vesuvian communities and towns and cities in other regions that will receive their inhabitants as evacuees.

Problems

Ever since it was first thought of, the Vesuvian emergency plan has been vigorously criticized in both a constructive and a negative sense. Detractors who have little that is positive to offer have simply argued that the attempt to evacuate between 600 000 and a million people from Europe's most congested metropolitan area would inevitably be doomed to end in chaos.

Some volcanologists have argued that the reference scenario is wrong. Rather than the 1631 sub-Plinian eruption, they argue, we should be preparing for a repeat of the AD 79 Plinian event, which produced a vertical blast up to 50 per cent higher than that of the later eruption. They also question the prediction of wind directions and strengths, which are vital to hazards associated with pyroclastic flows and ashfalls. Engineers have suggested that the threshold value for roof collapse during ash accumulation should be lowered to 100 kg per m^2.

Whatever the volcanological reality of the situation, there is no doubt that the circum-Vesuvian municipalities have been reprehensibly slow to produce emergency plans. These therefore cannot yet be worked up into a coordinated whole that will ensure the emergency is managed efficiently at the local level. Moreover, little has been done to improve the area's infrastructure with a view to facilitating quick evacuations. This is worrying, as volcanologists have suggested that a failed evacuation could lead to 15 000–20 000 deaths, mostly in the densely populated communities on the southern and eastern flanks of the volcano where eruption hazards are greatest.

The 1999 simulated emergency at Somma Vesuviana was a groundbreaking (if you will excuse the pun) but modest development. Although successful, most aspects of the plan remain hypothetical and untested. Indeed, such is the magnitude of operations that it is hard to see how they could be tested in anything other than a genuine red alert. Many people have grave doubts about their efficacy.

Notwithstanding these reflections, it is clear that emergency planning is absolutely necessary on Vesuvius and that it must be pursued vigorously whatever the chances of success are. It is important that the plan be flexible enough to accommodate changes in scientific, logistical, social and economic factors that might affect it. To its credit, the Italian Agency for Civil Protection, which masterminded the plan, is keeping an open mind about the whole problem.

Source: Various articles in *Nature* and the *Bulletin of Volcanology and Geothermal Research*. DPC Informa no. 19, 1999. Newsletter of the Department (now Agency) of Civil Protection, Rome.

process of interactive modelling, in which chance plays an important role. Games methods have to be designed for the situations they are intended to represent and are primarily used as a means of teaching roles to participants. First, a basic scenario is postulated. Next, roles are assigned to participants, who will act as disaster managers, firemen, doctors, and so on. Then, resources are allocated to the players, and constraints are placed on their activities. Objectives are defined (e.g. to rescue and care for all living survivors). Play begins and the game develops much as a scenario would (and for this it must be monitored as it progresses). Random numbers or throws of dice can be used to simulate chance occurrences (the numbers can be equated to specific circumstances). Alternatively, unexpected developments can be described on cards, which are then selected randomly from a pack. Figure 3.3 shows a map used in a **tabletop** simulation game in which participants manage an emergency caused when a landslide that dams a river in an Alpine valley causes a severe risk of flooding in a large town down stream. The example is hypothetical but based on various real examples of such emergencies.

Whereas in most forms of game the objective is to compete, in games simulation for emergency management it is to achieve the best sort of outcome by collaboration. The exercise teaches participants how to do this and also improves their ability to negotiate constructively. It helps the organizers to define participating groups, mediators, working relations, rules, goals, roles,

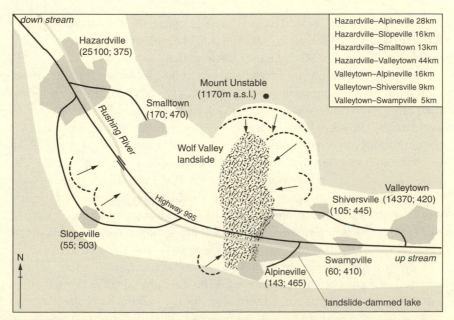

Figure 3.3 Map of a scenario for a tabletop emergency simulation game involving a landslide in an Alpine valley, which dams a river and creates a serious flood risk down stream.

strategies, policies and tactics. At appropriate stages in the game, the monitors can stop play and take stock of the progress of the scenario that it represents. In this, the role of chance can be evaluated. Playing the game repeatedly will not only help participants familiarize themselves with roles and relations, and the scenarios that the game represents, but will also permit the limits of chance factors to be established. This will make it easier to plan for the unexpected.

At this stage the emergency planner may feel that this may be all very well, but does games simulation actually offer any advantages as a support method for disaster planning; and, in any case, is it feasible? The answer may depend on the planner's ingenuity. There are remarkably few examples of prepackaged commercial games that can be used in emergency training. Hence, it is up to the individual to devise one. Recreational games such as Monopoly™ offer some inspiration. Indeed, computer strategy-games may one day inspire someone to produce a problem-solving CD-ROM game for emergency managers. In the meantime, a couple of dice, some blank cards and a board may be all that is necessary to start organizing players into groups (e.g. firemen, ambulance drivers, and so on) and get them competing against time and many difficulties in order to save lives in a hypothetical disaster.

3.1.2 The Delphi technique and expert systems

In games simulation the emphasis is usually placed on training novices, but there are also formalized ways of making use of experts. One is the Delphi technique, whose name comes from the Oracle at Delphi in ancient Greece.

Although discussion is a very useful aid to understanding, at times it can inhibit thinking and constrain less assertive discussants to contribute less than their more loquacious colleagues. The Delphi technique has been designed to avoid this by eliciting written information from a panel of experts who must not communicate between themselves. Each member of the panel is given a questionnaire and a limited amount of time to write answers to the questions. The scripts are collected and analyzed on the spot. The panel is then given a digest of the results and a new questionnaire that builds upon the results of the first one. The procedure is repeated until sufficient information has been elicited from respondents. Information supplied to the group can be controlled and guided in order to make the response efficient and avoid going off at a tangent.

It is fair to say that the Delphi technique has been underutilized in disaster planning and emergency management. With the right mix of experts, it offers a powerful means of gathering information for use in the emergency-planning process. Moreover, it could be given a new and more efficient emphasis via the Internet. This means that there is no longer a compelling reason to gather together a panel of experts in the physical sense if they can be convened via computer networks. The rapidity and relative efficiency of electronic mail and chat procedures mean that the whole exercise can be accomplished with the members of the panel sat at their computer workstations and connected to the convenor by modem.

Whether the methods used are traditional or are based on computer networks, the questionnaire will refer to specific needs connected with the local emergency-planning situation. However, in general it is appropriate to seek information from the panel on such matters as the following:

- What are the most significant hazards or combinations of them?
- What scope is there for monitoring and predicting the local hazards and how could this best be done?
- What are the principal sources of vulnerability in the local community and where are they located?
- What are the principal emergency-management needs?
- What are the inadequacies of the current system for managing emergencies and what is the best way to tackle them?
- What new resources are needed in order to manage emergencies better and how might they be acquired?

In many cases the panel should be chosen in such a way as to represent a broad range of approaches and opinions, both physical and social, as is needed. In this manner, the Delphi technique can be used to achieve a consensus on the local problems of emergency management and the best way to solve them. Successive questionnaires will become progressively more focused on the key issues and will elicit more and more detail. Furthermore, the method will often highlight unexpected aspects that can be followed up and investigated. If the initiative proves to be highly successful and the panel of experts is sufficiently authoritative, the Delphi technique can be combined with scenario modelling. The experts can be fed a risk or disaster scenario and a series of questions on how it should be managed, what resources and what type of organization are needed to manage it, and so on.

A second means of utilizing high-level knowledge is to devise an **expert system**. This is a computer method based on the management of information by means of formal logic. The expert system consists of a knowledge base, a set of rules and a series of decisions, accompanied by appropriate explanations. In simple written form, the knowledge base sets out all vital information about the situation at hand. It poses a series of problems with answers that will depend on particular combinations of conditions. Appropriate combinations are set out in the rules component of the system, which uses logical operators (IF, THEN, AND, etc.) to link the various strands of reasoning. The user works his or her way through the questions, chooses from among the options posed by the conditions, and is supplied with a series of suggested answers to them, with appropriate explanations.

In very simple terms, a small part of an expert system for the management of earthquake damage might run as follows:

- Question 1: Should a house damaged by earthquake be demolished?
- Question 2: Is the house damaged beyond repair?
- Question 3: Is repair economically and technically feasible?
- Decision 1: Do not demolish.
- Answers: [1] Yes, [2] No
- Rule 1: IF q1a2 AND q2a2 AND q3a2 THEN decision 1

Explanation: Demolition is not appropriate when rehabilitation is technically feasible and can be accomplished at moderate cost.

In reality, expert systems tend to be much more complex than this because they create webs of interdependent decisions. If well formulated, they offer an invaluable support to complex decision-making processes. However, expert systems for emergency planning are not commercially available and must therefore be developed by the user if they are required. A possible first step would be to download an expert-system construction program from the Internet. For example, WinExp, a small expert system program, is available from various sites as shareware. It is designed to teach the user how to construct the rules that will serve in making an expert system.*

3.1.3 Computer methods

Computers are playing a rapidly increasing role in the process of emergency planning and management. We have already seen how computer cartography and spatial data manipulation have come to the fore. Nowadays, several kinds of commercial and non-commercial software are available to help manage disasters. Given their utility, it is opportune to allow for their use when planning how emergencies will be managed.

Typical emergency-management support software will include the following components:

- *Databases* These include descriptions, lists, numerical data, charts, diagrams, maps and tables of information selected for its usefulness to the local emergency manager. Typically there will be:
  - basic, local information on different kinds of hazards

---

* See *Building models of conservation and wildlife*, A. Starfield & A. Bleloch (Edina, Minnesota: Burgess, 1991). This text explains the expert-systems model and is compatible with the WinExp program.

- route maps, street maps and building layouts, possibly a GIS
- a contact list with communications procedures, perhaps for the modem-based dialling of emergency personnel, with names, addresses, numbers and competencies
- lists of acronyms and organizations relevant to disaster management
- explanations of procedures for particular emergency situations.

- *Duty log* A **duty log** is a schema for tracking events and activities during the emergency period. It can be studied after the event in order to expose the weaknesses and show the strengths of the management response. For example, a record can be kept of telephone conversations and periods of duty.
- *Spreadsheets* These can be used to keep inventories of available resources such as fuel stocks, food, water, blankets, camp beds, heavy earth-moving equipment, medical supplies and so on.
- *Word processing* A means of writing messages and importing supporting documentation from disks and diskettes.
- *Data interchange* A modem- or network-based module allows for messages and other information (including graphics) to be sent and received within the confines of the program.

Basic versions of such programs generally do not supply all of the necessary information, which must therefore be provided by the user. What the program does is to offer a framework for the accumulation, classification, exchange and management of information. It will be useful for keeping a running inventory of resources and personnel, and as a manual for procedures and responsibilities. It will also help make decisions about the availability of routeways.

Computer programs can be integrated with emergency plans to the extent that they provide vital support for each plan, help develop it, and enable it to be diffused to all offices or organizations that form part of the computer network, onto which the plan is loaded (Fig. 3.4). However, it is important that the processes of planning be constrained neither by the limitations of the program nor of any hardware platform that runs it.

Several general principles underlie the use of computers in disaster planning and management. To begin with, the disaster-response environment supplies information for the emergency planner or disaster manager to interpret and utilize (Fig. 3.5). Information must be accumulated (as inputs to the computer model), updated whenever necessary, and edited to ensure accuracy, usefulness and conformability to specified norms. It must then be read into a data-analysis module, which will be connected to a module that allows the retrieval and display of the results of the analysis. The process of analyzing data is that which best connects the hazard or disaster environment with the planner or manager. The purpose of a computer analysis is often to collect, arrange and select data in such a way as to produce a simple, clear, easily interpreted image

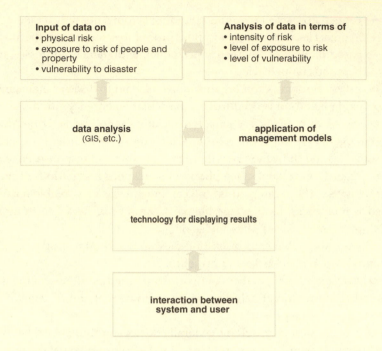

Figure 3.4 Using computers to help analyze and manage risk of disaster.

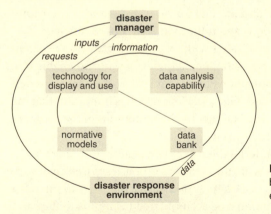

Figure 3.5 The role of computer-based information technology in emergency decision making.

of a particular situation – in short, a model. The planner manipulates the data by selecting a criterion: for example, "show me all functioning hospitals with **casualty** reception facilities in the area", or "select the best route from the scene of the disaster to the principle medical centre".

The advantages of computer methods include the ability to deal with and synthesize large amounts of datasets in minimum time; ability to update stored data rapidly and efficiently; ease and flexibility of display and printout; and portability, in that large volumes of data and complex programs can easily be transported from place to place

Nowadays, user-friendly interfaces make it easy to use the programs. The main disadvantage is that the computer environment is an artificial one and it can make the processes of decision making artificial by removing the decision maker from direct contact with the hazard or disaster environment. Nevertheless, in the future, computer methods will play an increasing role in disaster planning and management. Moreover, the models will increasingly be integrated with other facets of the information technology: the Internet and its modes of communication, satellite data, geographic information systems, online databases, and so on. Thus, the intelligent use of computers is now a very significant challenge to which the emergency planner must rise.

3.2 Risk analysis

As noted above, hazard, vulnerability and risk are complementary aspects of the same phenomenon: the interaction of physical forces with human or environmental systems. The concepts overlap somewhat and are often used in confusingly interchangeable or ambiguous ways.

Risk is essentially the product of:
- hazard (the danger or threat that a physical impact will occur)
- the vulnerability of items threatened by the hazard (people, buildings, etc.)
- their degree of exposure to danger (Fig. 2.6).

Aggregate risk is a function of the infrequency, but serious consequences of, high-magnitude events (Fig. 3.6). Specific risk is regarded as the likelihood of damage or loss multiplied by the number of items at risk. Exposure is the amount or proportion of time that an item is threatened by partial or total loss or damage. For example, the exposure level of tourists and residents in an area may differ, as the former are merely transient visitors.

Broadly speaking, risks can be classified as voluntarily or involuntarily assumed, and concentrated or diffuse. Although there is an apparent element of choice in those risks that are regarded as involuntary ones, it is much more restricted than in risks that are voluntarily assumed for the sake of amusement or recreation. Hence, the journey to work involves exposure to what are regarded as **involuntary risks**, whereas rock climbing and hang gliding are clearly voluntary risks. The classification is somewhat arbitrary, but it can be justified if the criteria on which it is based are fully spelled out. Among other risks, smoking and failure to wear seatbelts in cars are examples of diffuse risks, whereas tornadoes and building fires generally represent concentrated risks.

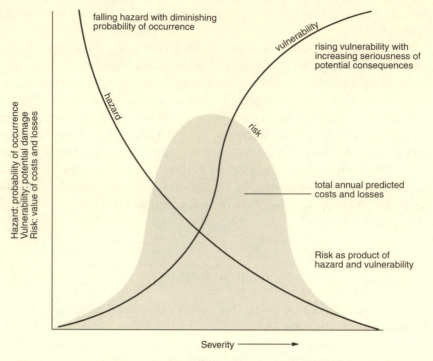

falling hazard with diminishing
probability of occurrence

vulnerability

rising vulnerability with
increasing seriousness of
potential consequences

hazard

risk

Hazard: probability of occurrence
Vulnerability: potential damage
Risk: value of costs and losses

total annual predicted
costs and losses

Risk as product of
hazard and vulnerability

Severity ⟶

Figure 3.6 How hazard and vulnerability combine to create aggregate annual risk.

Again, the classification is somewhat arbitrary. Rather surprisingly, this is also true of the distinction between natural and anthropogenic hazards. Current wisdom suggests that many natural hazards involve artificial risk, as in the case of an earthquake that can be survived in an anti-seismic building, which is proof against collapse, but not in an **aseismic** one, which is shaken into rubble. In this respect, anthropogenic vulnerability predominates over natural hazard. In addition, technological hazards such as dam failures or toxic spills may be directly caused by natural impacts, such as earthquakes or floods, which adds further weight to the argument against excessive differentiation of risk by typology of cause. The logical conclusion is that "natural" hazard is a convenience term rather than a reality.

Human perceptions of and attitudes to risk are far from complete and rational. It is as well to remember this in emergency planning, as the view of risks often governs how they can be managed and how far they can be reduced by the planning process. The perils of natural or technological disaster are part of a wide variety of risks that people must face up to every day. Indeed, no significant human activity is entirely free from risk. In terms of estimated or calculated levels, the risks of disaster are generally low to moderate. Hence, the risk of dying in a small road, rail, air or sea-voyage crash is generally somewhat higher than that of dying in a major transportation catastrophe. However, disasters tend to receive more dramatic publicity than the "background mortality" of minor accidents. As a result, there is a very high demand, for instance, for civil aviation safety, and the expenditures on it tend to exceed the benefits in terms of lives saved, at least when compared with how such money could have been spent to save lives by reducing other less spectacular risks. In any case, the period after disaster is one in which there is a brief increase in demand for safety. With good planning and timely expenditure, the risks can be reduced before public and political interest lapses. So predictable is this process that it may be fair to state that most advances in disaster mitigation stem from the impetus of particular catastrophes of the past, which exposed the inadequacies of efforts to mitigate and prevent them.

In more practical terms, many means are available to the emergency planner to tackle the problem of risk: estimation, analysis, management and communication. Let us consider a few of them individually.

The simplest way to estimate risk is to relate the distribution of physical hazard to that of human vulnerability and to give the resulting pattern a temporal framework. Risk then becomes a simple likelihood (e.g. high, medium or low – see Fig. 3.7) or numerical probability that casualties, damage or losses will occur with in a given interval of time or with a given return period. For example, one may conclude that there is a risk of severe damage to 25 per cent of housing stock at modified Mercalli intensity IX in an earthquake of magnitude

	Low	Medium	High
Hazard			
Vulnerability			
Exposure			

Severity	Probability of occurrence				
	Impossible	Improbable	Occasional	Probable	Frequent
Negligible					
Marginal					
Moderate					
Serious					
Catastrophic					

Risk level

Acceptable   Significant   Critical

Figure 3.7 Risk assessment matrix.

6.5 and recurrence interval of 50 years. In such a case, one uses the timescale of the physical hazard to describe the risk interval. Hence, **risk estimation** involves developing simple scenarios for hazards and vulnerabilities to them, and gauging the level of exposure of vulnerable items. A useful measure of this process is the aggregate risk of disease and fatal illness (Fig. 3.8). This represents the upper limit of **tolerable risk** for society, even though people would like to see it reduced significantly. Events such as a rare but cataclysmic volcanic eruption are a special case. As noted above, they constitute risk of high consequence and low probability – difficult to plan for because it is unlikely.

Risk analysis involves comparison of different risks, investigation of their causes and refinement of estimates with longer-term or more precise data. One assumption inherent in most analyses of the risks of disaster is that these must be viewed in the context of overall societal risks. An area may be susceptible to floods and landslides, but there are also risks of death in car accidents or aviation crashes, or from specific diseases and conditions. There are risks of loss, not only from the natural hazards but also from unemployment or crime. Comparison may thus reveal that certain risks are much less significant than others, no matter how important they may seem when viewed singly. Causal investigations may help clarify why risks exist and indicate the methods by which they can be reduced. Finally, the analysis of data on risk levels may transform a vague qualitative idea of risk into a more precise quantitative probabilistic one. The probability approach may involve notions of **release** rate at which the hazard strikes), **exposure** (vulnerability of populations per unit time), **dose rate** (impact per person) and **background levels** (inherent natural risk levels).

The applied side of risk analysis involves **risk management**. In this, a

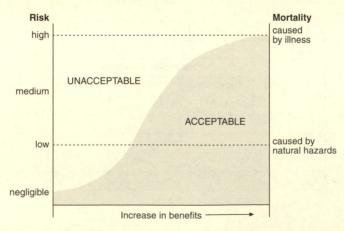

Figure 3.8 The acceptability of risk of death is proportional to the benefits to be gained from particular activities. High risks require substantial benefits to be acceptable. The risk of premature death from illness is a marker for the upper limit of tolerable risk levels.

strategy is developed in order to reduce and cope with societal risks in some rational explicit way. In one sense, that is what emergency planning is all about; hence, risk analysis and management are its fundamental underpinnings, whether they are carried out in quantitative or qualitative ways.

Emergency planners who want to manage risks that fall within their purview in a rational manner must first ask the following questions:
- What is the primary objective of the exercise?
- Which risks are to be managed?
- What resources are available to manage them? This includes resources of data and information.
- What approaches are economically, socially and politically acceptable?
- For whom are risks to be managed and who will be the main beneficiaries?

One of four main criteria can be used to determine the strategy. First, the utility condition causes one to try to maximize the aggregate welfare of all members of the community; that is, to spread the risk-reduction measures around so that their benefits are shared as widely as possible. Secondly, the ability condition leads one to reduce risks in relation to people's ability to bear them. This often results in the concentration of risk-reduction efforts among the poor and disadvantaged members of society. Thirdly, the compensation condition requires one to compensate for imbalances among the relative strengths of risks. This also necessitates the adoption of welfare measures among the poor and disadvantaged. Finally, the consent condition invites one to reduce risks only where there is support for the measures among the people who bear the risks. It is clear that these four criteria are not fully and mutually exclusive but represent different angles on the same problem.

In effect, the four conditions determine the ideology of risk management. Now for the practical strategy. One approach is to leave matters to market forces, which will impose unsustainably high prices upon risks that are in serious need of reduction. But such non-intervention is usually considered to be both inefficient and unfair to those who, through lack of resources, have little choice in the risks they face, which usually means the poor, who, paradoxically, may end up paying over the odds for improved safety. Beyond that, there are several forms of comparative approach. A simple one is to concentrate on reducing the risks that seem to be the most important ones: for example, floods but not earthquakes in an area of low seismicity but intense rainfall. In any case, emergency planners seldom have the remit to reduce risks that are not related to disasters and so the choices will be limited.

In this context, **cost–benefit analysis** is one of the most powerful tools for defining risk-reduction strategies, and for convincing politicians to fund them. Past catastrophes may furnish enough detail to be able to cost the future impact of given types and magnitudes of disaster. Estimates can be made, first of the

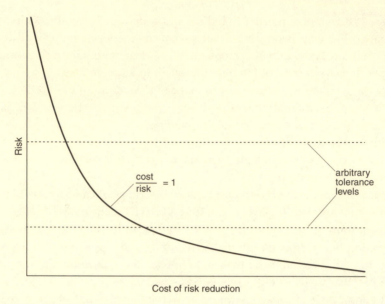

Figure 3.9 Economic rationality dictates that investment in risk reduction should cease at the point where the costs of reducing risk equal the value of losses avoided. However, society tends to impose arbitrary tolerance levels.

total cost of structural and non-structural mitigation measures, and secondly of the losses that would be avoided by using these to reduce the risks of the given impacts. With respect to natural disasters, the relationship between the magnitude of physical impacts and that of damage and destruction is non-linear: greater hazards involve disproportionately large losses. As a result, the ratio of mitigation costs to benefits in terms of damages avoided almost always favours the latter in ratios that vary from 1:1.25 to 1:1.9, assuming, of course, that the money is spent wisely and mitigation is successful. Cost–benefit analyses can also define the marginal point of diminishing returns, at which further expenditure produces less and less mitigation per unit of money spent (Fig. 3.9).

However, it is worth noting that economically rational methods such as these do not take into account what the public and authorities want by way of hazard management. There are several ways that this can be incorporated into risk management. For example, one can use social survey methods to reveal the public's preferences for risk reduction. The results are unlikely to be perfectly rational in terms of cost–benefit or other such criteria. Consider the following statistics and what they tell us about risk management preferences. In the USA as a whole the risk of death by leukaemia is 36 times higher than that of death by flood, 152 times the tornado death risk and 318 times higher than the risk of dying in a hurricane. Moreover, it is 47 times higher than the risk of dying in an earthquake in California. However, these figures are mere averages and they reflect assumptions about anyone's attempts to protect himself or herself against the risks, or about spatial variations in any of the risks. The data suggest that we should concentrate on reducing leukaemia risk rather than natural-hazard risk, and in general terms that is what has happened: more is spent on research on systemic cancers than on natural-hazard research, although it would be difficult to regard the sums as proportional to the relative risk. Moreover, leukaemia risks are probably much more evenly distributed than natural-hazard risks, which tend to concentrate where there are fault lines, floodplains, subtropical coasts, and so on. Hence, the average tornado risk is more than four times higher in the Midwest, where such storms are particularly common and devastating, than it is for the USA as a whole.

In the same vein, the overall US risk of dying in an automobile accident is 23 times higher than the flood risk, 95 times the tornado risk, nearly 200 times the hurricane death risk and 29 times higher than the likelihood that one would die in California in an earthquake. Compare this with the figures for aviation accidents in the USA: the flood risk is 22 times higher, the tornado risk is 5 times higher, the hurricane risk 3 times higher and the California earthquake risk 17 times higher than the likelihood of dying in an aeroplane crash. In relation to natural-hazard risks, there is thus a massive discrepancy between the high risks associated with automobile usage and the low risks of air travel. Of course, the statistics can be manipulated to give different results: for example, air travel looks much less safe in relation to road travel if conceived in terms of numbers of departures rather than distances travelled. Yet the expenditure on saving lives is not proportional to the risk, as the public perceives air travel as needing extra safety measures at almost any cost.

If one cannot design risk-reduction measures on a comparative basis, there are other rational procedures. For example, one can use professional standards or technical procedures to identify cases and situations in which risks are morally, ethically and technically too high to be tolerated. In synthesis, whatever the strategy used to manage risks, it is as well to make as explicit as possible the steps by which it is arrived at, and to list all criteria and assumptions that are utilized. One should bear in mind not only that risk analysis is a probabilistic and inexact technique but also that it involves uncertainties in both the input data and how they are interpreted. Thus, it is usual to have several outcomes based on different scenarios.

To conclude, in many cases emergency planning will not require a sophisticated analysis but merely a comparison of hazard and vulnerability maps and a modest amount of scenario construction. Nevertheless, it is a good idea to spend time examining as thoroughly as possible the assumptions on which it is based and developing a simple strategy to tackle different risks equitably. This will give justification and direction to the emergency plan.

Box 3.2 Analyzing the tornado risk in Massachusetts

Tornadoes are among the most destructive and common natural hazards in the USA, where about 1100 are recorded each year, including perhaps half a dozen that cause disasters. Nationally, the average annual mortality is about 60 per year, but the overall cost of damage is highly variable depending on how many major tornadoes go through large urban areas. It is generally assumed that Texas, Oklahoma, Tennessee and neighbouring states are most at risk, in part because these states have suffered major tornado strikes more often than other states, and in part because meteorological conditions tend to be more favourable to tornado formation in the lower Midwest and northern Gulf of Mexico area.

The New England state of Massachusetts is, of course, well outside this area. Compared with other US states, it ranks 35th for frequency of tornadoes, 16th for number of tornado-related deaths, 21st for injuries and 12th for cost of damages. Based on data from the period 1950–99 and in terms of numbers per square kilometre, it ranks first for fatalities, injuries and damage costs. The tornado risk is therefore deemed to be concentrated more powerfully in Massachusetts than in any other state, despite the reputation of Oklahoma and Texas as the leading states of "Tornado Alley".

On June 9, 1953, one of the most powerful tornadoes ever recorded in Massachusetts struck Worcester, the state's second largest city, and killed 94 people. The many who died in this single tornado and the small size of Massachusetts explain why it is top state in the tornado risk list. Although one could dismiss these data as a statistical anomaly, on a per-km^2 basis Massachusetts ranks 14th for frequency of tornadoes per unit area. This means that Massachusetts experiences more tornadoes per unit area than such states as Arkansas, South Dakota, and Wisconsin.

But how serious *is* the tornado threat in Massachusetts? Only a few deaths occurred in such storms in the 1990s and most tornadoes in the western part of the state were weak enough not to do any serious damage. Therefore, how seriously should the Massachusetts state authorities take the tornado threat?

George Bernard Shaw was not entirely wrong when he said that "there are lies, damn lies, and statistics". At the very least, the Massachusetts example suggests, first, that perhaps we can select ranks or ratios to fit whatever hypothesis we wish to promote, and secondly that risk analysis is as much a matter of judgement as it is of numbers.

The magnitude of the tornado threat in Massachusetts hinges on what interpretation one gives the 1953 event, the only major tornado disaster that the state has experienced since records began. On the one hand it is clearly an anomaly in the record, but on the other it could certainly happen again, although one hopes that 50 years of progress in meteorological forecasting and monitoring would allow a warning to be disseminated and people to evacuate vulnerable structures (no significant warning was broadcast in 1953).

In synthesis, it is clear that the tornado threat should be taken seriously in Massachusetts. The return period and future consequences of a disaster like the 1953 Worcester event are at present incalculable. Thus, it is probably prudent neither to overestimate nor to understate the tornado threat, but to regard it as generally moderate with a small chance of major disaster. Alarmism would not be warranted, but neither would complete lack of preparedness.

Sources: The Tornado Project On-line (http://tornadoproject.com); Anthony F. C. Wallace, 1956. *Tornado in Worcester.* Disaster Study 3, National Academy of Sciences/National Research Council, Washington DC.

The last aspect of risk to be considered is its communication to the public or other users of emergency plans. As noted, people tend not to have accurate perceptions of risk and often prefer not to think about the problem very seriously. People have difficulty in comparing widely different risks, and very few individuals have a clear and accurate idea of the range of risks to which they are subjected. When hazards manifest themselves, people tend to search for reassuring information and give preference to it over unsettling news, the so-called normalcy bias. However, responsible people do appear to want to be involved in the process of choosing risk-reduction measures. In fact, it is unlikely that these will succeed if they lack public and political support.

Research on **risk communication** suggests that ordinary people have difficulty in appreciating the meaning of probability in relation to danger. They also have a tendency to want to leave risk-reduction measures to qualified experts. However, if they are given simple concrete information on how to reduce their personal risks, they can often be induced to follow such advice. Hence, risk-communication measures should be made simple, although not oversimplified, should avoid complex scientific information, although not treat lay people as if they were unable to comprehend basic data, and not give the impression that vital information is being held back. Comparative risk data are of limited value, but supplying information on single risks can be an effective way of garnering support for their reduction, especially if publicity campaigns are sustained and forthright. However, monitoring the impact of risk-communication efforts and correcting misapprehensions that arise from them is essential. The first hurdle is to sensitize the public to the issue; the second is to supply information that ordinary people can comprehend and the third is to ensure that the information is sufficient and appropriate to enable people to make informed choices about the risks that they face.

Part of this process of risk communication can be incorporated into the section of the emergency plan that deals with public participation in emergency preparedness and mitigation. Thought may also need to be devoted to how to communicate the sense of risks to be reduced to the political and other authorities whose support will be needed in order to make the plan work. Scenario building (see §3.1.1), cartography and zonation will all help create a sense of risks to be communicated, but the planner's judgement will obviously be fundamental in deciding how to convey to the appropriate audience the sense of risk and need for mitigation.

3.3 Loss estimation

There are several reasons why loss estimation should be regarded as an integral part of emergency planning. To begin with it is necessary to have some idea of the potential scope of particular disasters in order to know what to plan for (e.g. how many casualties and collapsed buildings). Secondly, it helps gain an idea of both the need for emergency resources during a disaster aftermath and of the probable availability of local resources, given the damage and destruction that will occur. Thirdly, the estimate of losses may be alarming enough to convince politicians, administrators and the general public to support the formulation of the emergency plan.

Clearly, it is not easy to make a reliable estimate of what will be lost in a future catastrophe. Many uncertainties exist. However, it can be a process that is worthwhile and not obstructed by insurmountable technical barriers.

The first task is to construct one or more hazard scenarios that represent the probable pattern of impact in the local area for a given agent of disaster. For example, one may hypothesize a magnitude 6 earthquake, with 30 s duration of shaking, a focal depth of 15–20 km, maximum acceleration 0.3 g and the propensity to cause damage to intensity IX on the modified Mercalli scale. One may fix the epicentral location, say, about 20 km to the north of the principal town in the area. Of course, such a scenario must be backed by cogent reasons why it is likely, and thus it is common to base loss estimates on a repeat of some previous disasters: although the physical variables remain the same, the losses will vary in relation to changes in vulnerability that have occurred in the area since the last damaging event.

When loss must be estimated, it is usually considered in three ways: casualties, structural damage and economic (or productivity) losses. The last of these is partly dependent on the first two, as it can be obtained not only by looking at direct loss of income, sales, and so on, but also by costing damage and loss of life. More or less sophisticated procedures have been developed for making estimates, using complex forms of engineering, statistical and economic analyses, but the present work is intended to provide a more basic and simple approach, and hence such complexities are skipped over or generalized.

First, let us consider the estimation of structural damage. It requires that buildings be classified into a series of characteristic typologies, so that damage levels can be predicted for each example of each type. Fortunately, this is not an endless task, as a given area is likely to have a limited range of building types determined by local materials, indigenous building practices and traditions, economic considerations, and the uses to which the buildings are put. Typical building typologies can usually be assembled on the basis of the following sorts of information:

- construction type: load-bearing walls (mudbrick, random rubble, dressed stone, bonded brick, etc.)
- nature of vertical and horizontal load-bearing members (e.g. brick wall, steel beam)
- size of building: number of floors, number of wings, square metres of space occupied or cubic metres of capacity
- regularity of plan and elevation
- degree to which construction materials and methods are mixed in the building, or alternatively whether there is a single technique and set of materials
- age category of the building; this can be generalized according to the dominant building material of the period (usually this involves categories such as pre-1900, 1900–1940s, 1950s–1965, 1966 onwards)
- the building's state of maintenance (excellent, good, mediocre, bad).

In a small town it may be that there are only two or three main types of vernacular housing, and hence the typology will be simple.

On the basis of these typologies, and the prevailing hazard scenario, it will be necessary to estimate the approximate amount of damage sustained by buildings. One can generalize this into either four categories (none, slight, serious, collapse) or seven, which are as follows:

- *None*   The building has sustained no significant damage.
- *Slight*   There is non-structural damage, but not to the extent that the cost of repairing it will represent a significant proportion of the building's value.
- *Moderate*   There is significant non-structural damage, and light to medium structural damage. The stability and functionality of the building are not compromised, although it may need to be evacuated to facilitate repairs. Buttressing and ties can be used for short-term stability.
- *Serious*   The building has sustained major non-structural damage and very significant structural damage. Evacuation is warranted in the interests of personal safety, but repair is possible, although it may be costly and complex.
- *Very serious*   The building has sustained major structural damage and is unsafe for all forms of use. It must be evacuated immediately and either demolished or substantially buttressed to prevent it from collapsing.
- *Partial collapse*   Portions of the building have fallen down. Usually these will be cornices, angles, parts of the roof, or suspended structures such as staircases. Reconstruction will be expensive and technically demanding, so that demolition of the remaining parts will be the preferred option.
- *Total collapse*   The site will have to be cleared of rubble. A few very important buildings may need to be reconstructed (usually for cultural reasons), even though they have collapsed totally, but most will not be rebuilt.

The application of the hazard scenario to each building typology will generate a distribution of damage, by category, if a damage function is used

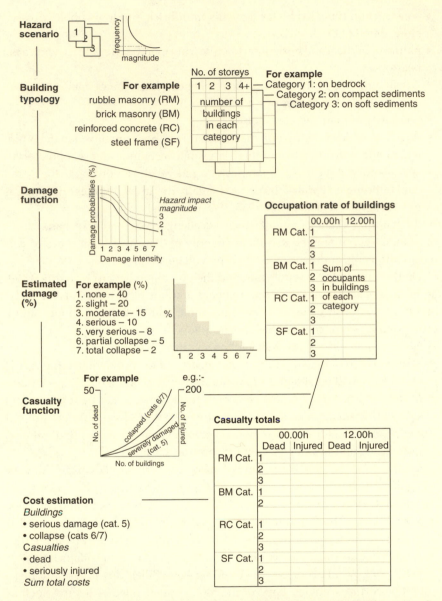

Figure 3.10 Procedure for making quantitative estimates of damage and casualties for emergency scenario building.

(see Fig. 3.10). Damage functions can be worked out by complex engineering equations that relate forces to resistances in particular structural and physico-dynamic situations. In an earthquake, damage-estimation spectral-response functions are widely used. These work on the basis of two facts. First, buildings are primarily (and often exclusively) constructed to resist static forces, such as internal loads and gravity. To do this they need merely to remain standing without significant structural deformation. But spectral-response functions relate to the transient dynamic forces that affect the buildings during an earthquake. A building must have enough stiffness to resist excessive deformation and enough flexibility to absorb part of it without sustaining major damage. Secondly, when subject to sudden dynamic loading, a building will resonate with a period and frequencies determined by its height and weight. The interaction between this resonance and the pattern (frequencies, amplitude, the velocities, acceleration and duration) of seismic waves will determine the maximum forces to which the building is subjected. If these exceed the resistance of materials (e.g. the compressive strength of brick, or the sheer strength of concrete), structural failure may occur. Many laboratory tests and field observations have determined the parameters of spectral response equations for a wide variety of seismic inputs and structural typologies. Despite this, the results are still only approximations. In less sophisticated analyses it is enough to use simple nomograms that relate hazard-impact magnitude to damage-intensity categories and produce average probability values. These can be derived from observations made during previous disasters, if statistics have been collected on damage and if hazard magnitudes are known.

Casualty estimation presents a thornier problem. This has been fairly thoroughly investigated for earthquake disasters (Fig. 3.11), although with somewhat indeterminate results. But for most other types of disaster, information is decidedly sparse. Again, the first task is to design a series of categories for injury. These may be as follows:

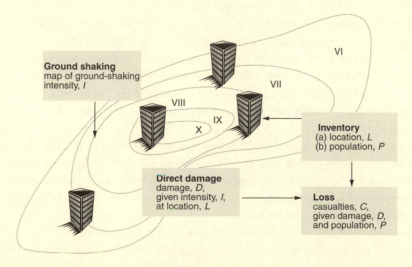

Figure 3.11 Estimation of earthquake impact potential using maps of the intensity distribution of past seismic events.

- No injury: the person is unharmed.
- Slight injury: minor medical attention will solve the problem and transport to hospital is not required; medical assistance is not urgently required.
- Serious injury that does not require immediate priority treatment. The patient will be taken to hospital but will not be among the first to be treated. The injuries are not life-threatening and will not lead to a significant worsening of the patient's condition if he or she must wait to receive treatment.
- Serious injury that requires priority treatment in order to produce some significant improvement in the patient's long-term prognosis, or simply to avoid deterioration in the patient's condition.
- Instantaneous death, which is a direct consequence of the disaster impact.[*]

In addition, there is a further category that covers death as a result of stress caused by the disaster, which provokes fatal heart attacks or catastrophic deterioration of pre-existing medical conditions, or death from complications arising from initially non-fatal injuries.

Again, a methodology is required in order to estimate the number and distribution of casualties in disaster. To begin with, it will be assumed that the vast majority of casualties will result from damage to buildings (and their contents) and that, the more severe the damage, the greater the number of casualties and the more severe their injuries. In this respect, assumptions will have to be made about the pattern of occupants of buildings. For this reason, estimates of casualties will vary with aggregate patterns of human activity and, of course, with the presence or absence of the evacuation that removes people from harm's way. The worst-case situation is probably that of a large, sudden, unpredicted impact that affects a densely settled area, in which vernacular housing (as opposed to offices or factories) is at particular risk of structural collapse, at night when people are sleeping and unable to react promptly. Casualty numbers may differ by an order of magnitude between the best- and worst-case scenario in a given area. Under such circumstances, it may be necessary to carry out some fieldwork in order to estimate the likely number of occupants in each type of building (residences, schools and other public buildings, shops, factories, etc.) at different times of day.

A relationship is then needed between damage and injury. This involves determining or hypothesizing the rate of injury or fatalities for each building type and each category of damage, and multiplying the results by the number of people at risk. In practice this is easier than it sounds because only the higher levels of damage will cause significant numbers of injuries, fatal or otherwise. Moreover, injury is likely to be concentrated in certain types of building – such as vernacular housing built of load-bearing brick masonry. Generally, the main

---

* These categories can be compared with those used in triage – see §6.1.2.

basis for hypothesizing a relationship between damage and injury will be the lessons of past destructive events. In this respect, if there are no such data it may be very hard to derive an appropriate relationship. The very least that will be required is a relationship, be it hypothetical or empirical, between the number of buildings that collapsed and the number of fatalities, tempered by the building occupancy rate at the time the disaster strikes.

The third aspect of loss estimation is the financial side. This involves calculating the following set of values:

- The direct cost of damage can be assessed either in terms of the proportion of the value of the building or structure that has been lost or in relation to the probable cost of reconstruction and restoration.
- Economic losses include the value of stock and inventory that has been damaged, the value of services that cannot be rendered, and goods that cannot be sold as a result of the disaster, and the social costs of temporary or permanent unemployment.
- Costs are associated with search-and-rescue operations, restoration of essential services and reconstruction of damaged buildings.
- Further costs are associated with injury and loss of life.

It is a routine procedure to estimate the financial cost a personal injury and of death. The procedure involves no consideration of the moral question of how much a human life is worth, for it is well known that in moral terms it cannot be costed. Instead, calculations are based on earnings forgone as a result of casualties. Losses are therefore considered as an estimate of financial productivity forgone. Studies in the USA in the mid-1990s gave the following average values: US$200 for a slight injury, $5000 for a moderate injuries and $2 200 000 for an injury that proves fatal or for instant death. These are average values based on average earning power and medium age. They will differ for specific groups in terms of social factors, gender and age.

Financial losses caused by a disaster can be considered in two ways. The first is in terms of instantaneous loss, with respect to the short-term impact of the disaster. The second refers to the long-term consequences, which include the continuing cost of lost transactions and income, and of reconstruction. Here, it is probable that inflation and re-estimation of the cost of reconstruction will increase the sums involved. So will any attempts to reconstruct to higher standards than those of what has been lost, or to combine reconstruction with development. The simplest way to tackle this is to include a fixed percentage increase in estimates in order to take account of probable future inflation.

The preamble to the good disaster plan should include a section on estimated losses. In most cases this need not require a full-scale house-to-house study of lossestimation. A simplified procedure would be as follows. First, accumulate statistics on the types of building (e.g. the number of **unreinforced** two-storey **masonry dwellings**) and demographic factors of land use and building occupancy (the average number of residents per room or per building at given times of day). Next, design some simple hazard scenarios that relate to what has happened in the past, but update them to take account of changes in vulnerability and improved knowledge of hazards and their impacts, then estimate the probable extent of damage or number of buildings lost. From this, and again with reference to past disaster losses in the area, estimate the number of casualties in the deaths and serious-injuries categories with respect to the estimated numbers of buildings that collapsed or were damaged in the disaster, and the estimated rate of occupation of buildings at particular times of day. Total costs can then be estimated approximately by summing the value of damage to buildings, the nominal cost associated with death and serious injuries, and, where it can be calculated, the costs of lost production, services not utilized and civil-protection services rendered during the disaster.

With respect to casualties, the following points should be borne in mind:

- In many instances, building collapse is the main cause of death and injury, but in specific cases toxicity, blast effects, burns, drowning or **trauma** caused by transportation crashes or by malfunctioning machinery may be the principal causes.
- The casualty totals and injury typologies vary markedly from one disaster to another.
- Casualty totals tend to be more variable in small impacts than they are in large disasters.
- Where loss of life and injury are not large, they tend to be concentrated at specific locations within the disaster area, in response to specific conditions such as the collapse of particular buildings.

As an aid to emergency planning, Table 3.2 offers a simple checklist for hazards and their consequences. Obviously, however, the factors and categories on the list require more thorough investigation, and the form is merely intended to help orientate the planner at the preliminary stage of deciding which hazards are capable of generating significant losses.

3.4 Resource analysis and inventory

Accurate knowledge of resources is important to plan their efficient deployment and use. It also enables the disaster manager to know what extra resources are required during the disaster, in order to supplement those that are available locally, and, conversely, what offers of help should be refused to avoid duplication of effort.

Table 3.2 Hazard and disaster impact checklist.

Hazard...

Brief description of scenario

....

....

Location...

Characterization
natural ❑   technological ❑   social ❑   secondary ❑   complex or multiple ❑

Probability of occurrence
calculable ❑   hypothesized ❑   unknown ❑   independent of past events ❑
dependent on past events ❑

Frequency
regular (e.g. seasonal) ❑   some regularity ❑   apparently random ❑

Pattern of impact
sudden catastrophe ❑   rapid build-up (<24 h) ❑   slow build-up ❑
imperceptibly slow build-up ❑

Duration
seconds ❑   minutes ❑   hours ❑   days ❑   weeks ❑   months ❑   years ❑

Spatial distribution of impact
widespread (50–1000 km^2) ❑   local (1–50 km^2) ❑   site-specific ($\leq$1 km^2) ❑

Short-term predictability (forecast capability):
Location   predictable ❑   variable but generally known ❑   unpredictable ❑
Timing   highly predictable ❑   quite predictable ❑   somewhat unpredictable ❑
highly unpredictable ❑

Warning capability
very high ❑   high ❑   moderate ❑   low ❑   very low ❑

Controllability (can physical processes be abated?)
definitely ❑   probably ❑   possibly ❑   no ❑

General assessments
Vulnerability   very high ❑   high ❑   moderate ❑   low ❑   very low ❑
Risk levels   very high ❑   high ❑   moderate ❑   low ❑   very low ❑

Preparedness measures
very effective ❑   effective ❑   uncertain ❑   ineffective ❑   completely lacking ❑

Structural and semi-structural mitigation measures
very effective ❑   effective ❑   uncertain ❑   ineffective ❑   completely lacking ❑

Non-structural mitigation measures
very effective ❑   effective ❑   uncertain ❑   ineffective ❑   completely lacking ❑

Probable future impact levels
very effective ❑   effective ❑   uncertain ❑   ineffective ❑   completely lacking ❑

Public awareness of hazard
very effective ❑   effective ❑   uncertain ❑   ineffective ❑   completely lacking ❑

Public support for mitigation and preparedness measures
very effective ❑   effective ❑   uncertain ❑   ineffective ❑   completely lacking ❑

General assessment of mitigation situation for this hazard
very effective ❑   effective ❑   uncertain ❑   ineffective ❑   completely lacking ❑

At the most fundamental level, many geophysical and technological phenomena are considered as resources until they become extreme, at which point they are viewed as hazards. Thus, rainfall is beneficial to crops and water supplies unless it is abundant enough to cause flooding or scarce enough to cause drought. But in terms of disaster planning, one can divide available resources into four categories:

- human resources (personnel): officials, technicians, volunteers, etc.
- capital goods: movable plant and other vehicles, buildings, heavy equipment, and so on
- renewable supplies, such as food, water, medicines and clothing
- information resources in terms of communications hardware, software and databanks.

Each category of resources can be considered in a static sense in terms of the starting inventory, and in a dynamic sense in terms of the running tally of resources during the disaster, as an input–output system. However, note that, as emergency management has become more sophisticated, so the emphasis has shifted from merely accounting for local resources to attempting to match what is available locally with what will be needed during the next disaster.

It is a good idea to begin the process of planning for emergencies with an audit of all available resources. With respect to supplies, categories are given in Table 3.3 and examples of goods utilized in disaster response are shown in Table 3.4. Table 3.5 gives a set of forms that can be used as templates for designing a resource-tracking system. Once the list is complete, one needs to design a system to ensure that it is kept up to date and another to ensure that the list can be adjusted rapidly during the emergency (which to all intents and purposes is a period of extremely rapid consumption of resources). The first of these may involve a monthly review, in which resources are re-surveyed according to a detailed set of categories; new additions are noted, as are obsolete or depleted items, and the functionality of the resource is subject to a quick assessment. The latter system may involve circulating a daily accounting sheet to all people in charge of supplies or commanding personnel.

The preliminary audit will serve the dual function of making known what resources are available to manage a disaster and identifying and prioritizing deficiencies in personnel, supplies, equipment and information. One fundamental question is whether the plant and equipment can be converted on the spur of the moment from non-disaster to disaster uses. Just as most of the disaster-management personnel will have dual roles split between regular work and emergencies, much equipment will normally be used for tasks that are unconnected with disaster. It may work in times of quiescence, but will it do what one expects it to do during the fully fledged catastrophe? Similarly, the inventory maintained during a disaster should be accompanied by notes on

Table 3.3 Categories and subcategories for the classification of relief supplies.

Medicines*
Analgesics
Anaesthetics
Antibiotics
Cardiovascular drugs
Steroids
Vaccines

Health supplies and equipment*
Human resources
Medical and dental
Surgical
Blood bank and laboratory
General anaesthesia and X-ray
Patient transport
Others

Water and environmental health
Human resources
Water treatment
Water distribution
Vector and pest control
Human and other waste disposal
Others

Food and beverages
Human resources
Cereals, vegetables and grain
Oils and fats
Dairy and meat products
Water and other beverages
Others

Shelter, housing, electrical, construction
Human resources
Shelter and housing
Electrical
Construction
Others

Logistics and administration
Human resources
Logistics and administration
Transport
Radio communication
Others

Personal needs
Human resources
Clothing
Sleeping gear and blankets
Personal hygiene
Cooking
Others

Unsorted
Clothing
Medical drugs
Food
Mixed
Other
Unknown

* International colour coding for medical supplies: red – cardiocirculatory; blue – respiratory; orange – paediatric; green – various.

Table 3.4 List of relief items supplied or requested in internationally declared disasters, 1994–9.

General items
Air-conditioning units
Arc lamps (floodlights)
Canvas
Chain saws
Collapsible tables
Diesel and electric pumps
Diesel-driven electrical generators
  (110–240 volts, 60 Hz, 1.5–60 kW)
Gas cylinders (liquid propane gas)
Jerry cans (for fuel and water)
Manual pumps
Plastic sheeting

Polyethylene pellicle
Reels of cord
Rope
School equipment and stationery (pens,
  pencils, chalk, workbooks, textbooks)
School furniture (desks, chairs, blackboards)
Submersible mud pumps
Tarpaulins
Vacuum pumps
Voltage transformers (e.g. 35 kV)
Voltage controllers (e.g. 10 kV)

Table 3.4 (continued)   List of relief items supplied internationally.

Communications and connectivity

Cellular telephones
Computers, printers, modems and computer
  media
Field telephones
Metal girder bridges (Bailey or Hamilton type,
  length 15–20 m)
Pedestrian bridges
Pontoons

Radios (AM/FM)
Replacement telephones
Satellite telephone units
Telefax machines
Two-way radios
VHF radios (36–162 MHz, 16 channels,
  antennae, cables)

Personal needs

Fibreglass latrines
Household tools and utensils (e.g. brooms,
  screwdrivers)
Kerosene lamps
Latrine pans
Overalls
Personal hygiene items (women and infants)
Petromax lamps
Plastic pails

Portable showers
Portable toilets
Soap
Sturdy footwear
Toilet and kitchen paper rolls
Torches and batteries
Towels
Warm clothing
Washing powder

Cooking and consuming food

Aluminium cooking pots
Cutlery
Detergents
Kitchen utensils

Mobile kitchens
Plastic cups, plates, bowls and jugs
Stoves

Food and water

Baby formula
Beans and other pulses
Blended and concentrated foods (Unimix,
  Spermix, etc.)
Canned foods
Chlorine for water purification
Concentrated milk
Fish and fish products
Flexible water containers
Flour (various grains)
Food-grains (wheat, maize, rice, oats, barley,
  rye, etc.)
Fresh water
High-protein biscuits (BP5, etc.)

Meat
Milk
Onions
Pasta
Potatoes
Powdered milk (*not* for consumption by
  infants)
Salt
Sugar
Tea
Vegetable oils
Water purification filters
Water purification tablets and potions
Water collection pots

Shelter and sleeping needs

Bed linen
Blankets
Caravans (trailers)
Cots
Ground sheets
Hammocks

Mattresses
Mosquito nets
Sleeping mats
Sleeping bags
Tents (military style)

Table 3.4 (continued)   List of relief items supplied internationally.

Transportation

Aluminium boats
Ambulances
Aviation fuel
Barges
Buses and minibuses
Diesel fuel
Dump trucks
Fixed-wing aeroplanes (cargo,
  reconnaissance, troop-carrier, etc.)
General haulage trucks
Helicopters (light, two-rotor, etc.)
Milk trucks
Off-the-road vehicles (four-wheel drive)
Outboard motors
Petrol (gasoline)
Railway sleepers, rails, ballast
Rubber boats
Spare vehicle tyres
Truck and other vehicle spare parts
Water-tank trucks

Agricultural needs

Agricultural implements
Cane knives
Chemicals and bags for disposing of animal
  corpses
Combine harvesters
Concentrated livestock feeds
Electroseparators
Fertilizers
Fishing nets, hooks, lines
Forage
Grain seeds (wheat, rice, etc.)
Handcarts
Hoes
Insecticides and sprayers
Irrigation pumps, hoses, pipes, sprays and
  sprinklers
Machetes
Mowing machines
Pesticides
Pickaxes
Potato seeds
Poultry feed
Shovels and spades
Tractors
Vacuum milking machines
Veterinary biocompounds

Construction materials, plant and tools

Aggregate
Bamboo poles
Bituminized board
Bolts
Bricks
Builders' tools
Bulldozers
Carpenters' tools
Cement
Concrete, ceramic and metal pipes (various
  diameters)
Copper plumbing pipes
Corrugated, galvanized steel sheets (3 mm)
Electrical insulators
Electrodes
Equipment to produce cement or composite
  blocks
Excavators
External PVA wiring
Face masks
Gas pipes
Gravel
Heating boilers
High-voltage cables
Industrial helmets
Internal VPP wiring
Lime for mortar
Locking steel scaffolding
Mobile cranes and winches
Nails
Plywood sheets
Rocks
Roof tiles
Roofing insulation material
Spigots, connectors, braces and stays
Steel bars for concrete reinforcement
Steel cables
Steel-toed boots
Valves
Varnishes and paints
Veneers
Welding cables
Wheelbarrows
Window glass
Wooden baulks, planks, laths
Work gloves

Table 3.4 (continued)   List of relief items supplied internationally.

Firefighting equipment
Breathing equipment
Cargo nets
Cistern aeroplanes
Fire hoses and long lines
Firefighting tanker trucks
Fireproof clothing

Portable water tanks
Scooper planes and helicopters
Storz couplings for fire hoses
Swivel hoses
Water cannons

Medical equipment
Hospital-style camp beds, mattresses and
 bedding
Kidney dialysis units
Portable bacteriological analysis kits
Stretchers (regular and all-terrain types)

Surgical apparatus, instruments, kits
Transfusion apparatus
X-ray machines and film

Medical supplies
Acids and alkalis
Adrenomimetics
Alimentation media
Amino-acids of sugar
Amoxycillin
Analeptics
Analgesics
Anti-diarrhoea drugs
Anti-fever medicines
Anti-infection vaccines
Anti-inflamation medicines
Anti-rabies vaccines
Antibiotics for pneumonia treatment
Bacteriofages
Band-aids
Bandages and dressings
Blood and blood products
Blood coagulation stimulants
Cardiological medicines
Chlorine
Chloroquine and other anti-malaria drugs
Co-trimoxazole
Dermatological compounds
Disinfectants
Emetics
Erythromycin
Fermentation compounds
General anaesthetics
Glass slides
Histamines, anti-histamine compounds

Hypotensives
Kidney medicines
Levomicitin
Local anaesthetics
Measles vaccines
Metronidazole
Neuroleptics
Paracetamol
Penicillin
Phloroglucinol
Pituitary gland compounds
Plasmo-replacement solutions
Polyvalent sera
Promethazine
Protein compounds
Rehydration salts
Sedatives
Sleeping pills
Sodium, potassium and calcium compounds
Spasmolytics
Sterile drips and tubes
Steroids
Sulphanilamidic compounds
Swabs
Syringes
Tetracycline
Tongue depressors
UNICEF emergency medical kits
UNIPAC medical kits
Vitamins and their analogues

Table 3.5 Sample database structures for emergency services.

EMERGENCY-WORKER DATABASE
Record no.   of
Date record created: __/__/__   Most recent update: __/__/__
Municipality   Census district code:
Province/County/Region/Department/State:
EOC name: Identification code:
EOC address:
Tel.:   Fax: Telex:
Telegrams: E-mail: Internet http:
Map reference: Latitude: Longitude: Metres a.s.l.:
Principal disaster manager (name, address, contact numbers):
Alternative principal disaster manager (name, address, contact numbers):
Emergency worker name (forename, FAMILY NAME):
Title: Identification code:
Home address:
Tel.:   Fax: E-mail:
Cellular:
Work address:
Tel.:   Fax: E-mail:
Alternative address:
Tel.:   Fax: E-mail:
Other means of contacting the subject (e.g. pager, relatives' or colleagues' phone nos):
Operations group, squad or unit:
Specific competence:
Title of job:
Description:
Time required to report to base:
Usual working hours:
Programmed absences:
Notes and annotations:

Table 3.5 (continued)   Sample database stuctures for emergency services.

VEHICLE DATABASE
Record no.   of
Date record created: __/__/__   Most recent update: __/__/__
Vehicle type:
Make, model, year:
Registration (licence plate) number:
Key numbers:
Vehicle is property of (organization, name, address, contact numbers):
Engine size and type:
Fuel type and tank capacity:
Is there a reserve fuel tank? (capacity?):
Estimated range on full tank of fuel (km):
Transmission (manual/automatic, high–low ratio, four-wheel drive, etc.):
Seating capacity:
Does vehicle have:
Two-way radio Y/N   Car telephone? Y/N   AM/FM radio? Y/N
First-aid kit (type):   Review contents every ____ months
Trailer? Y/N   Details:
Tax details: Date of renewal: __/__/__
Insurance details: Date of renewal: __/__/__
Any other financial information (e.g. leasing)?
Maintenance interval (km or months) Vehicle:
Components (e.g. engine oil change, tyres substituted):
Equipment carried aboard:
Maintenance normally carried out by (name, address, numbers):
Maintenance contract details (if any):
Vehicle normally parked at (address, contact numbers):
Keys held at (address, contact numbers):
Registered drivers and operators (name, address, contact numbers: 1. 2. 3.

Table 3.5 (continued) Sample database structures for emergency services.

COMMUNICATIONS DATABASE		
Record no. of		
Date record __/__/__ created:	Most recent update: __/__/_	
Municipality	Census district code:	
Province/county/region/department/state:		
EOC name:	Identification code:	
Name:		
Address:		
Telephone:	Cellphone:	Fax:
E-mail:	Internet http:	
Location:		
Organization:		
Position (title, job description):		
Radio operator:		
Radio ham:		
Internet operator:		
Telephone operator:		
Journalist: Radio	Television	Print media
Title of program or publication:		
Equipment:		
Capacity:		
Coverage:		

usability, functionality and efficiency, otherwise there is likely to be a serious difference between what is available on paper and what can actually be used. Hence, the resource tally requires a mechanism for assessing the quality of items: there is no value in listing rancid milk or rusted-out machinery.

The preliminary audit can divide manpower into a series of categories. These include disaster managers, emergency responders and other decision makers; emergency-service personnel connected with fire, ambulance and public-order corps; **volunteer** specialists, such as paramedical workers and volunteer doctors and registered nurses; general volunteers, who are usually associated with voluntary associations; military personnel; and technicians. The last group includes those who are expert at search and rescue to recover victims from wreckage of rubble or to locate missing persons, and those who work on restoring and maintaining essential services, such as electricity, gas,

Table 3.5 (continued) Sample database structures for emergency services.

WAREHOUSE AND STORAGE DATABASE
Record no.   of
Date record   __/__/__   Most recent update: __/__/__ created:
Municipality   Census district code:
Province/County/Region/Department/State: .
EOC name: Identification code:
EOC address and contact numbers:
Warehouse name: Identification code:
Address and contact numbers: .... 24-hour telephone?
Map reference: Latitude: Longitude: Metres a.s.l.:
Directions for reaching the facility:
Total area (m^2):   Cubic capacity area (m^3):   Area under cover (m^2):
Loading bay height (m, cm):
Largest entrance dimensions (height ´ width, in metres):
No. of floors: No. of rooms:
Floor and pavement type: Reinforced? Y/N
Cranes, forklift trucks, lifting gear?
Principal administrator (and alternative p.a.):
Emergency-operations units assigned to warehouse:
Alternative address:
Personnel: Number?   Who?
Lighting: Heating:
Water supply: Toilets/washrooms:
Beds and bedding (if any):
Emergency electricity generator? Y/N: Details:

Special facilities:
  Secure storage:
  Cold store:
  Chemical response equipment:

Fire-suppression facilities:
What is the warehouse or storage facility usually used for?
Requisition procedure (if appropriate):

Table 3.5 (continued)   Sample database structures for emergency services.

EMERGENCY MEDICAL SERVICES DATABASE
Record no.   of
Date record created: __/__/__   Most recent update: __/__/__
Municipality   Census district code:
Province/County/Region/Department/State:
Name of facility: Identification code:
Category: hospital ❏   clinic ❏   medical centre ❏   doctor's surgery ❏   other
Address:
Tel.:   Fax: E-mail:
Map reference: Latitude: Longitude: Metres a.s.l.:
Number of doctors: Number of nurses:
Paramedics: Medical orderly: Administrative staff:
Number of beds: Number of bathrooms: Number of showers:
Emergency medical facilities: With receiving bay? Y/N Details:
Area of indoor facilities (m^2):
Burns unit? Y/N   Details:
Number of intensive-care beds: Number of kidney dialysis units:
Cardiac resuscitation facilities:
Diagnostic facilities: X-ray facilities?   CAT scanner?
Magnetic spin resonance?   Haematological lab.?   Bacteriological lab.?
Surgical facilities?   No. of operating theatres: Description:
Surgeons: Surgical teams: Anaesthetists:
Night-time staffing levels: Doctors: Nurses
Number of ambulances available at this site:
Auxiliary electricity generating facilities:
Special facilities for disabled patients:
Average stocks of blood and blood products (by type and Rh±):
Autonomous EMS plan? Y/N   Participation in general EMS plan for area? Y/N
Participation in municipal emergency plan? Y/N In other plans:

telephone, water, sewerage and public-health services. It also includes specialized categories such as geotechnical engineers, divers and hazardous-materials responders. The audit should include names, full addresses (work, home and alternatives), contact numbers and notes on specific competences and experience. A key feature of resource inventory is substitution. If the particular personnel (or any other resources) are not available at a particular time, arrangements should have been made to substitute with others. Hence, lists of personnel require much cross referencing to ensure that, when a disaster unexpectedly strikes, there are no critical gaps because of absence or because personnel fall victim to the event. **Call-up** procedures are discussed in §4.1.2; they have a critical relationship with manpower inventories.

Capital goods include the following categories: heavy plant, such as earth-moving equipment, snow ploughs and cranes; emergency and other vehicles, including all-terrain vehicles and helicopters; arc lights for working at night and robust electricity generators for field operation; buildings to be used as stores, emergency-operations centres, evacuation centres, garages for vehicles, by nurses and doctors for relief work, and so on; beds and re-usable bedding; temporary buildings such as mobile trailers and **prefabricated huts**; and specialized equipment, such as hydraulic rams and jacks, all-terrain stretchers, and temporary road signs. Particular attention needs to be given to sources of fuel for vehicles, as it is easy for these to run out. When disaster strikes and vehicles are operated around the clock, alternative and back-up sources should be identified, along with means of rapid re-supply by the key sources. Strictly speaking, fuel stocks come under the heading of renewable supplies, which include food, water, bedding, clothes, footwear, disinfectants, water-sterilization preparations, medicines, medical supplies and some very basic materials such as plastic sacks and rolls of plastic sheeting. There are, of course, many links between capital goods and consumable supplies. For instance, shelters require beds, which require sheets, blankets, pillows and pillowcases, not to mention towels, bars of soap, toilet paper and disinfectants. Perishable supplies, such as fresh fruits and live vaccines, require climatically controlled storage (so do dead bodies awaiting last rites or autopsies). As there is a risk that supplies will overwhelm storage facilities, it may be helpful to conduct a census of local cold-storage facilities, from mortuaries to food warehouses and refrigerators.

Medical supplies constitute a special category of consumable supplies. In mass-casualty disasters there will be a very high demand for anaesthetics, analgesics, sedatives, antibiotics, antiseptic preparations, anti-tetanus sera, splints, sterile dressings, syringes, needles, water-sterilization preparations and X-ray film. To the extent that the demand for these items can be estimated in advance, they can be stockpiled specifically for disasters. However, one has to ensure that storage conditions do not allow them to decay and that all supplies are replaced, irrespective of condition, when they reach their use-by dates. The alternative to this is to ensure that normal usage is very efficiently counterbalanced by re-supplies, so that an adequate surplus is always maintained. Inventories of medicines may require a system for identifying drugs from brand names, storing them and dispensing them as needed. This can be a complex procedure, which should be assigned to a qualified pharmacist.

In effect, a separate disaster plan is required for information resources in order to ensure that during disasters these are protected, functional and available to people who need them. There are several aspects to this. Computers must have surge protection sufficient to ensure they will not break down during variations in the electricity supplies. They must also have alternative sources of the electricity, either from generators (and transformers, when necessary) or from batteries whose charge will last as long as the equipment is needed. Back-up copies of programs and data must be made and kept up to date. They should be conserved at separate locations away from the primary sources. As far as is possible, computers and all communications equipment should be protected from flooding, blasts, seismic shaking or other impacts, and should be supplemented by a functional back-up system that will act as an alternative in the case of failure at critical moments during the emergency.

Once again, the initial audit of resources should be supplemented by an input–output system for registering consignments and disbursements. This should also include a means of assessing needs as they arise, and matching them to resources, and one for deflecting the unwanted supplies and assistance. These latter items can be seriously debilitating to a relief effort because they take up time, space and resources that should be devoted to managing the disaster. On the other hand, resources can be supplemented by mutual aid agreements with neighbouring jurisdictions, although one must ask to what extent these will function if both authorities are badly affected by the same disaster. Lastly, supplies must be managed to balance the ability to estimate warehouse capacity in relation to what is in stock, which may vary considerably in little time as supplies are received and dispersed (Fig. 3.12).

The estimate of resources needed will be a function of vulnerability analysis and hazard scenarios, as discussed above, and of the capabilities of the disaster plan, as described in Chapter 4. It can be a complex and painstaking process to match available resources to needs, but it is surely a worthwhile exercise.

3.5 General and organizational systems analysis

A system is a set of attributes, together with their interrelationships in terms of mutual influences and the processes and flows that connect them. Systems

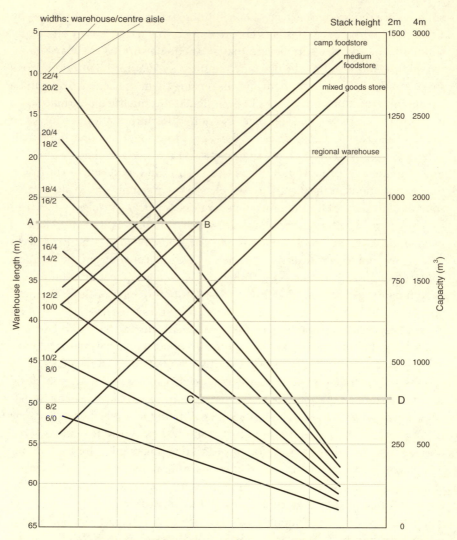

Figure 3.12 Nomogram for calculating the capacity of warehouses. In the example, a 28 m-long warehouse (A) is used as a store for mixed food and non-food items (B). It is 12 m wide and has a 2 m centre aisle (C). This yields a storage space of just under 400 m; for a 2 m-high stack or just under 800 m; for a 4 m-high stack (D).

analysis examines the components and functions of systems in order to illuminate either their input–output relationships (black-box analysis) or their inner workings (grey-box analysis). It is a form of modelling, a simplification of reality designed to make complex situations comprehensible. Systems analysis may be a useful approach to help understand the functions of complicated sets of organizations that must work together in emergency management and to illuminate their interrelationships.

Although formal systems analysis is not based on graphical methods, it helps to begin by conceptualizing a system in terms of graphic relationships, especially as a flowchart. The first stage is to define the boundaries of the system. If these are physical entities, they may have a geographical expression that is mappable; if they are social, one may have to decide which individuals or groups of people are to be included and which are to be excluded.

A simple black-box system works on the principle that

output = input ± change in storage within the system

The input is termed a forcing function and the output a response; the two are linked by a transfer function, which is usually a simple empirical relationship that reliably transforms starting values into ending ones. This is appropriate where one observes that a given action, or change, produces a predictable reaction (or outcome) and when one is not particularly concerned to understand how the change comes about, but merely to replicate it for predictive purposes.

A more complex grey-box system may be divided into several interconnected but semi-autonomous components termed subsystems. For example, where several organizations must work together in bringing relief after disaster, each can be considered as a distinct subsystem. They may be connected by regulators – mechanisms that govern the flow of information, commodities or other subsystem inputs. Where subsystem outputs form inputs to other subsystems, the system is set to become canonical or cascading – flows of information or other commodities in line directly from one subsystem to another. This is the case, for example, in a hierarchical chain of **command** as a linear system composed of command units deployed as linked subsystems.

A simple illustration of subsystem linkage is that of information flows in the immediate aftermath of a localized disaster. Operations at the site are directed from an **incident-command post** (ICP) located on its periphery, which is under the jurisdiction of the area's main emergency-operations centre (EOC), located at some distance from the disaster (Fig. 3.13). The EOC receives some information inputs from the site of the disaster, from the ICP, and from other sources such as central government. It outputs instructions to the ICP, which sends them as commands to the workers at the disaster site. Thus, a systems approach enables one to focus attention on the character and quality of information flows and to assess how they cross the boundaries between subsystems.

Most functional entities are open systems. If they are physical systems, this means that they exchange mass and energy with their surrounding environment; if they are social systems it means that they exchange information and personnel. The description of a system at any single point in time is its state, and any change in state represents its trajectory. Systems that tend to be stable are described as homeostatic, meaning they use the servo-mechanisms of negative

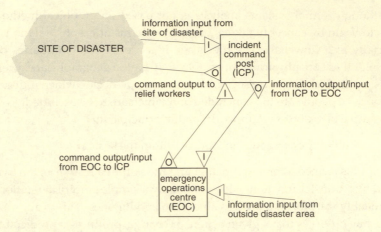

Figure 3.13 The functional relationship between the emergency-operations centre and incident-command post.

feedback to maintain dynamic equilibrium – in other words that their internal mechanisms predispose them towards stability. This means that changes that take the system's trajectory away from its average values are counteracted by forces that bring it back again. Self-regulation of this kind is characteristic of systems that are resistant to shocks and that maintain their form and function under duress. In contrast, systems that utilize positive feedback undergo transient responses. Change provokes more change, as is common in disasters with both physical and social systems that must react to events by adapting themselves to rapidly changing circumstances. However, rapid change in system properties cannot continue for long and is usually followed by negative feedback, which establishes a new equilibrium.

Systems characterized by bursts of rapid change separated by periods of stability are termed metastable. The stable behaviour is separated by thresholds, which the system must cross in order to regain stability. A threshold is a form of barrier or dividing line: when the variables that control the system reach certain values, then there is a change of behaviour, perhaps a change of system state. In physical terms, the top of the river levee signals a threshold between water flow in the channel and overbank spillage associated with flooding. In social systems, thresholds may be manifest in certain staffing levels, above which the organization must be run in a different way that takes account of the form and level of interaction among a larger group of people.

In the physical sciences, many attributes and aspects of systems have been discovered and analyzed using mathematical and statistical methods. One or two common characteristics deserve a brief mention. All systems, physical and social, are subject to entropy when deprived of constant re-energization. This means that they tend towards ever more disorganization. In the gain in entropy the system "runs down" by distributing its energy at ever wider radii and lower concentrations, which tends to make it ever more formless. For example, a group of search-and-rescue volunteers will tire and become inefficient unless they are substituted after many hours of work or are allowed to rest and regain their strength. In thermodynamic terms, or their social equivalent, entropy maximization is a form of homeostasis, that trends towards low and steady performance. Often, it can be achieved in various ways and the system is set to be equifinal, that is, the same outcome will result from a variety of possible combinations of processes and controlling values.

The systems approach is of value to the emergency planner for various reasons. One is that it can be applied in similar ways to both physical and social systems, so that it provides a common structure that helps bridge the gap between the two. Another is that it helps to formalize ideas about the structure of groups and organizations, to define the boundaries where these are unclear and to break them down into semi-autonomous subsystems. It facilitates the analysis of inputs and outputs (e.g. resources consumed and resources obtained) and of the interrelationships between the component parts of organizations, landscapes, social communities and so on.

Systems analysis can be applied at various stages in the course of the emergency-planning process. It has been widely used in hazard studies by geophysical scientists. On the social side, vulnerability can be analyzed in terms of a system's propensity to change under duress. The resources of manpower, equipment and supplies that are available for disaster relief can be thought of as a management system that must be integrated and tuned in order to improve its overall performance. Relief units such as volunteer groups and emergency task forces respond in certain ways to external stimuli. However, it is as well to ensure that the over-enthusiastic application of the systems approach does not lead to artificiality and loss of the vital connection with reality, as this has been a recurring criticism of **general systems theory** ever since its founder, Ludwig von Bertalanffy, first presented the methodology in 1950.

3.6 Field exercises

It is a very good idea to build a requirement for detailed **field exercises** into the emergency plan as it is written. The exercises are one of the most important ways of testing the plan to ensure that it works efficiently when disaster strikes.

A field exercise is a simulated emergency that puts the disaster-relief community to work but without the dangers and losses that arise in a real emergency. As it can be expensive and time-consuming, it must be carried out with the maximum of seriousness and dedication. However, it should be clearly identified as an exercise, so that no-one mistakes it for the real thing. In this respect one needs to plan for the hopefully remote possibility that the exercise will be overtaken by a real disaster and participants will find themselves working in earnest in the roles in which they had hitherto only been acting. In one instance in which I was involved, three days of simulated earthquake and volcanic-eruption exercises ended quite unexpectedly with a real-life explosion and fire in a factory that had the ambulance crews rescuing real victims rather than pretend ones. At this point it was necessary to remove the signs from the emergency vehicles that read "Don't worry – it's a civil-protection exercise". One hopes that alerting the public to the simulated emergency did not lead them to shrug off the real one that followed it.

Besides sensitizing the public to matters of **civil protection**, field exercises have several other important purposes (Table 3.6). They allow emergency managers to study the practical **logistics** of mounting a relief operation. Alternative routes can be tested for emergency vehicles. Response times can be measured under different sets of conditions. The details of a relief effort can be worked out as the exercise progresses, and participants will gain knowledge and experience of their roles and effectively learn on the job. Routines can be established and tasks apportioned. Shortcomings in planning can be identified, as can needs for particular types of equipment and supplies. A typical conclusion might be that, in the event of a building catching fire in an historic city centre, large fire appliances will not be able to manoeuvre into the narrow streets

Table 3.6 Some objectives of field exercises in civil protection.

- Demonstrate the ability to staff the main emergency-operations centre to an operational level and initiate primary communications within 60 minutes of a disaster and keep it operational for 36 h by a shift system of duty.

- Test and evaluate the allocation and coordination of regional resources for local usage.

- Demonstrate the ability to alert the public in a timely manner.

- Test alternative communication systems to be activated and functional within 90 minutes of a disaster.

- Test the ability to provide an initial assessment of the emergency within 4 h of time zero.

- Demonstrate the ability to provide food, quarters, communications and administrative support to emergency-operations centre personnel throughout the exercise.

- Demonstrate the ability to verify the existence of reported events and hazards.

- Assess the ability of participating agencies to coordinate emergency response operations, maintain communications, and inform the public.

around it and an alternative strategy must be worked out. Fire services usually evaluate these conditions on a routine basis, but their strategies to overcome such difficulties may need to be taken into consideration in the plan.

Field exercises serve to work out the timing and practical functionality of relief systems, to familiarize personnel with tasks and routines, to test the degree of cooperation between participants, and to identify shortcomings in planning. Hence, the emergency plan should not merely make provision for field exercises, it should *interact* with them; that is to say, improvements suggested by the exercise should be incorporated into the plan. As on-the-job training is an important component of the exercises, it is often useful to combine them with a formal training session, workshop, short course or civil-protection conference to broaden the horizons of participants further.

The first step in setting up a field exercise is to decide whether to test the whole plan or only a part of it. Logistics and available resources will probably determine this and, moreover, it is expensive to mount a major civil-protection exercise. In a similar vein, one must decide whether to cover a large geographical area (the broad picture) or a small one (the detailed approach). Next, a detailed programme of operations needs to be worked out. This will be based on a hazard and impact scenario. Updated versions of past events can be used if one fears that a particular hazard event will recur in the future. At the appointed time, personnel, vehicles, equipment and supplies are assembled at the exercise's designated **base camp**, which must be spacious, central to the places where the exercises will take place, and equipped with adequate communications. Before starting on the exercise, personnel are briefed and the coordinator must ensure that all details of the plan are fully known to participants. Parties who are not involved in the exercise must be briefed in advance to avoid misunderstandings and conflicts when the simulation takes place.

Many different aspects of emergencies can be simulated. Residential areas, archives, works of art, schools, factories or hospitals can be evacuated. Medical emergencies can be staged, with selected relief workers posing as injured victims. Complex search-and-rescue operations can be staged. Emergency-service personnel can learn to work with volunteers and members of the armed forces.

One final point is that in field exercises it is permissible to make mistakes, provided that one learns from them. I recall a simulated medical emergency in which an enthusiastic squad of young, newly trained paramedical volunteers stabilized the condition of one of their companions, who was acting as a simulated victim, and gently loaded him onto a stretcher. When the overseeing doctor arrived, he took one look and said, "Good work, lads: you have made a perfect job of stabilizing a burns victim. The only trouble is, this one was supposed to be a heart-attack case!" Mistakes should be made before, not during, the real emergency.

3.7 Use of information technology (IT)

Information is one of the prime resources in the planning and management of disasters. Its quantity, quality, flow and utilization can effectively determine the level of success achieved in mitigating catastrophes or dealing with emergencies. In the modern world, an information revolution is in progress, with profound technological, social and cultural implications. It is hardly surprising that this is spreading to the field of civil protection. This opens up a potentially vast array of new opportunities, and it also throws out a challenge to the disaster planner to make use – good intelligent use – of the emerging technologies.

The issue of information is a complex one for the following reasons:

- Remarkably few standards exist that define the quality and reliability of information on hazards, emergencies and disasters.
- Information must be evaluated in cost–benefit terms just like any other commodity. If the costs of collecting it, or paying royalties to intellectual property owners, exceed the value of the information to a particular project, it may not be worthwhile proceeding.
- Sharing information can lead to problems of rights and ownership, or to the need for reciprocal arrangements to distribute the burden of collecting data and maintaining databases. It is common that one **agency** bears these costs while others reap the benefits.
- Laws, copyrights, conventions and other "intellectual property" restrictions can render information difficult of access. In fact, some countries have national legislation that prohibits the use of "sensitive" information, including that which is directly useful for emergency preparedness.

Despite these issues, the amount of information that is available to emergency planners has increased exponentially in recent years and it requires that new techniques and standards of efficiency be developed to cope with it. This is the world of information technology, which is gradually revolutionizing the way in which disasters are managed.

Table 3.7 lists a range of information technologies, all of which have applications in civil protection. Many of them are a direct result of the development of the computer microchip; others in their modern form are controlled by computers. One problem of dealing with computer technology in a work like this is that the treatment can easily become out of date, such is the pace of change. Hence, a significant part of the process of using the new technology is being flexible enough to adapt to changes as they occur. It is also important to recognize that a substantial portion of the investment will go into periodically updating obsolescent equipment.

As a general rule, the disaster planner should make as much use of computer technology as possible, but note that it will not solve many of the problems that

Table 3.7   Information technology resources.

Word-based communications	Computer facilities
Internet resources	**Computer**
• bulletin boards	• computer cartography
• electronic mail	• computer graphics
• hypertext links	• geographic information system
• modem	• microcomputer
• newsgroups	• network computer
• World Wide Web pages	• supercomputer
Facsimile transmission (telefax)	• workstation
Postal services	**Data storage**
• courier	• Compact disk (CD-ROM, CD-RW, etc.)
• regular post	• Databanks
Radio	• Hard copy
• AM/FM, long-wave	• Magnetic disk
• citizen's band	• • floppy diskette
• microwave	• • hard disk
• short-wave	• Magnetic tape
• transceiver	• • cassette
• transponder	• Microform
Telegraph	
Telephone	**Image-based communications**
• computer-generated messages	**Motion picture photography**
• cordless (field, cellular, etc.)	• developed film
• hard wire and fibre-optic	• digital images (MPEG, etc.)
• voice mail	• television and video
Telex (teletype)	• • cable television
	• • closed circuit television
Satellite	• • digital television
Communications	• • normal television transmissions
• global	• • satellite television
• regional	• • videocassette
Earth resources	**Still picture photography**
Geophysical	• developed film
• volcanological	• digital images
• Meteorological	• • aerial images
	• • GIF, JPEG, TIFF, etc.
	• • ground-based remote sensing
	• • space-based (false colour, infrared, panchromatic)

Source: Updated and adapted from "Information technology in real-time for monitoring and managing natural disasters", D. E. Alexander, *Progress in Physical Geography* **15**(3), 238–60, 1991.

he or she will encounter; it will only contribute to their solution. Moreover, technology is a double-edged sword: it can *create*, as well as reduce, vulnerability to disasters. Hence, every time a form of new technology is incorporated into processes of disaster planning, mitigation or management, some fundamental questions need to be asked:

- Is the system of disaster reduction or management over-dependent on the technology in question?
- Will a back-up system be available if the technology fails?
- Will unintended side effects or complications result from the technology?
- Is the technology adequately understood (at least in terms of its mode of use

and eventual products) by those who will use it or benefit from it?

- Will the technology deliver a product that is genuinely valuable or will it be merely showy and insubstantial?
- Lastly, is the investment needed to acquire and use the technology justified by the results, or will the technology become obsolete before it has delivered a good return on the initial investment?

Despite these reservations, information technology has a bright future in disaster planning. One of the fastest growing areas here is the use of the Internet and associated growth of **intranets**. In addition, when the latter have substantial, but selective, external connections, they may be termed extranets. Here, we will concentrate on the use of the Internet in disaster planning, as, at the time of writing, this is the most developed aspect. As space limitations preclude the review of many of the details, the reader who is not familiar with the Internet should consult an appropriate text, CD-ROM or website for basic instruction.

For disaster planners and managers, the Internet offers several advantages over more traditional means of communication. To begin with, it is a robust, versatile and fast-growing means of mass communication. The software that sends messages will repeatedly seek the most efficient means of getting them to their destinations. As it is dealing with a highly ramified network, routes that are blocked by breakages or excesses of information traffic will be abandoned in favour of those that are open. This is particularly important during emergencies, when cables may be severed, antennae felled and equipment damaged. However, the extent to which the local network is robust during a disaster may depend on the level of investment in it and the measures taken to ensure that it can survive damage. Nevertheless, as much Internet traffic goes by satellite, it is possible rapidly to substitute some local facilities with mobile ones when damage occurs.

The following are some of the potential uses of the Internet in disaster planning and management:

- *The World Wide Web* (WWW)   The World Wide Web is an excellent place to display a disaster plan. Hypertext links enable one to do this efficiently in a hierarchical manner that sends the reader directly to particular chapters or annexes of the plan. Detailed data, photographs and maps can be presented, along with the text, and colour can be used for greater clarity. The plan or any of its parts can be downloaded and printed out by any Internet user who has a modem or Ethernet connection to a server, a computer and a printer. The electronic version of the plan can be substituted frequently and easily as the plan is updated and revised. Newsgroups, discussion groups, electronic mailing and electronic newsletters can be used to publicize the plan.
- *Information resources*   Although the quality varies considerably from site to site, very many sources of information are accessible on the Internet.

Published disaster plans can be used as models by aspiring disaster planners. Databanks and archives may provide useful information on hazards. Bibliographic searches can be conducted at many different sites. Equipment and commercial products can be investigated and often ordered over the Internet. Finally, some academic research is accessible over the net and there are on-line international conferences on disasters and their mitigation.

- *Information for the public*   In the developed world, public use of the Internet has increased dramatically in recent years. It is therefore an important place to provide non-specialist information about hazards. Besides the disaster plan itself, this may include instructions on how to prepare in the home or workplace for disasters. It should also contain a directory of local resources that the ordinary citizen can call upon in times of disaster, including addresses and contact numbers. Experience from one of the world's most successful civil-protection sites on the Internet, that of the US **Federal Emergency Management Agency** (www.fema.gov), shows that such sites can become important points of reference for the general public: during Hurricane Floyd in 1999 the FEMA site logged more than two million hits (i.e. occasions when members of the public logged onto the site) in 36 h.

- *Situation reports*   If full use is to be made of the Internet during a disaster, the plan should include provision to publish frequent situation reports ("sitreps", as they are known) that outline developments in hazard impacts and relief efforts over the previous, say, 12 h. The Internet is one of the best places to disseminate situation reports, including those aimed at the news media.

- *Virtual systems*   A new development is the invention of "virtual" emergency management systems. The ability of the Internet to transmit data rapidly in real time or near real time means that disaster managers can participate in an emergency from distant sites. Although this does not obviate the need for an on-site presence, it does enable one to call upon different sources of expertise where it is available. The ability to transmit messages, diagrams, maps, photographs and film clips ensures that a copious two-way transfer of information is possible. Plans are being laid to connect hazard-monitoring instruments via satellite to the Internet, which will broaden and quicken the range of real-time analyses to which data can be subjected.

- *Advertising*   Products and services can be publicized and sold over the Internet. These include consultancy services associated with disaster planning and hazard monitoring, and the sale of equipment for civil-protection activities. This is expected to be a major growth area in the near future.

Obviously, information-technology resources are not restricted to the Internet, as the discussion of satellite imagery and geographic information systems illustrates (see §2.3.1). The trick is to achieve a good level of integration between the various facets of IT. These should not only connect but also provide enough overlap (termed "redundancy") to ensure an unbroken flow of information when one part of the system fails. Generally, over-reliance on a single channel of communication is dangerous. But ability to direct multiple channels presupposes adequate familiarity with both the technology and the consequences of using it. The present section deals primarily with the former, but the latter must not be underestimated. Like all information systems, communications can easily suffer from the "garbage in, garbage out" (GIGO) syndrome. A Californian expert in this field, Art Bottrell, has identified four layers in the process of communicating emergency information ("Networks in emergency management" Internet discussion group, 1996). The first, technology, is purely mechanistic in character. It would be considered neutral were it not for the fact that the choice of communications technology has a great deal to do with what can be communicated by using it (to take that one stage further, "The medium is the message", as Marshall McLuhan noted). The second layer is that of procedures, the rules and conventions used to communicate information. These include interfaces between users and technology. Thirdly, there are human factors, which include the ways in which both messages and the technologies by which they are conveyed are perceived. Finally, communications occur in the context of organizations that have different targets and priorities.

The result is undeniably complex. The correct formula, or happy medium, for the benign and productive use of information technology thus depends on the judgement of the emergency planner. In order to design a functional IT system for emergency usage, a close analysis is required of the four layers of communication, the cultural matrix in which they are set, and the needs generated by emergencies.

CHAPTER 4

The emergency plan and its activation

Having considered the basic aims of emergency planning, and some of the tools and methods that are available to achieve it, we will now examine the writing, revising and implementing of an emergency plan. In this process, technology must mesh with geophysical, social and cultural factors to produce a multi-faceted document flexible and detailed enough to direct operations in a wide variety of circumstances, but simple and well structured enough to be understood by a wide variety of users.

As noted elsewhere in this book, a basic choice must be made between single-hazard and multiple-hazard (or all-hazards) planning. The latter is preferable, as it achieves greater levels of protection and economies of scale. However, it is more expensive and time-consuming than the former.

4.1 The process of planning

Emergency planning is indeed a process, more than it is a static sort of goal. It should be practised continuously and progressively, as plans should be allowed to grow and adapt to changing circumstances. In the following sections, a model of generic emergency planning is presented. In this, the plan offers a set of procedures for coping with eventualities that have been predicted by making a detailed assessment of hazards, vulnerabilities and risks. The methodology offered here is most suited to comprehensive (all hazards) planning at the level of municipalities or other local jurisdictions. However, much of it can be adapted to other situations, such as planning for industrial sites or medical facilities. In most cases the basic principles are the same. To begin with, Table 4.1 lists the basic ingredients of an emergency plan and Table 4.2 offers a model for the structure of a generic plan.

Table 4.1   The basic ingredients of an emergency plan.

Context
• legislative framework
• participating organizations
Scenarios
• hazard
• vulnerability
• risk
• impact
Emergency needs
• search and rescue
• medical care
• public safety
• food and shelter
• damage prevention and limitation
Available resources (structure, items, competencies)
• manpower (personnel)
• equipment
• vehicles
• buildings and facilities
Resource utilization
• application of resources to problems posed by scenario
• dissemination of plan
• testing, revising and use of plan

Table 4.2   Structure of a typical emergency plan.

Title

Preface: jurisdictions, sponsorship, authorities involved, dates

1. Introduction
- policy statement on disaster planning by chief executive officer
- legislative authority for the design of the plan and for the steps it contains
- aims and general purpose of the plan
- conditions under which the plan comes into force
- national and regional or provincial framework of local emergency planning

2. Local hazards
- nature of local risks
- introduction to local area, its characteristics, resources and hazards
- historical description of impacts

3. Vulnerability and risk analysis of local population, built environment, economic activities, social and cultural systems
- assessment of community disaster probabilities
- risk analysis (hazard × vulnerability × exposure)
- risk and disaster scenarios for the local area
- risk-management strategies

4. Legal and jurisdictional responsibilities for emergency management
- including warning, evacuation, search and rescue, healthcare and sanitation

5. Introduction to local emergency management resources
- personnel, equipment, supplies, communications, etc.

Table 4.2 (continued)   Structure of a typical emergency plan.

6. Structure of local emergency command system
- authority organization chart
- hierarchies, organizational relationships, coordination and command structures, reciprocal aid agreements and links with other jurisdictions
- relationships with other levels of government, particularly emergency-related agencies

7. General all-hazards plan *OR* plans for specific types of emergency
- roles, relationships and tasks

Operation of warning systems
- types of warnings, how they will be distributed, obligations on receiving warnings

Pre-impact preparations
- relationships between type of disaster agent and necessary preparations
- responsibilities of different agencies
- location of sites of greatest risk

Emergency evacuation procedures
- conditions under which evacuation is authorized
- routes to be followed and destinations
- how the special needs of the elderly, ill, or institutionalized will be accommodated
- locations and facilities of emergency shelters

Emergency-operations centres and incident-command centres
- locations, equipment, operation, staffing

Communications

Search and rescue
- responsibilities, equipment, places most likely to require SAR work

Public order

Public information
- management of the mass media

Medical facilities and morgues
- location, transportation, capacity, facilities

Restoration of basic services: order of priorities, responsibilities
- protection against continuing threat:
- the search for secondary threats, actions to be taken if discovered
- continuing assessment of the total situation: responsibilities, distribution

8. Plans for specific sections
- aviation, hospitals, art treasures, secure institutions, archives, tourism, factories, nuclear reactors, etc.

9. Arrangements for testing, disseminating and updating the plan
- plan distribution and publicity
- field exercises and their evaluation
- standard or automatic procedures for updating the plan

Appendices
- tables of data
- maps and photographs
- lists of names, addresses and contact numbers of all relevant agencies, their heads and deputies
- detailed descriptions and strategies

4.1.1 The basic ingredients of the plan

The plan must make provision for hazards that are locally important. The first task, then, is to rank hazards in terms of their magnitude, frequency and, above all, impact on the local area. Some aspects to be considered when characterizing local hazards are shown in Table 4.3. They should be introduced at the start of the plan, with a brief description of past impacts and potential future scenarios. Measures taken to mitigate the hazards, assess vulnerability and reduce risks should be summarized. Anticipated consequences should be discussed, of both future disasters and efforts to provide relief. Although there are many chance factors that cannot be anticipated in prior planning, a brief description of some less likely hypotheses and scenarios may be provided as long as they are clearly identified as outside chances. Scenarios should in any case allow for secondary, compound or multiple hazards, or for multiple or repeated impacts, and the plan should outline procedures for tackling these.

The bulk of the plan should provide details of resources, structures, networks, procedures and competences (i.e. assigned tasks) that will be brought to bear on disasters of different types. If there is relatively little difference between the responses required for each hazard (and the same agencies will tackle it each time), or if the area is dominated by a single risk type, then the

Table 4.3 Aspects of hazards and disasters of relevance to planning and management.

Physical occurrence
Probability
Frequency
Transience (duration)
Physical magnitude
Energy expenditure
Physical effects: direct, indirect and secondary
Area affected: directly and indirectly
Degree of spatial concentration or ubiquity
Volume of products (e.g. lava, floodwater)

Predictability
Short-term (for avoiding action)
Long-term (for structural and non-structural adjustment)

Controllability
Can physical processes be modified?
Can physical energy expenditure be reduced?
Can effects be mitigated?
Can effects be modified?

Socio-cultural factors
Belief systems inherent in societies
Degree of knowledge of risk
Complexity of social system and its constituent groups

Ecological factors
Environmental damage propensity
Environmental compatibility of mitigation measures

arrangements can be dealt with in a single sequence of descriptions. If, instead, the hazards and their potential consequences are varied, then it may be necessary to divide the plan into a series of sections, with some overlap and repetition (or at least heavy cross referencing) in what they contain.

The purpose of the plan is to inform, instruct and direct participants. It should therefore tell them what to expect, what to do and what emergency resources to employ. It is vital that participants know not only their own roles but also those of their colleagues, in order to facilitate interaction, increase the level of collaboration and reduce confusion. Hazard scenarios must therefore be matched by descriptions of arrangements made to tackle the hazards. Thus, it is opportune to include a list and description of local emergency resources categorized personnel, vehicles, equipment, consumable supplies, services, institutions (e.g. hospitals) and organizations (e.g. the local **Red Cross** or **Red Crescent** chapter). There are three ways to present this description:

- as an early chapter of the plan, and therefore as a comprehensive survey of the resources in question
- in relation to each hazard scenario for which resources will be needed
- as an appendix to the plan.

In any case, large disasters will require more resources than those that are available locally. Hence, the plan must describe what regional (i.e. county, province, region or state-level) and national resources can be expected to be supplied in a local disaster, bearing in mind that impacts covering a wide area will necessitate sharing resources among neighbouring jurisdictions. It helps to be frank about anticipated shortfalls of relief and assistance. In some cases, these can be overcome by concluding reciprocal aid agreements (as part of the planning process, that is; see §4.1.6) or by launching international aid appeals. Where possible, procedures and expectations should be outlined in the plan. Special attention should be given to how local arrangements will intermesh with regional, national and international ones, as the duplication of aid is more than merely disadvantageous, because it debilitates the relief effort.

Some of the most basic forms of planning follow a linear progression through the stages of the emergency (Fig. 4.1). Forms of preparation and training during "peace time" (i.e. between impacts) can be outlined in the plan. Next come procedures for turning scientific forecasts and predictions of impending impact into **alert** procedures and warnings. In this context, the response to predictions needs to be considered with care in the plan, as it involves transforming scientific information, which may be obtuse, inconclusive or difficult to comprehend, into a set of definite actions. These will include procedures to warn and call up emergency responders, and those to alert members of the public and, where appropriate, evacuate them. Except in cases where there is no forewarning of disaster, alerts are often conducted in stages

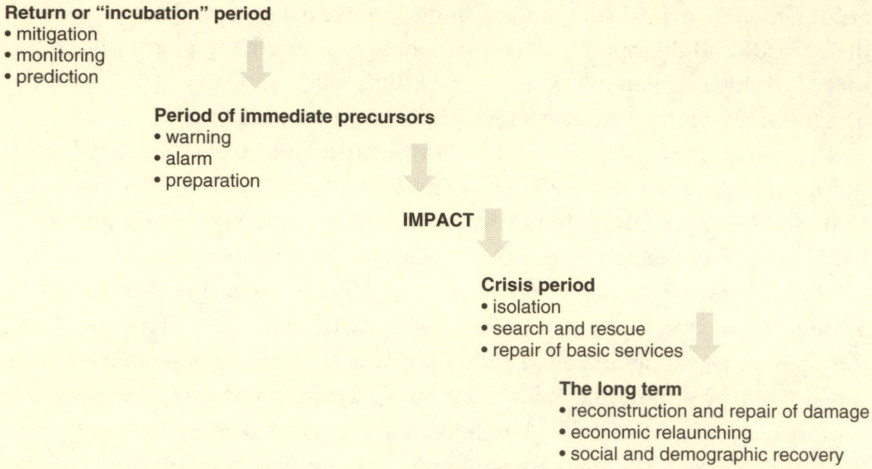

Return or "incubation" period
• mitigation
• monitoring
• prediction

Period of immediate precursors
• warning
• alarm
• preparation

IMPACT

Crisis period
• isolation
• search and rescue
• repair of basic services

The long term
• reconstruction and repair of damage
• economic relaunching
• social and demographic recovery

Figure 4.1 The temporal phases of disaster and major planning and management tasks associated with them.

(see §5.2), including stand-down and end-of-emergency procedures when the danger recedes. Each stage and each action connected with it require the planner to design mechanisms to verify the action that has taken place as planned and, if it has not, to adapt the broadcast message, or the **control** structure.

In terms of structures, the plan should outline participatory organizations, communications channels, response procedures and the chain (or web) of command. The description should begin with an outline of call-up procedures (see §4.1.2) and progress to a description of individual responsibilities and the arrangement of operating units and organizations, plus their interconnections. It is often appropriate to use diagrams and flowcharts to illustrate the command structure and it is vital that, whatever the structure, all participants know to whom they are responsible and who is under their charge. The consequences of each command action should be outlined in terms of their repercussions for public safety and public order. This will include provisions for evacuating vulnerable neighbourhoods, and special provisions for the aged, the very young, the handicapped, the sick and prisoners (see §5.10). Evacuation routes and centres must be designated and preliminary shelter provided for. Clothing, bedding and food may need to be distributed during the short-term aftermath. There must be provisions for traffic control, maintenance of public order, and restriction of access to dangerous areas or places where emergency personnel are working. Plans that make provision for the long-term aftermath of emergencies will have to designate specific areas for setting provisional shelter, such as prefabricated structures or commandeered public buildings. This will require specification of the types of shelter to be provided, sources of materials or buildings, forms of urbanization, and service provisions at the site.

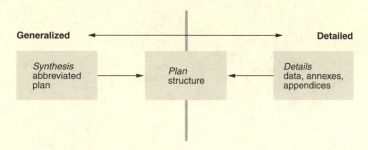

Figure 4.2 Tripartite structure of an emergency plan.

Two sections of the plan may require special attention. The first refers to the treatment of information flows. For emergency workers, communications channels, **protocols** and priorities will need to be specified. For the general public, a system of information dissemination will be required. This will include an information service, message-relaying service, situation reports, and press liaison facilities (e.g. press releases, press conferences and interview facilities). The second special section refers to the more urgent matters of search and rescue and primary medical care. These form a plexus of problems that requires an integrated solution in terms of dispositions for the rapid retrieval of injured victims, their triage on site, and their transportation to treatment centres (see §6.1). Procedures may differ between hazard types, as the expected kinds of entrapment and bodily injury may vary.

The presence of special sections in the plan calls attention to the question of how much detail an emergency plan should contain. On the one hand, it is necessary to ensure that nothing fundamental is left out, but, on the other, any excess of detail may lead to confusion and inability to comprehend the essential details of the plan. A happy medium must be struck. One way to do this is to make the body of the text a "grand plan" that outlines the basic elements and relegates much of the detail to technical annexes and appendices (Fig. 4.2). An abbreviated digest of the plan can be produced as an aide memoire for use in the stressful period at the start of an emergency (Fig. 4.3).

4.1.2 Command structures and call-up procedures

One of the most important and fundamental parts of the disaster plan is its command structure, the backbone of emergency operations. A weak or inappropriate structure may lead to confusion and inefficiency during the provision of emergency relief. In contrast, a well designed system of command will operate smoothly and without ambiguity. The first principle is that all participants in the emergency should know to whom they must answer and exactly for whom

101

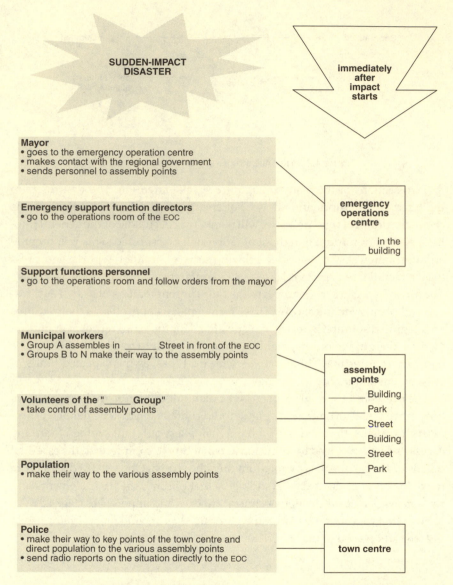

Figure 4.3 The emergency plan reduced to a single page. Simplified procedure for municipal operations. Original diagram courtesy of the Italian Department of Civil Protection.

and for what they are responsible. In many countries and regions, some of these responsibilities are specified in civil-protection laws; for example, a mayor, president, state governor or cabinet minister may be the figurehead at the upper end of the command structure. But the emergency plan must map out the full structure as completely and unambiguously as is needed to make it work.

A basic choice must be made between the more traditional hierarchical structure, or chain-of-command, and the newer and more avant garde structure known as the **incident-command system** (ICS). There are advantages and disadvantages in both methods of control. A typical chain of command will be headed by someone with ultimate responsibility for managing disasters. This is usually a senior politician, although he or she may delegate responsibility to a senior civil servant, perhaps with the title of disaster manager or emergency coordinator. On rare but happy occasions, the politician who has the civil-protection brief is also an expert in the field, perhaps an academic or former member of the emergency services. As the head of operations is usually responsible for signing legally enforceable orders or directives, or at least for devising them, his or her nomination and position are usually determined by existing laws, not by the formulator of the plan.

Commonly, the working head and the nominal head of civil protection (whether or not they are the same person) are backed up by a committee of leaders and experts who help determine policy, strategy and tactics. These may include senior police chiefs, engineers, emergency-service directors, and politicians with responsibility for such departments as public works and healthcare. There may be separate advisory committees, made up, for example, of academics and technicians with responsibility for monitoring hazards. In more complex structures the policy and executive committees are separate, the former devising broad approaches to situations and the latter formulating orders to put them into effect. In much simpler cases, the civil-protection director may do without a formal strategic committee and simply garner advice where it can be obtained.

Further down the hierarchy come a series of operational units that represent the part of the command structure that directs and carries out the various emergency functions. The precise nature of these depends on the forces that must be managed and the government departments of which the emergency structure is, at least in part, an outgrowth. Each department will be responsible for a range of units composed of chiefs (or principals) and fieldworkers. These will represent various ranks and competencies, such as pilots, drivers, paramedics, volunteers and so on. Whatever the configuration chosen, in the end it should take at least some account of the following categories:
- search and rescue, rapid response and initial medical aid
- telecommunications and other forms of communications channel
- vehicles and ground transportation
- airborne operations, including air-traffic control
- hazardous-materials response
- civil and structural engineering
- health and sanitation, including direction of medical services and facilities

- emergency shelter and housing
- physical welfare, including distribution of food, clothing and bedding
- social and psychological services
- materials, supplies and warehousing
- public order and public safety, including traffic control
- military forces
- public-information services and mass-media liaison
- accounting, finance and personnel, including distribution of emergency assistance funds.

These are the **emergency-support functions**. By way of an example, Table 4.4 compares the actual categories used in two countries, the USA and Italy.

Participation of military personnel in civil disasters is a somewhat thorny issue. In some countries they are the principal, perhaps the *only*, source of disaster relief, although this is a rather outmoded situation and one that stems from the time when civil *protection* was thought of as **civil *defence*** – the protection of the civilian population against the effects of bellicose attack. In any case, a duplication may exist between civil and military command structures. In the latter, the command system is rigid and unalterable, as is appropriate to warfare. Military forces have the advantage of being well organized, efficient and self-sufficient in the field. Their main drawbacks are a possible tendency towards authoritarianism, in situations that require delicate handling from the social point of view, a certain slowness of collective movement, and the difficulty of suborning the military command structure to civilian leadership. It is axiomatic that command systems and relief efforts should not be duplicated (except in so far as is necessary to ensure that no parts are missing when disaster strikes). There is therefore a risk that military forces will work at cross purposes with civilian ones. However, it can be overcome by sensitive planning, close collaboration and constant communication. In some cases a possible solution is to convince the military to lend personnel and equipment to the civilian authorities for the duration of the emergency. The soldiers, sailors or airmen will, of course, be able to use many of their characteristic skills in the new civilian context. In general, military forces work best in civilian disaster when they are assigned specific tasks and allowed freedom of action to perform them.

In the case of military forces, a hierarchical command structure is inevitable, but in civilian ones it is not. After experiencing problems with the organization of operational units to fight **wildfires** in the 1960s, in 1970 the Californian authorities devised one of the first non-hierarchical incident-command systems. These have steadily become more popular with the onset of the information-technology revolution, as their viability depends upon the maintenance of a copious and constant flow of emergency information, which modern technology facilitates. In an ICS the hierarchical chain of command is limited, and

Table 4.4 A comparison of US and Italian emergency support functions.

Emergency Support Functions – US Federal Government

ESF	Agency
1: Transportation	DOT; USDA, DOT, DOE, DOS, GSA, ICC, TVA, USPS
2. Communications	NCS; USDA, DOC, DOD, DOI, DOT, FCC, FEMA, GSA
3. Public works and engineering	DOD; USDA, DAC, DOE, DHHS, DOI, DOL, DOT, VA, EPA, GSA, TVA
4. Firefighting	USDA; DOC, DOD, DOI, EPA, FEMA
5. Information and planning	FEMA; USDA, DOC, DOD, DOED, DOE, DHHS, DOI, DOJ, DOT, TREAS, ARC, EPA, GSA, NASA, NCS, NRC
6. Mass care	ARC; USDA, DOC, DOD, DHHS, DOE, DHHS, DHUD, DOT, AID, FEMA, GSA, USPS
7. Resource support	GSA; USDA, DOC, DOD, DOE, DHHS, DOL, DOT, AID, FEMA, NCS, OPM
8. Health and medical services	DHHS; USDA, DOD, DOJ, DOT, AID, ARC, EPA, FEMA, GSA, NCS
9. Urban search and rescue	DOD; USDA, DHHS, DOL, DOT, ARC, FEMA, GSA
10. Hazardous materials	EPA; USDA, DOC, DOD, DOE, DHHS, DOI, DOJ, DOL, DOS, DOT, FEMA
11. Food	USDA; DOD, DHHS, DOT, ARC, EPA, FEMA
12. Energy	DOE; USDA, DOD, DOS, DOT, GSA, NCS, NRC, TVA

Full capitals: primary agency responsible for management of the ESR. *Small capitals*: secondary agencies.

US Federal Agencies	
USDA	Department of Agriculture
DOC	Department of Commerce
DOD	Department of Defense
DOEd	Department of Education
DOE	Department of Energy
DHHS	Department of Health and Human Services
DHUD	Department of Housing and Urban Development
DOI	Department of Industry
DOJ	Department of Justice
DOL	Department of Labor
DOS	Department of State
DOT	Department of Transportation
TREAS	Treasury
VA	Veterans' Administration
AID	Agency for International Development
ARC	American Red Cross
EPA	Environmental Protection Agency
FCC	Federal Communications Commission
FEMA	Federal Emergency Management Agency
GSA	General Services Administration
ICC	Interstate Commerce Commission
NASA	National Aeronautics and Space Administration
NCS	National Communication System
NRC	National Research Council
OPM	Office of Personnel Management
TVA	Tennessee Valley Authority
USPS	Postal Service

Table 4.4 (continued)   A comparison of US and Italian emergency support functions.

Italian National Emergency Support Functions ("Augustus" method)

1. **Scientific/technical – planning**
   National Research Council; national scientific research groups; National Geophysical Institute; regional departments or agencies of civil protection; National Technical Services.

2. **Health and social services**
   Ministry of Health; local and regional health authorities; Italian Red Cross; social and health sector volunteer associations (NGOs).

3. **Mass media and information**
   RAI – state television and radio services; national and local private television and radio stations; the press.

4. **Voluntary associations**
   National Agency for Civil Protection; local, provincial, regional and national associations.

5. **Materials and equipment**
   CAPI; Ministry of the Interior; Mercury Information System; armed forces; Italian Red Cross; public and private companies; voluntary associations.

6. **Transportation**
   Italian State Railways; road, sea and air transportation; National Roads Consortium (ANAS); autostrada companies; provinces; municipalities.

7. **Telecommunications**
   Telecom Italy; Ministry of Posts and Telecommunications; INMARSAT; COSPAS/SARSAT; amateur radio operators.

8. **Essential services**
   Electricity, gas and water companies; banking system; fuel distribution services.

9. **Census of damage, people and commodities**
   Industrial, artisans' and commercial associations; Ministry of Public Works; Ministry of Cultural Heritage; infrastructure; private sector.

10. **Search and rescue**
    National Agency for Civil Protection; fire brigades; armed forces; Italian Red Cross; Carabinieri; Revenue Police; Forestry Corps; police forces; volunteer organizations; National Mountain Rescue Corps (Italian Alpine Club).

11. **Local authorities**
    Regions, provinces, municipalities; mountain intermunicipal associations.

12. **Hazardous materials**
    Fire brigades; National Research Council; storage areas and industries at risk.

13. **Evacuation logistics and reception centres**
    Armed forces; Ministry of the Interior; Italian Red Cross; voluntary associations; regions, provinces, municipalities.

14. **Coordination of emergency operations**
    Connections to intermunicipal emergency-operations centres (COMS); resource management; computer technology and methods.

Box 4.1 A case of inefficient planning

The following example illustrates the problems associated with having an emergency plan – two of them in this case – that does not work. It offers an opportunity to reflect on what a plan should consist of, and what pitfalls to avoid when compiling and implementing one. As the example is both a real and somewhat controversial, professional discretion precludes revealing all the details.

The example

This case history involves a seaside town with an international reputation for expensive high-fashion tourism. Its average January population, largely composed of year-round residents, is 62 000, and in August this swells to 150 000 or more, as holidaymakers arrive *en masse*. Prices are high in town, real estate is very expensive there and hence the municipality is flush with cash (which thus endows it with adequate means to promote emergency preparedness). An international cultural festival and a major sports event are held in the town each year and these add to the revenue base, as does a large gambling casino on the waterfront.

The town is situated on a rocky coast in a Mediterranean environment, with mild winters and hot dry summers. Steep slopes forested with pine trees come down to within a few metres of the narrow beaches and small yachting marinas situated along the coast. A branch railway and a narrow main road wind their way between the backslope and the beaches. The urban area extends about 4 km linearly along the coast; the centre of town runs up to 2 km inland along three narrow valleys in the mountain front. An international freeway runs on a series of viaducts and in tunnels through the mountains about 3 km inland from the coast, whose direction it follows.

The town has two emergency plans, both commissioned by the previous municipal council from recognized professionals in the emergencies field. Neither plan has ever been activated, nor have any steps been taken to implement it. Both are huge, unwieldy documents, very theoretical in character and old fashioned in concept. They remain on the bookshelves of the town's planning office.

On a hot July day, forest fires break out in the hills around town and considerable effort is required on the part of firefighting crews, who are mostly unpaid volunteers, to bring them under control and avoid damage to the periphery of town. Later in the month a sudden heavy rainstorm generates a flash-flood that sweeps through town late one afternoon. It runs down the central boulevard and, although the water is only a few tens of centimetres deep, the current is strong. A woman is swept off her feet while trying to get out of her car, which has been immobilized by the floodwaters. She is swept under a parked vehicle and within two minutes she has drowned, despite the efforts of onlookers to form a human chain and rescue her. During the following winter, sustained rainfall gives rise to large land-slides that block through-roads (including the freeway). A quarry worker and two motorists are killed by landslides. Several houses have to be evacuated for fear that debris flows may destroy them.

The diagnosis

I was asked by the municipal administration to examine the emergency plans and suggest ways of making them functional, in line with a general reorganization of the town's civil-protection services.

Both plans offered considerable detail about hazard scenarios, which had been investigated quite thoroughly. One plan was limited to flood propensity; the other largely concentrated on seismic hazards. In essence, they stated the following:
- Six ephemeral streams run from the mountains to the coast through the urban area. All of them are susceptible to flashflooding, which would be worst where culverts under roads and the railway line restrict the flow of water. In each case water would back up over an area of up to 300 m by 500 m and inundate up to 50 buildings.

Box 4.1 (continued)   A case of inefficient planning

- A late nineteenth-century earthquake (magnitude 6.0) was used as a reference event for seismic studies. Its epicentre was located off the coast and it generated a small tsunami. Predictions based on site surveys suggest that the probable maximum seismic event could damage up to 20 000 buildings. Most of them would be in the older neighbourhoods at the centre of town, where nineteenth-century buildings, although elegant, have not been seismically retrofitted. A tsunami would seriously threaten holidaymakers if it were to occur when the beach was in use. There could be as many as 300 deaths in town when buildings collapsed.
- The forest-fire hazard remains highly significant, especially in summer months, when hot dry winds blow across the slopes. The continuing extension of urbanization into the backslopes has added to the risk. Many fires are deliberately set by arsonists.
- The section of coast that includes the town is subject to occasional fierce storms that can drive tall waves onto beach-front properties. There is also a significant tornado and waterspout risk.

Although the emergency plans were detailed with respect to hazards, they were deficient in most other respects. To begin with, vulnerability, risk and impact studies were inadequate for flood, land-slide and storm events, which in this area are all more common than earthquakes. Secondly, the plans did not employ the same methodology as the regional emergency plans with which they would have to interface. It is obviously essential that plans interact efficiently, rather than conflict with one another or simply fail to connect. It is often said that the only thing worse than having no emergency plan is having two of them. There was no justification in this case for not having a single integrated plan for all hazards.

Put together, the plans approached 900 pages of documentation, but to what effect? It was striking, for instance, that one of the plans identified the floodable areas, road by road and building by building, on large maps, but said nothing about the probable consequences of flooding. It is one thing to flood an empty building and another to flood, for instance, a fully occupied old people's home.

One of the authors of the two plans clearly thought it was important to educate the reader about the nature of seismic hazards. He therefore devoted a large amount of space to a general, introductory review of the causes and consequences of earthquakes. This would have been ideal in a university course on natural hazards but was utterly out of place in an emergency plan. One therefore had to wade through reams of material on magnitudes, epicentres and so on, in order to reach the operative part of the plan. This included remarkably few prescriptions about what to do next time an earthquake actually strikes the town.

Both plans suffered from a common ailment of such documents: they prescribed remedies to haz-ards and risks on the basis of what would be appropriate under ideal conditions. Realism is obviously an essential ingredient of an emergency plan. Hence, it is inappropriate to state what one would *like* to see in the way of resources, structures, manpower and organization. For example, one of the two plans specified an emergency-operations centre that would have required a staff of about 20 and considerable investment in equipment, furnishings and indoor space. Yet it was several years before the municipality was able to insert a line into its budget to fund an EOC at all, and space never became available to the extent that the plan required it.

In synthesis, this example illustrates the following:

- It is important to maintain a balance between the elements of a plan, including the definition of haz-ards, vulnerabilities and risks, the construction of impact scenarios, the assessment of resources, and the design of measures to tackle hazards and impacts.
- A plan should not be encumbered with masses of irrelevant or semi-relevant details.

Box 4.1 (continued)   A case of inefficient planning

- Plans that rely more on bulk of documentation than realistic provisions for tackling specific kinds of eventuality are unworkable.
- There is a need for plans that can be fully integrated with the overarching regional emergency-support programs and facilities that will furnish extra resources and direction when municipal capabilities are overwhelmed by an exceptionally large event.
- Plans should be realistic, pragmatic and operationally orientated, not theoretical, hypothetical and didactic. They *must* specify exactly what the emergency services are to do in an emergency.

In short, the two plans constituted a classic case of what other authors have termed the "paper-planning syndrome", in which the plan exists, and it may even contain useful material, but in the present form it cannot be implemented.

The solution

The first aspect of the remedy was obviously to combine elements of the two documents into a single instrument that would act as a workable, generic, all-hazards emergency plan. Data on flood and earthquake hazards could be extracted, synthesized for the new plan in order to give a clear picture of the basis of risks, and placed in supporting appendices. However, groundwork still needed to be done, as the picture on hazards, particularly landslides and forest fires, was incomplete.

The second stage was to make a realistic assessment of resources available under the jurisdiction of the municipality and ascertain how far these could be supplemented in an emergency by the regional authorities and other forces. Once the hazard, risk and impact scenarios had been refined and updated, a start could be made on specifying how the available resources could be matched with predicted needs during an emergency. This would be written up using current national and international methodologies. Steps would need to be taken to ensure that the resulting plan was (a) simple enough to be implemented rapidly, but detailed enough to cover all the main eventualities, and (b) flexible and robust enough to be applicable to hazards that could not be foreseen adequately (a crash involving radioactive materials on the freeway or railway, for instance).

A very important aspect that the two existing plans had not tackled effectively concerned how to manage threats to the transient population while not giving the town, always conscious of its role as a major tourist resort, a negative image. This would involve launching a public-awareness campaign that encouraged tourists and visitors to consider their own safety, but by introducing the problem in a way that emphasized the positive aspects of personal security, not the negative ones of hazards. This required the dissemination of information in hotels, restaurants and other public places, and also organizing very public emergency exercises "in the holiday spirit", as it were, to encourage full participation.

Note: Sources of information in this example, and precise details of the locality discussed, must remain confidential.

parallel operational units are managed by a coordinator, who has few powers of coercion. Tasks are apportioned and shared out by mutual consultation; overlap and duplication of effort are minimized by ensuring a continuous multilateral flow of information about what needs to be done and what is already being done in a particular emergency. Special problems are tackled by forming task forces; regular operational units spread out to tackle the more routine emergency tasks.

The main advantages of an incident-command system lie in its flexibility in operation and its ability to mutate or expand to meet the requirements of the situation. It is particularly valuable when the extent, seriousness and complexity of an incident are not clear to the first responders. It is also most appropriate when the incident needs to be directed largely, or entirely, from the field.

In a typical small-scale application of ICS, the leader of the first emergency response team to reach the scene of the incident assumes command. Typically, this would be a fire chief. He or she pulls on a reflective tabard with INCIDENT COMMANDER written on it in large letters and begins directing operations from the back of the emergency vehicle, while keeping in constant contact with base (the fire station, dispatch room, or emergency-operations centre). If reinforcements arrive (either because the **incident commander** calls for them or because they are sent by dispatchers), they are successively integrated into the evolving command plan.

It is reckoned that a single commander can personally direct between three and seven operations or workers. This is the **span of control**. Once the situation becomes more complex than this, it is time to form separate teams, or task forces, and to delegate command. The incident commander keeps a running record of the architecture of operations as these grow. He or she may cede the role of commander to another operations chief if a more senior person arrives, if he or she is needed elsewhere, or if exhaustion points to the need for a rest.

The larger an incident-command system becomes, the more it involves division of competencies (Fig. 4.4). Major events tackled by ICS involve branches of finance, logistical support, planning, operations and relations with the public and mass media.

One particular problem with ICS is that its impromptu architecture is easily compromised by the "freelance" activities of personnel who decide to work alone. In order for operations to run smoothly, all participants must be part of the command system. If any work separately, or are not accounted for, this can make it difficult to assign tasks effectively and avoid duplication of effort. If this is a significant risk, then a more rigid, authoritarian command-and-control structure may be more effective. To ensure proper integration of forces in a small incident-command system, a system of identification tags can be used. The tag is handed to the incident commander as the holder (or the team he or she supervises) starts an assignment. The commander keeps the tag secured to a chart of operations to remind him or her that the holder is at work.

Another disadvantage of an ICS is that it is difficult to control from above, although advantages are reaped in terms of flexibility and promptness of action, as the individual units do not have to wait for orders to be formulated from the general leadership before they can carry out tasks. There are, of course, risks of lack of coordination and of failure to address all the needs generated by

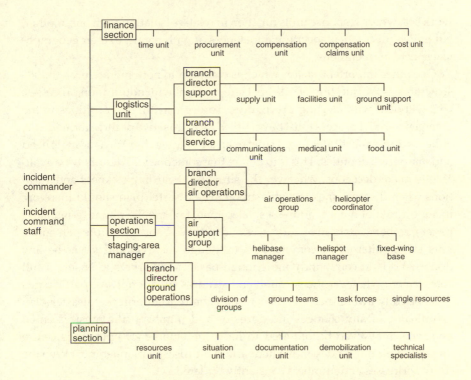

Figure 4.4 The organizational structure of a large incident-command system.

the emergency. However, most of these can be solved by ensuring that there are no gaps and inadequacies in either the decision-making process or the flow of information between teams in the field. In synthesis, incident-command systems are most appropriate to the management of large heterogeneous emergencies that need to be handled with great flexibility, and also to situations where the institutional culture does not demand a rigid structure of command. They are much less useful in small concentrated emergencies and places where the chain of command is an ingrained idea, without which emergency workers would be lost.

The emergency planner who must design a command structure first needs to evaluate the legal requirements of emergency response in his or her jurisdiction. These may specify all or part of the structure or may apportion specific emergency powers to particular figures. In a case in which there are few such constraints, a choice can be made between the chain-of-command and ICS options. Cultural or organizational factors may point to one or the other, for example, to a hierarchical structure where people are accustomed to giving or receiving orders or where emergency response is the preserve of organizations (such as a civil-guard corps) that are already very hierarchical. ICSs tend to

111

work best where response units function in relative isolation from one another, either in terms of their specific competences or with regard to their geographical sphere of operations.

One other important factor, which is common to both hierarchical and ICS structures, is the question of call-up. Here, we deal with notification processes; §5.2 will deal with alerting procedures. To begin with, key positions in the emergency-relief process must be occupied by at least two or three individuals, either as direct substitutes (e.g. a chief or an alternative chief) or as a single head and one or two deputies. This is to assure that someone will definitely be available to make decisions, whatever the situation regarding sickness and vacations when disaster unexpectedly strikes. The disaster plan should therefore include a rota of names, addresses, telephone, fax, mobile phone, e-mail and pager numbers, alternative addresses and numbers, and substitute personnel. This means alternative commanders (although there should never be any doubt who is in command) and enough basic fieldworkers to ensure a full complement whatever the vacation and sickness situation is. All participants must be fully aware that it is their responsibility to inform the disaster-plan coordinator of any changes in addresses and numbers (the most essential personnel may even be required to keep the emergency-operations centre aware of their day-to-day movements), and the disaster planner must keep the list of addresses and numbers constantly updated.

When an alert is imminent, or disaster unexpectedly strikes, one encounters the problem of how to ensure that all participants in the plan report for duty. There is no perfect system for alerting them, and each alternative has advantages and disadvantages. A single operative may contact each person on the list by telephone, radio or pager. This can be time consuming, complex and inefficient, especially if the list is long, although computerized dialling and telephone-message broadcast can be used. In addition, verification can be complex if responses are slow or people are absent. Alternatively, a **fan-out system** can be used in which person A contacts persons B and C, B contacts D and E, C contacts F and G, in a "telephone tree" or equivalent (Fig. 4.5). This shortens the time required but is liable to break down when people on the contact list are absent. If electronic communication is impracticable, a courier can be sent to the home or workplace of each participant. This tends to be ponderously slow and may put the courier at risk if conditions are dangerous. The final option is to expect all emergency personnel to report to pre-arranged **rendezvous points** as soon as an alert is broadcast or disaster strikes. Of course, one has to be sure that the rendezvous points are the right place to be (and are not, for example, submerged by floodwaters or carried away by landslides). One must also be reasonably sure that the participants are likely to hear a broadcast alert. In short, there is no perfect call-up procedure and hence it is wise to

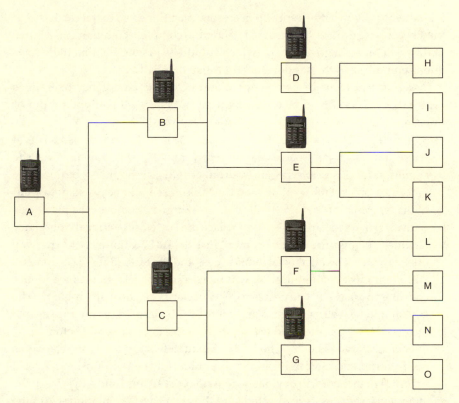

Figure 4.5 Emergency alert and call-up using a telephone fan-out procedure. Person A calls persons B and C, B calls D and E, etc. Cross checking is required.

design a fail-safe system in which one method can be substituted by another if it shows signs of failing. In any case, the emergency planner can take comfort from the fact that, if signs of the disaster are self-evident, emergency personnel will probably report spontaneously for work without needing to be contacted.

In conclusion, the command structure and the call-up procedures must be designed on the basis of an analysis of how to integrate and make functional the various organizations and participants in the emergency. These are described in the next section.

4.1.3 Emergency coordination structures

Good emergency management requires a well organized command centre to act as the hub of the command and communications structure. In some instances, different levels of government have been known to have separate command centres, but this may lead to overlap, confusion or inefficiency and

113

it is better to have a single centre, if necessary partitioned so that it can be occupied by diverse groups. These may include the complete command structures of different levels of government (which should be planned so that their operations enmesh) and the support functions listed in §4.1.2.

The physical focus of emergency planning is the emergency-operations centre (EOC). This should be a well thought out and well equipped room or building with the following characteristics:

- *Site location* The EOC needs to occupy a location well connected in terms of telecommunications, roads, and possibly rail or air terminals. The site should, as far as possible, be safe from hazards (e.g. it should not be floodable or located on unstable ground) and the building should be resistant to wind damage, water infiltration, earthquake shaking, or whatever the plan deals with in terms of hazards. It is wise to place the EOC close to critical emergency facilities, such as a fire station or ambulance depot. Location close to a railway line or main highway may facilitate access and the offloading of equipment. Although most communication will be by radio or telephone, face-to-face consultation may still be a necessary way of clarifying difficult problems. The EOC should be well signposted and identified by large signs on its principal façades. However, it should *not* be easily accessible to casual visitors.

- *Communications* The EOC should be adequately supplied with telephone, fax, Internet, and radio-transmitter equipment and connections. The level of investment in these will obviously depend on available funds, the size of the emergency services to be directed, and the extent and population size of the area to be covered. It is as well to have some redundancy (i.e. duplication) in communications, especially in terms of having different means of sending the same messages: damage or overloading can render some channels useless in emergencies, especially public telephone services.

- *Other equipment and facilities* The EOC will require various other facilities, according to the scale of operation. These include computing equipment and software (as described in §2.3.1 and §3.7), possibly including computer-display equipment for large-scale communal display of data and maps, a GIS for the analysis of local site conditions, and emergency-management and communications programs. Television and radio receivers provide news-media treatment of evolving disasters, as often the general public will be managed on the basis of what messages have been broadcast, and it will be necessary to monitor both these and the public's reactions to them. Well equipped EOCs have stocks of food and drink, limited cooking and sleeping facilities, and supplies of tools and protective clothing. These are not for the benefit of victims or emergency workers, but are intended to help guarantee the EOC a level of autonomy when disaster strikes and workshifts are greatly prolonged. Finally, it is useful to have plenty of parking space in the vicinity of the EOC so that emergency vehicles and supplies can be assembled there during exercises and real disasters.

- *Media-briefing facilities*   Many EOCs have a conference room where information can be given to journalists, and interviews conducted for radio and television. This room will require audiovisual equipment and other appropriate aids, such as a lectern and a backdrop with the agency's logo on it (this confers an official air on television interviews conducted in front of it). Office space for reporters, and communications facilities for the media, may also be provided in some of the largest EOCs.
- *Instructional facilities*   Some EOCs also have lecture halls with blackboards, slide projectors, and overhead projectors. This means that they can conveniently be used for training sessions, seminars and planning meetings. At the least it is helpful to have a small committee room in which heads of emergency services, scientists, and political and community leaders can get together and confer on tactics as conditions change during emergencies.

In reality, EOCs vary widely in terms of size, equipment and staffing levels. Small facilities may consist of a desk in a room attached to, for example, a municipal technical office, staffed by a single emergency manager whose regular employment is usually in activities such as landownership registration or urban planning. Equipment is limited to a single computer, radio transmitter, fax machine and a couple of telephone lines. Large EOCs may have many workstations and dozens of satellite-based telephone links. Commonly, the larger ones are divided into a conference room, in which tactics are worked out, and an operations room, in which emergencies are managed using telecommunications equipment (Fig. 4.6). In addition, EOCs are usually supported by the operations rooms of organizations such as fire and ambulance services.

Although it is logistically risky to have multiple EOCs within a single administrative jurisdiction, there is an exception to this rule. It can be helpful to extend the command structure with the aid of a mobile or portable incident-command post (ICP), or **field-command post (FCP)**, as it is sometimes known. This is a small facility that can conveniently be located in a camper or mobile trailer, which can be driven or towed to the site of the disaster and parked in a safe place adjacent to the main scene of search, rescue and triage operations. Emergency policy is formulated in the EOC and relayed as a series of instructions and directions to the ICP. The latter directs field observations on the basis of the EOC's instructions and it relays information back to the EOC on the state of the emergency. Hence, the ICP needs to be equipped with robust and dedicated communications channels – usually radio transmitter–receivers – to the EOC. It also needs to be staffed by one or more field-operations directors. However, it is important not to confuse the roles of the EOC and ICP; they must be complementary facilities.

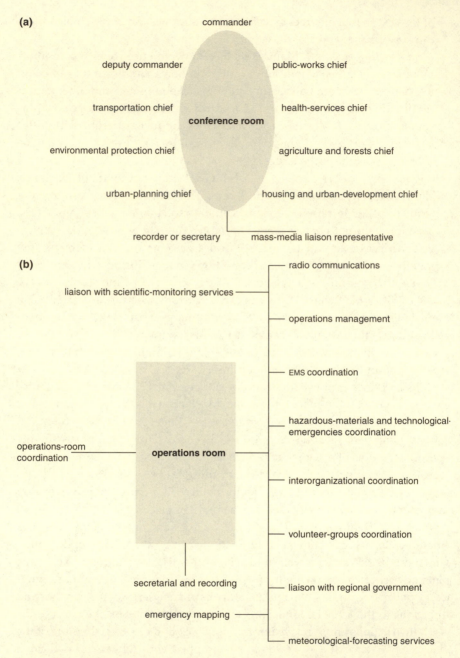

(a)

commander

deputy commander   public-works chief

transportation chief   health-services chief

conference room

environmental protection chief   agriculture and forests chief

urban-planning chief   housing and urban-development chief

recorder or secretary   mass-media liaison representative

(b)

radio communications

liaison with scientific-monitoring services

operations management

EMS coordination

hazardous-materials and technological-emergencies coordination

operations-room coordination   **operations room**   interorganizational coordination

volunteer-groups coordination

liaison with regional government

secretarial and recording

emergency mapping

meteorological-forecasting services

Figure 4.6 A bicameral emergency-operations centre consisting of a conference room, where strategy and tactics are worked out, and an operations room, from which emergencies are directed.

4.1.4 Participants and their organization

Although there are key individuals who participate in emergencies, such as a mayor, cabinet minister, emergency-plan coordinator or state governor, the plan will mostly coordinate groups and organizations, classified as follows:

- bureaucratic and legislative structures: parliaments, councils and political administrations, with emergency and relief committees where appropriate
- coordinating agencies such as emergency offices, emergency-management agencies and interdepartmental councils
- scientific bodies, such as geological surveys, meteorological offices, and hydrological-research bodies, with responsibility for monitoring and predicting, and for warning and advising the civil authorities
- normal emergency services: police, fire, ambulance and paramedics
- parastatal or non-governmental organizations, including accredited associations of volunteer workers
- teams and task forces, for example, for search and rescue or hazardous-materials response
- other types of organization, including news-media teams and associations of survivors.

Sociologists[*] have classified organizations that operate in disaster as follows (see also Table 4.5):

- *Adapting organizations* retain their original structure and complement of personnel, but adapt their operations to the needs of the disaster; thus, a local-government council may form a relief committee.
- *Expanding organizations* increase their complement in order to cope with the disaster, perhaps by taking on volunteer workers, cancelling leave of permanent personnel, calling in consultants, or increasing the hours of part-time workers.

Table 4.5 A classification of organizations in disaster.

Type of organization	Functions	Relationships	Example
Established	Regular	Old	Emergency-operations centre
Extending	Non-regular	Old	Red Cross absorbs volunteers
Adapting	Regular	New	Local government relief committee
Emerging	Non-regular	New	*Ad hoc* volunteer group
Redundant	None	None	None

Source: After "Organizational response to disaster: a typology of adaptation and change", J. W. Bardo, *Mass Emergencies* **3**, 87–104, 1978.

---

[*] See "Coming to terms with community disaster", R. R. Dynes, p. 115 in *What is disaster? Perspectives on the question*, E. L. Quarantelli (ed.) (London: Routledge, 1998). The sociological classification of organizations has also benefited greatly from the work of Enrico L. Quaranelli and Gary A. Kreps.

- *Extending organizations* increase the range of their activities to cover needs generated by the disaster; thus, a construction company may be involved in **structural mitigation** and **urban search-and-rescue** activities.
- *Emerging organizations* are born out of the situation created by the disaster and the emergence of people with latent gifts of organization and leadership; for example, victims and survivors may form an association to represent their needs more effectively. (Thus, what sociologists call a disaster sub-culture is born among the affected group.)
- *Redundant organizations* have no role to play during disaster and are usually abandoned by their members for the duration of the emergency. These may include sporting or cultural societies, although occasionally they adapt their functions and find a role in the emergency.

Participants in the disaster-relief effort, and the organizations that give them structure, method and purpose, must tackle various exigencies. Sociologists have classified these into **agent-generated demands**, which are created by the hazard and its impact (e.g. the need for sandbagging of levees in order to reduce flooding), and **response-generated demands**, which are a function of the relief effort (e.g. the need to house and feed relief workers). A worksheet for assessing the principal demands is given in Table 4.6.

In many respects, the art of emergency planning is one of creating adequate

Table 4.6 Agent- and response-generated demands in emergency situations.

Agent-generated demands	What is the demand?	Who will assume responsibility?	How is the demand to be met?
Warning			
Pre-impact preparations			
Search and rescue			
Care of the injured and dead			
Welfare			
Restoration of basic services			
Protection against continuing threat			
Community order			
Response-generated demands:			
Communication			
Continuing assessment of threats			
Mobilization			
Coordination			
Control and authority			

Source: A perspective on disaster planning, R. R. Dynes, E. L. Quarantelli, G. A Kreps (Disaster Research Centre, Ohio State University, Columbus, 1972).

connections between the various groups and organizations in order to ensure that they work together efficiently. The planner should therefore make a list of agent-generated and response-generated demands, and assign particular organizations the task of dealing with them. If the geographical area to be covered is large, populous or complex, this may need to be done region by region in order to assign full coverage. In this case, special attention should be given to the urban/rural dichotomy: search-and-rescue and relief needs and methods of delivery are likely to be different between cities and more sparsely populated or less accessible rural areas. Arrangements should be made to divide resources and responsibilities in both time and space, in order to take account of this fact (see §5.3). If the dichotomy is very strong (e.g. between a large city and a remote mountainous hinterland), it may be worthwhile trying to organize the disaster plan along dual lines, with separate agencies for the two environments, although they must be linked by high level of coordination and exchange of information and resources.

So far in this account it has been assumed that all relevant organizations will already exist, but of course there are situations in which some of them will not (the emerging organizations of the classification described above). The planner must review the twin lists of demands created by the emergency and ask whether the forces that can be marshalled are, in the event, likely to be adequate in terms of organizations and their staffs. If not, can new organizations be created and new members enrolled in them? This may involve inviting non-governmental organizations to set up local chapters, requesting more funds for emergency services, or concluding mutual aid agreements with neighbouring jurisdictions (see §4.1.6). In any case it is likely to be a long and possibly frustrating process. As an important general rule, emergency management should be conducted through existing tried-and-tested local organizations as much as possible, as these will have the contacts, expertise and experience to function well in the local area.

The emergency plan, as written, must clearly and concisely spell out the tasks to be accomplished by each organization. In addition, the planner should ensure, first, that each organization is fully aware of its role in future relief efforts (bearing in mind also that roles may vary from one kind of an emergency to another) and, secondly, that it has made all of its members fully aware of their personal roles in the emergency. No participant in the plan should be left without a well defined role; all participants should know both the substance and the limits of the actions they will be required to perform. However, the body of the plan should not be encumbered by the minutiae of details about roles and actions. These should be left to the participating organizations or put in an appendix to the plan.

Awareness of roles is also a matter for training and exercises. In these, each

119

individual participant learns about his or her own role, but also about the roles of others with whom he or she will interact during the emergency. Participants also need to be appraised of their legal responsibilities, either through the plan or through training sessions. At the top of the hierarchy, failure to declare an emergency when one occurs may be considered to be criminal negligence, as it will inhibit the flow of relief to the affected area. At all levels, some countries legally require citizens to furnish aid to people whose lives are threatened by accident or disaster, a fact that has important implications in terms of who gives assistance to whom and under what circumstances. It is worth studying these aspects of the law before formulating the disaster plan.

4.1.5 Volunteer groups and religious organizations

Major disasters require many relief workers. It is seldom possible for these to be permanent full-time salaried personnel, hence the important role of volunteers. Volunteers are by definition self-motivated occasional workers in the given field, who receive no remuneration for the work that they do. Indeed, lack of any form of conscription and payment together determine the *sine qua non* of volunteer status. There are essentially three types of volunteer worker: those who in times of need swell the ranks of regular permanent organizations, such as volunteer firemen who work alongside permanent members of the brigades; those who pertain to specific volunteer organizations, such as rescue squads or charities; and single unattached volunteers or small *ad hoc* groups. It is probably reasonable to suggest that the last type of volunteer not be given a role to play in the aftermath of a disaster unless there is both an overwhelming need for untrained manpower and ample facilities available to organize, direct, feed and house such workers. Usually, this will not be the case. Therefore, this section will concentrate on official volunteer groups with at least some degree of autonomy and professionalism.

In the aftermath of disaster, a mechanism may be needed politely but firmly to refuse the help of unorganized volunteers and to ensure that they do not get in the way of relief operations. That being said, there are nevertheless cases in which untrained and unorganized people will help in disaster, by searching through rubble for survivors, distributing food supplies, giving first aid, transporting the injured to hospital, and so on. In the emergency plan, such assistance cannot be specified or assigned, and should not be relied on. However, emergency managers should anticipate such forms of spontaneous help in disasters and be able to integrate them as far as possible into official efforts.

The two main questions regarding volunteer groups concern accreditation and integration into the disaster plan. In some cases accreditation is a matter of national or regional legislation. Statutes determine the criteria for recognizing the bona fide status of volunteer groups, what they can do, and the extent and limitations of volunteer workers' liabilities with respect to the people they assist, to other relief workers, and to themselves. Official status is usually highly prized by volunteer groups and may be a key factor in determining whether or not they can be utilized in plans. But where status is not conferred by statutory powers, or by national accreditation bodies, the disaster planner may need to make his or her own evaluation of each group.

There are various criteria to take into account when assessing the usefulness and reliability of a volunteer organization:

- Is the organization purely local or does it have a national or international structure to back up local chapters?
- Is the organization new and untried or well established? An objective, impartial review is needed of the group's reputation, admissions criteria, level of experience, history of success or failure, quality of leadership, esprit de corps, and organizational philosophy and policies.
- On a practical level, before an organization can be made an integral part of the disaster plan, an assessment of its training, equipment and autonomy levels is needed. Can it operate for prolonged periods and in hazardous conditions without draining resources from other disaster-relief operations?
- Specifically, does the organization provide its own food and shelter, protective clothing, transportation and materials needed to do the work at hand (e.g. water pumps for putting out fires, or jacks for lifting wreckage, or stretchers for moving the injured), in order to guarantee that it will function in the field?
- What sort of training are its members given, to what level and how often (including refresher courses)? In other words, how is the competence of each individual volunteer certified or attested?
- How easy is it to identify the group and its members: do they have recognizable uniforms, logos and liveries?

If there is no national or regional programme of accreditation, the disaster manager may wish to consider other means of certifying groups of volunteers, such as through professional associations. Groups without adequate guarantees of professionality should not be included in the disaster plan, nor encouraged to participate in relief efforts independently. However, it may be worth encouraging them to seek appropriate recognition.

Among recognized groups, some, such as the Red Cross and Red Crescent societies, have remarkable expertise in the disasters field. They are able to provide basic services to survivors and victims efficiently and autonomously. They therefore merit special inclusion in the disaster plan, as their services can be judged to be highly reliable. Other associations act as coordinators for

volunteer agencies (these include associations of radio operators, rescue workers and ambulance services). It may be important to consult these bodies both
when formulating the disaster plan and when putting it into operation, as they
can act as overseers for the volunteer effort.

Volunteer groups can fulfil several kinds of role in disaster situations. These
include acting as general auxiliaries to relief (lifting and carrying, cooking, running errands, doing repetitive manual work, and so on), providing medical and
healthcare services (volunteer doctors and nurses are needed here), conducting
search-and-rescue operations, looking after the handicapped or other special
groups, and operating radio-transmission services. It is as well to remember
that various professions, such as medical doctors and engineers, have volunteer organizations that are composed almost entirely of highly qualified people. Overall, however, the skills and professionalism of volunteers vary widely.

Once a profile of usable volunteer resources has been built up, the disaster
planner needs to consider the extent to which volunteer organizations are to be
utilized in disaster relief, and how to integrate them into response plans. Keeping in mind the evaluation of each organization's capabilities, a list of tasks that
it can perform should be compiled and an estimate should be made of the forces
that it can muster for each task. Limitations should be noted, for example, in
equipment or training. Ways should be sought to strengthen the organization,
perhaps by sharing resources with it or offering training courses.

Once the fundamental decision has been made to incorporate volunteer
groups into the disaster plan, they should not be kept at arm's length. Instead,
they should be encouraged to participate fully, not merely in the emergency-
relief effort but also in the planning process. This means consulting with the
groups and their leaders over planning questions, keeping them fully informed
about the plans and having them participate fully in courses, discussions and
field exercises. Two pitfalls to avoid are jealousy between professional and
volunteer groups, and rivalry among the latter. The emphasis must be placed
firmly on collaboration at all stages of the proceedings.

Radio enthusiasts form a particular group of volunteers who benefit in
special ways from organization and some training to participate in disaster
work. They operate as a very extensive network and often have considerable
technical knowledge of communications procedures. However, if they are not
formally involved in disaster work and are completely untrained, they can end
up passing on false, inaccurate or misleading information, even though they
may not wish to. For this reason, they need to be involved in preparations for
disaster and accredited; they will also need a detailed code of conduct. Many
areas have associations of amateur radio operators ("hams"), which can be an
appropriate point of contact for the disaster planner.

One additional aspect of planning for disaster that is often overlooked concerns the role of churches and other religious organizations. This is very important for several reasons. In many communities, churches (in the widest sense of the term) are a strong focal point and source of local identity. Religious functionaries provide moral and spiritual leadership, which can be very important in times of disaster, when there is great social stress. Many religious institutions are accustomed to providing charity to those in need, or can easily be prevailed upon to do so after disaster has struck. Moreover, the local parish priest or imam may have an unrivalled knowledge of the community, especially of the whereabouts of people in need (the sick and the elderly, for example), whom the emergency services might easily overlook.

Except through missionary work and their own charitable organizations, religious leaders are often reluctant to become directly involved in civil protection. If such figures prove to be particularly receptive to the idea of disaster planning, they should be allowed to become part of it. Otherwise, attempts should be made to consult with them and use their expertise wherever it would benefit relief efforts. Apart from being caring people who are deeply involved in the community, local pastors and parish priests often possess a fund of useful knowledge about their parishioners, which can be very useful in an emergency.

4.1.6 Reciprocal aid agreements

Response to disasters and emergencies can sometimes be improved by concluding formal agreements with neighbouring jurisdictions to furnish mutual or reciprocal aid, and then taking account of these in the disaster plan. However, such agreements are not easy to formulate and implement, as there are pitfalls and obstacles.

It is likely that aid will be forthcoming from outside the disaster area in any case as part of the **convergence reaction** of rescuers and others (see §5.4 and Glossary). But casual assistance cannot be planned for, and hence mutual aid agreements are important. Several key questions must be asked when such agreements are negotiated. These mostly refer to possible imbalances of resources. Will the requirement to provide aid to a neighbouring area leave the originating jurisdiction short of resources, materials or manpower at a critical moment? Who will pay for the aid supplied? Is the burden of supplying aid likely to fall unequally upon one or two neighbouring areas? Problems such as these may lead to a certain reluctance to conclude agreements. In any case, reciprocal aid should never be treated as a substitute for what can be generated locally or what is due from central government.

A reciprocal aid agreement should specify several things very clearly and, if necessary, in separate form for each of the jurisdictions involved. Exactly what is to be provided in given circumstances should be spelled out in terms of manpower, equipment, vehicles and supplies, as appropriate. The duration of such external assistance should be specified, along with any limitations to be placed on it. Unless the financial burden of supplying reciprocal aid is deemed to be roughly equal between the parties, arrangements may have to be spelled out for financial compensation. It may also be appropriate to state the conditions in which mutual aid is not expected to be furnished. Finally, there are cases in which mutual aid is best mapped out at a conference attended by various jurisdictions, in order to ensure that the assistance is efficiently planned, rather than provided for in a series of bilateral agreements that tend to duplicate resources or lead to imbalances.

A corollary of the mutual assistance pact between neighbouring jurisdictions is the question of how much aid can be expected to be supplied during an emergency and with little time available, by higher levels of government, such as regions, provinces or federated states. Although such aid does not necessarily fall under the heading of mutual assistance, it is wise to ascertain its probable extent before concluding other bilateral or multilateral assistance. If these intermediate levels of government can be persuaded to join in the process, they can be a very useful catalyst to mutual aid agreements, by providing the framework, and perhaps also the financial incentives, to construct them.

In all probability, mutual aid functions best when disasters are localized, of fairly short duration and simple in terms of their impacts. If this is the scenario that the disaster planner is facing, then mutual assistance pacts can be a valuable reinforcement of existing tools, strategies and resources.

4.1.7 The legal situation

Emergency planning must take place against a background of legal instruments that mandate, facilitate or regulate it. These include laws, ordinances, protocols, decrees, norms and requirements. A list of typical directives or ordinances used in emergency situations is given in Table 4.7. Like other legal provisions associated with disaster, they may be applicable at the national, regional (state, provincial, departmental, etc.) or local levels. In addition, there are super-national instruments, such as European Union directives, and international ones, such as the Law of the Sea. As this book is intended to be a general guide to emergency planning, it will not treat the legal situation very specifically, for laws and regulations vary considerably from one jurisdiction to another. Nevertheless, some general points can be made.

At a very early stage in the construction of an emergency plan, the planner should identify and study the legal situation as it pertains to his or her

Table 4.7 Examples of directives or ordinances for post-disaster administration.

Public safety, rubble clearance and demolition
- Mandatory closure of roads or restricted access where public safety would be at risk
- Work-order for clearance of road blockages
- Interdiction on access to buildings that have been seriously damaged or pose a risk to safety in other respects
- Mandatory demolition of buildings and structures designated as unrepairable and a threat to public safety

Evacuation orders
- To mandate general evacuation of populations at risk
- To mandate evacuation of specific groups of people
- To mandate evacuation of particular buildings which are at risk
- Forced evacuation of families in cases where there is a substantial risk to their safety but persuasion has failed

Requisition orders
- Requisition of buildings in order to provide temporary accommodation
- Requisition of lodgings for displaced families
- Requisition of buildings for emergency use
- Requisition of means of transport
- Requisition of workers with particular skills
- Requisition of materials or foodstuffs for temporary supply to groups of displaced people

Assuring continuity of public services
- To ensure that particular retail shops of critical importance to the public remain open during the disaster aftermath
- To ensure that particular gasoline stations remain open during the disaster aftermath

Environmental and public-health measures
- Interdiction on public consumption of non-potable water supplies
- Temporary suspension of environmental laws regarding refuse disposal and institution of alternative arrangements following damage to normal structures
- Precautionary measures to suspend production, distribution or consumption of food and drink that may be contaminated

---

N.B. An ordinance is a temporary measure that is legally binding; unless it is renewed, it remains valid for a set period limited by permanent laws. The validity of most ordinances should be maintained for the duration of the emergency.

jurisdiction and with respect to all applicable levels of government. It is also necessary to keep up to date with legislation, as new laws may be passed and old ones, or some of their articles, may be repealed. This can significantly change legal responsibilities over time.

Civil protection in different countries is governed to varying degrees by the legal framework behind it. There are essentially two situations. In the first, laws and regulations are complex and numerous enough basically to determine what can and must be done to protect people against disasters. In the second, laws merely provide a basic structure within which emergency planning can take place. In the latter case, individual laws may facilitate planning or may place limits on possible actions; very occasionally they succeed in doing both.

Upon reviewing the legal situation, the emergency planner should ask several important questions. These can be classified according to the four main aspects of legislation: to stipulate obligations, determine responsibilities,

125

guarantee rights and create the framework for civil protection. To begin with the first of these, is there a legal obligation to save people who are threatened or injured by disaster? If there is, then failure to honour it could be construed as criminal negligence and could open the way to prosecution. Secondly, what are the legal implications of failing to warn or evacuate people, mitigate hazards or manage risk of disaster? Again, claims of negligence must be avoided. As an extension of this, what are the legal implications of false alarms? Generally, forecasts of hazards are unlikely to be 100 per cent accurate and there is inevitably some risk of precautionary action being taken without disaster occurring at the end of it. However, evacuation and other forms of disruption of regular activities can lead to loss of income or production, and attempts may be made to seek costs from those who were responsible for the emergency actions. Some degree of indemnity is required, or otherwise no disaster manager would have the courage to declare an emergency before it is too late to take avoiding action such as evacuation. In this context, careful documentation of activities during an emergency serves to counter legal claims of negligence. To convince a judge and jury, a well documented and precise narrative of what happened is more likely to succeed than a vague and hopeful statement.

The legal situation can also be divided into that which governs the relationship between emergency workers and the public and that which regulates the emergency workers themselves. Are the activities described in the emergency plan legally permissible? Do they conflict with any laws or regulations that are not directly related to emergency management? These may include norms on safety in the workplace, and regulations that guarantee freedom of access to places, facilities or information. Are participants in the emergency plan aware of their legal responsibilities? Have frontline emergency workers signed a legally valid release document that absolves their superiors from liability (either partial or total) if they are injured while carrying out emergency work? Are emergency units insured against legal claims and does the budget provide for paying the premiums? These are obviously problems that should be sorted out before emergency workers are sent into the field, especially as many of them will have to carry out their duties in difficult and dangerous situations.

It is worth paying special attention to the applicable catalogue of laws and regulations in terms of how they divide up responsibilities, tasks and jurisdictions. Unfortunately, when the legal situation is complex, there is a certain likelihood of conflicts between legal provisions enacted at different levels or emanating from different organs of government. There is obviously little sense in planning for activities that are the legal responsibility of another branch or sector of government. In addition, the emergency plan needs to be examined to ensure that it conforms to legal requirements. It is essential to have it looked over by a legal expert with experience in this field of hazard legislation.

Various considerations pertain to any question of liability after a rescue mission has gone wrong. Rescuers who are injured may be able to claim worker's compensation (a situation that should be clarified with appropriate government departments). Questions of negligence, wilfulness, risk taking and causality may be complex and subtle, although courts generally tend to recognize that rescues are undertaken in good faith and in spite of personal risks and limitations. If rescue workers have signed release forms that absolve their leaders from liability in the case of personal injury, these documents should be scrutinized for legal validity. Likewise, policies for liability or malpractice insurance should be examined carefully to ensure that they are valid for disaster situations and do not contain exclusions that would make them useless under such circumstances. Vehicle insurance should also be checked to ensure that it is valid for disaster operations and is sufficient to protect drivers, operators and others who may be involved in any accident that could occur.

A few other legal questions deserve some attention. One is the bona fide status of participants in the emergency plan. Of course, not all will be subject to special regulations, but engineers, doctors and nurses, for example, will need to be licensed or certified. In this sense, the qualifications must be matched to the job; it is dangerous to assign to inadequately qualified people medical or engineering tasks that carry a great deal of responsibility. The emergency plan should therefore specify the minimum professional qualifications required of people who must perform certain tasks. This will help guard against legal action for negligence, incompetence or malpractice if anything goes wrong. In addition, volunteer units should be encouraged to seek non-profit tax-exempt status. Disaster planners can help with this by producing information sheets on the appropriate requirements and assisting with the process.

Finally, it is a good idea to design a mechanism to investigate any legal transgressions or claims that may arise out of future emergency-management activities that produce negative results. The ground rules for convening a commission of enquiry should be laid down and prescriptions made for the sort of experts who will sit on it and how they will be co-opted.

Most other legal aspects of disaster planning and management will be specific to given countries, regions and jurisdictions.

4.1.8 Avoiding discrimination

This section will briefly address two important questions of equity in emergency relief. Despite the brevity of the treatment, they are important issues that need to be tackled seriously and effectively.

To begin with, in disaster, women's rights need to be safeguarded and their concerns taken into account. In the chaotic period that follows the impact of a disaster, they can suffer, among other things, sex discrimination, increased rates of domestic violence, and disadvantage in the allocation of relief. They may be the first victims of loss of employment or reduction in wages when factories are damaged. Furthermore, it is important that their collective voice be heard in all post-disaster issues. Consequently, women who are sensitive to and knowledgeable about women's issues and rights should be involved at all stages of the emergency-planning process. Disaster should not be an opportunity for the further subordination of women. Moreover, a neutral stance may be insufficient, as what appears impartial to men, who are accustomed to their own hegemony, may well seem biased to women.

Secondly, disasters should not be allowed to be occasions for racial or ethnic insensitivity. Multi-ethnic and multicultural societies may require safeguards to be built into emergency plans to ensure that operations do not violate cultural norms or cause offence to minorities of any kind. Institutional racism is a form of discrimination inherent in the way that organizations function, although not necessarily in the actions or pronouncements of any individual within them. It needs to be guarded against in all aspects of planning and management, for it is an insidious phenomenon that can take hold in organizations while their leaders are unaware of its growing presence. It can be counteracted by ensuring that a multi-ethnic committee not merely reviews and monitors emergency preparedness but that its views are listened to and acted upon. In some cases, emergency-preparedness staff of all kinds, from firemen to emergency managers, may require racial-sensitivity training, which is well worth providing, as accusations of discrimination can destroy the goodwill of the public, which emergency planning requires in order to succeed.

4.2 Disseminating the plan

A well thought-out emergency plan will contain provisions for disseminating and updating its content. Here, we deal with the former, and the latter is the subject of §4.3.2. The emergency plan must be communicated to the two categories of people who will most need to know about it: those who will participate directly as emergency workers and those likely to be otherwise affected, either directly or indirectly, when it is put into operation.

The plan can be disseminated in various ways. Copies can be printed, reproduced and distributed. The document can be placed on a Website and the Internet address widely publicized in cyberspace bulletins, discussion groups and mailing lists. Copies of the plan can be appended to newsletters or inserted into other serial publications, such as magazines and newspapers. Posters and advertisements can be designed in order to publicize the plan. It can be launched at a local conference, with invited speakers to explain its functions and various provisions.

However, none of these initiatives can guarantee that the plan will be read and understood, especially by the general public, if they are to be involved in it. For example, in the late 1970s a six-month campaign to publicize the flood risk in central London resulted in widespread public misunderstanding of warning processes and required actions, such as not using the underground trains or cross-river buses when the **alarm** was broadcast. It was not clear from the subsequent inquiry whether the plan itself was at fault or merely the means of publicizing it, but it is clear that publicity needs to be designed very carefully.

With respect to emergency workers, who are the main direct participants in the plan, it is essential to ensure that they receive a copy and that they read and understand it. They must gain a full appreciation of both their own roles and those of any other participants with whom they will have to interact. The first task is to distribute copies of the plan by name and address. Its functioning can be explained in a series of briefings or seminars, which may best be conducted with individual groups (firemen, paramedics, police, etc.), although occasionally it is opportune to explain common elements of the plan to mixed groups, especially regarding how they are expected to work together. In rare instances, it may even be appropriate to test key participants in the plan to assess their knowledge of it.

In essence, public knowledge and understanding of the emergency plan can never be guaranteed to be adequate, such is the variety of beliefs and perceptions that the public holds. Hence, the best that can be hoped for is that many members of the public will grasp some simple principles (e.g. regarding alert stages and evacuation procedures) and that, when disaster strikes, the rest will follow their example. The plan thus needs to be given as much mass-media coverage as possible, but with emphasis on simplified aspects of the parts that most affect the general public.

An emergency plan is more a continuing process than a finished product. Hence, it is not good enough to rely on a single one-off attempt to disseminate it. Instead, this should be a periodic process of reminding people of the plan's existence and content, and instructing newcomers. It is a good idea for the plan itself to contain formal provisions for periodic dissemination and refresher courses. Knowledge of the plan can also be made a component of induction and training courses for local emergency workers and government employees.

4.3 Testing and revising the plan

In the sense that an emergency plan is a process rather than a finished product, it should be considered as an approximation that can be improved by making adjustments on the basis of new expertise and experience. Of course, this means that versions of the plan will become obsolete over time, but this is inevitable, given that conditions will change: equipment will wear out or be superseded, staff will come and go, risks alter with land-use changes and mitigation measures, and so on. It is therefore worth designing a process to update the plan periodically, perhaps by holding a six-monthly review and arranging for revisions to be incorporated once a year, in concert with renewed efforts to disseminate the plan among its users.

4.3.1 Assessing the plan's functionality

There are several ways in which the functionality of the plan can be assessed without having to revert to the supreme test of putting it into effect in a real disaster. To begin with, feedback can be obtained from participants in **briefing** sessions; indeed, it may be worth collecting it systematically. Emergency workers are a source of many kinds of practical expertise and can significantly enrich a plan, especially if their help and suggestions are actively sought in a spirit of collaboration.

Secondly, field exercises can be used. As noted in §3.6, these help work out the timing and logistics of the emergency-relief process and will reveal deficiencies in planning. They should end with a debriefing session in which participants can air their views on what worked well and what did not. Such opinions need to be recorded carefully and factored into the planning process.

Perhaps surprisingly, a real disaster should be considered as part of the planning process. The plan should designate one emergency worker as the chronicler and recorder of relief efforts. Logs should be kept of the timing of emergency actions (most emergency-management software allows for this to be done). Video films should be made of fieldworkers in action, and photographs should be taken by way of documentation. Overt recording of events and actions will help reveal deficiencies in the plan, for correction in future versions. Once again it is helpful to hold a debriefing session at the end of the emergency period and to take detailed notes on participants' reactions to the plan as executed.

In the event that none of these options is possible, the plan can be evaluated by the scenario method. In this, a model emergency is worked through in a round-table discussion by an invited panel of representatives of emergency workers. At each stage its probable ability to meet the demands posed by the emergency is evaluated, and suggestions for improvement are made. However, it usually takes a real emergency, or at least a major field exercise, to confirm what the scenario appears to have revealed. In all cases, the acid test of the plan's functionality is whether it saves lives, protects people and reduces damage to property.

4.3.2 Upgrading the plan

No emergency plan is likely to remain fully operational and completely efficient over time unless it is updated. Procedures for doing this need to be a formal part of the planning process.

Suggestions for improvement can be collected at any time, but especially when the plan is being disseminated or tested. Beyond this, it is necessary to monitor changes in the conditions on which the plan is based. Emergency planners, or their scientific advisers, must decide whether changes in hazards, vulnerability or risk levels are sufficient to require full-scale re-evaluation, including remapping. Disaster scenarios may need to be rewritten as a prelude to adjusting the emergency response. Next, emergency-resource levels need to be monitored. These include the quality, capability and functionality of equipment, staffing levels, and the inventories of stockpiled supplies. On this basis, adjustments should be made to the planned emergency actions.

It is also evident that scientific research will reveal more about hazards, and technological change will alter the opportunities to respond to them, as time goes on. If the original plan is based on very scarce information about local hazards and risks (as is often the case when emergency planning is started in an area), the perspective on them may be changed considerably by further research. This may include enhanced forecasting capabilities that make prior evacuation possible where before it was out of the question, and more detailed maps that help redefine the spatial location of hazards and vulnerability.

In sum, the assessment of an emergency plan's functionality and the process of updating it need to be considered as a cyclical process, perhaps a semi-annual one. The emergency planner needs to apply this process in such a way as to strike a balance between the extremes of a plan that is fossilized by lack of renewal and one that is changed so often and so radically that it confuses participants. In short, flexibility needs to be matched by continuity. Versions of the plan can be numbered or dated in order to identify them. On occasion it may be appropriate to draw attention to changes and additions by highlighting them or mentioning them in an introduction.

4.4 The integration of plans in theory and practice

Hitherto, we have discussed the emergency-planning process as if it were applied only at one level of government and via an all-purpose set of dispositions. In reality there will probably be not one but several emergency plans for a single event. It is a good idea to see what can be done to integrate them in order to avoid confusion, duplication and waste of resources.

In as much as there is an ideal spatial scale of emergency planning, it is probably that of a single large municipality or association of smaller ones. At this scale, problems can be resolved in detail and local self-sufficiency can best be encouraged (a typical municipal civil-protection organization is shown in Fig. 4.7). However, other levels of government and other types of organization have emergency plans. In general terms, the function of a national plan is to convert government policy into actions, provide leadership and coordination for lower levels of government, offer models for planning, mitigation and training at lower levels, and provide emergency resources where regional and local ones are insufficient. The main function of a regional (or state or provincial) plan is to provide coordination and continuity for local planning initiatives. The regional and national plans should offer a model that can be developed at the local level in a standardized way that helps make the quality of emergency planning uniform from one community to another (although in practice this

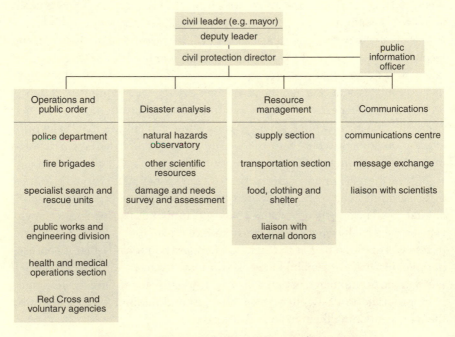

Figure 4.7 Structure of a local civil-protection system.

aim is seldom achieved and local planning tends to be thoroughly hetero-geneous). A regional plan should also identify the nature of the regional threat, including variations from place to place in hazard, vulnerability and risk. It should aim to channel emergency resources to where the risks are greatest.

At any level of government, emergency planning cannot be expected to tackle the full minutiae of detail that must be considered in order to direct operations at the institutional level. Hence, separate emergency plans may be needed for hospitals, airports, railway systems, factories, prisons and schools. To some extent, these will be smaller versions of the general emergency plans, but they will have to tackle specific problems with particular sets of resources.

Where possible, emergency plans should be written in such a way as to encourage as much local self-sufficiency as possible, but to promote integration of plans at various levels. This may require making arrangements to share the same emergency-operations centre, establishing special lines of communica-tion between disaster managers who work under different plans, dividing up or delegating emergency tasks, and ceding authority where necessary. In short, emergency planning should take place in full recognition of the environment and context created by other plans and planners.

CHAPTER 5

The plan in practice:
emergency management

Hitherto we have reviewed some of the general principles and elements of emergency planning. The following sections will add details that relate to the content of the plan and will do so by outlining the needs and exigencies of emergencies as they are encountered through its operation. This will serve to illustrate the sort of situation that planning must tackle and some of the practical details that need to be considered before disaster strikes.

Once again it is important to emphasize that good emergency management seeks to reduce improvisation to an essential minimum. Although some aspects of any emergency will have to be extemporized because there is no way that they can be foreseen, and flexibility will have to be demonstrated in the handling of the crisis, the aim is to anticipate conditions, as far as possible, and make advance provision for them. In terms of managerial capacity, it is important to divide decisions into those that can be made *before* the disaster strikes (e.g. about the allocation of resources) and those that must be made *during* the emergency. Procedures must be in place to streamline the latter, for delay can be fatal.

The following sections will outline some of the general and specific planning requirements for each distinct temporal phase of the emergency. The phases are outlined in Figures 1.2 and 4.1.

5.1 Management styles

The following section discusses problems and techniques of managing groups of people and single individuals, including oneself. It encompasses both the formal role of emergency coordinator and the less formal need to alleviate problems of stress, impaired judgement, and friction between co-workers.

Management has been studied systematically and scientifically since the mid-twentieth century, although remarkably few research findings have been applied to the coordination of disasters and emergencies. Indeed, throughout the world in this field of endeavour it seems to be considered normal to muddle through, managerially, rather than apply principles of human organization. This is a pity, as it reduces the efficiency of managerial operations. However, change is on the way. The usual profile of the disaster manager is being raised progressively in response to the need for a more professional figure, with a better scientific training, more political savvy, and greater computer literacy. This change towards a more precise and rigorous specification of job requirements provides an opportunity to re-assess the managerial status of the emergency coordinator and to suggest ways to improve his or her performance in the job during disasters.

5.1.1 Towards adaptive management of disasters

Good emergency management requires experience and training, but it also needs a certain degree of reorientation in respect of normal management practices. In management terms, emergencies are hardly normal situations, but techniques of managing them can successfully be learned. To understand what is different and what needs to be learned, we must first consider what is involved in the process of management in the absence of emergencies, in what disaster managers tend to refer to as "peace time".

Essentially, management consists of evaluating information, making decisions and ensuring that they are put into practice. First, the problem is characterized and objectives are defined in order to focus on how to tackle it. These are clarified by refining criteria for decision making that limit or constrain the range of solutions to a practicable level. In democratic situations, a preliminary decision is made and consensus is sought among those people who will be affected by the decision. If the consensus is strong enough, the decision is accepted or ratified. It is then put into operation: in order to meet the defined objectives, resources are marshalled and applied to the problem according to well defined procedures. People are directed to carry out specific tasks and supervised as they do so. The consequences of the decision are monitored and an evaluation is made as to whether it is succeeding in fulfilling the objectives. Over time, circumstances may change to the extent that the decision ceases to be valid (hence, in some conceptual manner a threshold is crossed), and a new decision must be made.

Criteria for decision making are often defined on the basis of listed or ranked priorities. Limitations may assume any of several forms. Manpower and equipment may be in short supply, as time may be also. There may be geographical constraints of distance or area that can be covered; these fall into a broader category of physical limitations. There may also be constraints on expenditures, affordable costs and the availability of funds. **Culture** and perception constitute a diffuse but important source of limitations, as problems may not be perceived accurately, or cultural factors may militate against change and innovation. All of these factors add up to a set of limitations on the management of complex and otherwise difficult situations. In this, lack of knowledge and information are often the key constraints upon action, especially when it is necessary to manage the situation prudently.

Despite the existence of many possible limitations on decision making and action, there is plenty of scope for managing emergencies well. However, such situations inevitably require rough compromises and give rise to inexact solutions; hence, there is no room for perfectionism in emergency management. But let us examine why that is so.

Emergencies are situations of great uncertainty in which there is usually a severe shortage of vital information. Circumstances can change with great rapidity and unexpectedness; hence, so can the factors that limit decision making, and often towards a more restricted range of options. At the same time, the impact of threatening or depressing circumstances, the weight of responsibility and the pressure to get things done and make decisions rapidly can all conspire to subject the emergency manager to great stress. The fixed parameters of management at such times are indeed few. Large fluxes of personnel and materials occur. Working relationships change and may accumulate strain. Lack of information reduces the efficiency of the emergency operation with respect to ideal levels.

In emergencies, the number of organizational goals is usually reduced, and sometimes drastically so. Lives often depend on the ability to make decisions rapidly but well. It is equally important for managers to communicate effectively with people, especially colleagues, and to maintain a sense of realism by recognizing the limitations of people, equipment and specific situations. Goods must be prioritized and knowledge of the developing situation must constantly be increased. All this adds up to the need for a clear pattern of authorization to make decisions. As there is no time to consult, obtain a consensus and ratify decisions, authorization must be simplified, speeded up and, wherever possible, given in advance (it is often inherent in a chain of command – see §4.1.2).

This leads us to a wider problem associated with the organizational continuum that extends from consultation and consensus (the slow process of democracy at work) to command and control (the swift hand of authoritarianism). Extraordinary powers may be needed to manage crises, but they can of course be abused. In some countries such risks have turned out to be a sticking point that has delayed the emergence of a fully fledged civil-protection structure; in other nations, it has led to the creation of undemocratic institutions that engage in repressive activities. Moreover, it is not merely a national problem, for variants and degrees of it can be found at all levels of government. However, it can be tackled in various ways. One is to impose strict and well defined limits on the powers of any individual or committee that directs emergency operations. Another is to "lend" powers by forms of delegation that can be revoked if things get out of hand. Monitoring-procedures and checks and balances are required here. Such controls can also be utilized when decision makers turn out to be ineffective by vacillating or sticking to decisions that are patently wrong. But again, checks and balances need to be built into the system.

Other risks are involved in the management of emergencies. Failure to delegate is one of these and it can lead unwise emergency managers to tackle problems that their skills are inadequate to solve. A corollary of this is overconfidence in one's ability to solve problems. Often this is based on misapprehension (i.e. misperception) of the fundamental facts of the case. A wrong analysis of the situation can mean that decisions turn out to be fundamentally wrong: unpopular, unworkable, equivocal, open to diverse interpretations, vague, incorrect, too risky, foolish, fatuous or arrogant. Worse still, colleagues and other people may be antagonized as a result. Such mistakes are codified in M. J. Smithson's classification of ignorance, which he applied to the (mis)management of disasters (Fig. 5.1).

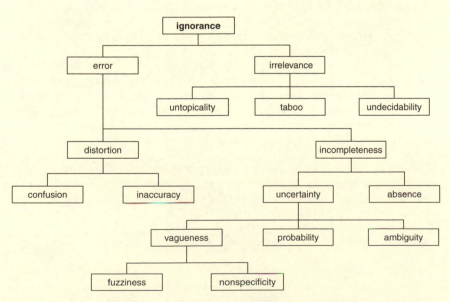

Figure 5.1 Smithson's classification of ignorance.

Box 5.1 Air-ambulance problems

Narrative of the event

One morning in May a public hospital in a major city asks a specialist commercial firm to send one of its air ambulances and a medical team to transport a woman in her thirties with liver failure to another public hospital that has specialized facilities for treating her. As the transfer has to be authorized by the regional ambulance service, a telephone call is also made to its head, who, however, refuses to allow the commercial helicopter to land at the hospital. Instead, he calls the airforce command and requests that a military helicopter be sent immediately. The only one that could be used is out on a search-and-rescue mission for two lost boys. It is a large aircraft, more suited to air-to-sea rescue than ambulance work.

Next, a member of the medical team associated with the commercial air ambulance calls airforce command and points out that the team's aircraft has had to land at the hospital, but the head of the regional ambulance service is preventing them from loading the patient. The military command, which regard the commercial company as perfectly competent to do the work, have trouble in understanding the head's motivation, as the commercial air ambulance is only 30 per cent of the cost of the airforce helicopter to use on this mission. Nevertheless, the regional ambulance-service head makes a special appeal, in rather intemperate language, to the military to send their helicopter. He describes the commercial air-ambulance service as a "cowboy outfit". In the meantime, the commercial vehicle runs out of legal flying time, so the military eventually suspend the search-and-rescue mission for ten hours and send their helicopter. The patient reaches the specialist hospital four hours later than necessary and dies shortly afterwards. Military and civil administrations initiate formal inquiry processes.

Significance of this case

The true significance of this case may never be made public, but it does illustrate an important problem associated with command structures and perception of needs. The head of the regional ambulance service had full unsupervised authority to order or block air transportation of a critically ill patient. Through prejudice and bad judgement he made the wrong decision and through inflexibility, or unwillingness to lose face, he stuck to it. Tempers flared and the level of stress rose, which did not contribute to rational debate. In the end, the head's insistence on using the military helicopter compromised two missions, one very seriously and the other fatally.

Besides subjecting the ambulance-service head to disciplinary proceedings on the basis of a formal inquiry into what happened, the command structure and decision-making processes involved in this unhappy incident needed to be reconsidered. Better training and supervision might have helped avoid the worst. At the least, this case highlights the importance of choosing the right people for certain jobs and ensuring that they remain capable of performing well and rationally. In all organizations that have a frontline role in emergencies, periodic review of decision making should be complemented by staff-evaluation procedures.

Additionally, the commercial air-ambulance team may have had a good case for defying the orders from the head of the regional ambulance service and taking off with the patient regardless. However, as easy as it is to argue this with the benefit of hindsight, it is not so easy to justify in the midst of the incident, when the outcome is uncertain.

Sources of information: internal reports and news bulletins; anonymity maintained.

The solution is to have good leaders. Without implying that every person in charge of emergency operations need be a hero and a saint, we may nevertheless outline some of the characteristics of ideal leaders, the innate qualities of leadership. By virtue of physical or intellectual presence, and other factors, they will be able to exert moral, institutional, psychological and perhaps even legal authority. They will have the knowledge, experience and training to do the job unselfconsciously. Good leaders are dependable, reliable and usually not given to unpredictable mood shifts (although there are, of course, some famous exceptions to this). An optimum combination of rigidity and flexibility of character is valuable, so as to stick to a chosen path, although not when it becomes untenable. The continuity of leadership style needs to be maintained, but with appropriate adaptations to circumstances. In other terms, this means an ability to maintain one's perspective, to balance detachment against involvement. An excess of the former can result in artificiality, whereas too much of the latter can lead one's judgement to be overwhelmed by circumstances.

Let us remember, however, that despite this counsel of perfection, good leaders can also have substantial defects or inadequacies, which are amply compensated for by their talents.

Part of the process of maintaining the balance is to recognize the abilities and limitations of colleagues, whether they be heads, peers or underlings. It also helps to convey to others the sense that the enterprise is going in a given direction, to inspire colleagues with the feeling that their own efforts are part of a significant collective endeavour. This involves giving attention to the individual needs of members of the management team. It also requires a firm but not insensitive hand and an absence of vacillation. Lastly, a vital quality of leadership is the ability to obtain results without incurring excessively high costs, including the social costs of making enemies and antagonizing people.

The foregoing may seem like a counsel of perfection and it is not to be expected that emergency managers will be bursting with all the qualities that are listed. Regardless of personal or personnel factors, there are some additional management techniques that can help to ensure that the emergency is well run. These involve adaptive management.

In traditional methods based on principles of command and control, procedures and methodologies learned beforehand are applied, usually somewhat rigidly, to problems as they occur during the disaster. The adaptive method requires one to change one's approach as new information arrives and qualifies the prevailing picture of the situation, and as the situation itself changes and demands adjustments in the way it is tackled. The key to the process of managing a disaster adaptively lies in the rapid interpretation of scientific, social and logistical information. The methodology learned beforehand must be reinterpreted in order to find a more flexible and efficient way of applying it.

At the same time, procedures must be developed to avoid confusion. This involves a process of constant self-monitoring to ensure that decisions made in any given moment do not conflict with those made previously. Adaptive management is not especially appropriate to the rigid structures of chain-of-command systems, but is particularly well suited to approaches based on incident-command systems. Although there must be no equivocation in the chain of responsibility, authority rests more in the effectiveness of particular solutions (i.e. in the way that they are seen to be effective) than in the ability to hand down orders.

This section ends with a word of warning. We have developed a profile of ideal emergency managers and have reviewed the formidable challenges that they face. In the end it is unlikely that all decisions will turn out well, but there are potential pitfalls in the way they are evaluated, for hindsight can radically change the criteria by which decisions, actions and performance are judged.

5.1.2 Stress management

In this section we shall examine how to recognize and manage problems of stress in oneself and one's colleagues. Stress and mental-health problems in the general public will be considered separately, although only in the context of emergency planning (see §6.10). There are various sources of stress in emergency situations. The weight of a disaster manager's responsibilities may become oppressive, or the risks involved in directing emergency operations may seem excessive, and so may the demands that are constantly made on his or her time and attention. The problems that need to be solved may seem enormous and, as the hours go by, tiredness may give way to exhaustion, which may tend to erode or distort the emergency manager's judgement. Stress may accumulate physically in muscular tenseness and mentally in the degradation of the quality of reactions. It may also be present in the form of shock as a result of witnessing extremely unpleasant events. This is **critical-incident stress** (CIS), a subset of **post-traumatic stress disorder** (PTSD), and it is common among emergency-relief personnel who have to work in situations of great risk, terrible destruction or very serious injuries, especially where there are dead and dismembered bodies.

Among disaster managers, stress can cause a series of negative reactions. Loss of perspective is the first of these. It amounts to a failure to maintain the necessary balance between detachment from the continuing situation and involvement in it. Irritability, irascibility or anger may follow. The magna mater and Jehovah complexes may also set in (see §6.10). With regard to emergency-operations centres, one solution to the stress problem is to have a quiet room to which staff can retire in order to rest, have a drink or a snack and regain their equilibrium.

Knowledge of the risks of accumulated stress should enable an experienced emergency manager to recognize the danger signals in himself or in others and to take appropriate action. Rest, relaxation and sleep, or at least some form of mental diversion, may be needed. Hot food and drink may help. A period of absence from duty may be needed, even if for only half an hour. Relief workers who operate in extreme conditions should be monitored for critical-incident stress and taken out of service if they show signs of it. These may vary from anger to irrational actions or spontaneous withdrawal into the inner self (frequently manifested as incommunicability and motionless gazing into space).

Stress can also be managed by socializing it. In extreme cases professional counselling may be needed, but less drastic remedies are usually sufficient. As stress if often greatest in young and inexperienced relief workers, they can be paired up with older, more experienced colleagues in a system of mutual support. Debriefings and post-disaster reunions can be organized in order to recount experiences and share one's grief, anger or exultation. Certain types of physical and mental exercise can be practised singly or in groups in order to disperse the tenseness that accompanies stress. Follow-up counselling can tackle the depression that often follows a period of highly stressful activities.

A modicum of stress in inevitable in the process of emergency management; it can even be beneficial by stimulating reactions. However, the chaotic aftermath of disasters and the accompanying high degrees of uncertainty act as catalysts for the accumulation of stress that can be harmful to good emergency management and must in turn be managed.

5.2 Alert procedures, warnings and evacuation

Disasters differ widely in the amount of warning and lead time that they give before impact occurs. Some, such as explosions and earthquakes, may give no warning at all; for example, although there may be precursory signs of seismic activity, they tend to be difficult to identify and interpret and are not always present. Others, such as slow-rising floods and developing volcanic eruptions, may give weeks of forewarning, if the signs can be interpreted properly.

The monitoring of precursory phenomena (Table 5.1) is the preserve of scientific and technical staff. It should, of course, be strongly encouraged wherever it is likely to produce results. But it is equally important to design a robust and reliable method of communication between scientists, who are responsible for predicting, forecasting and monitoring events, and public

Table 5.1 Monitoring technologies for some of the principal natural hazards.

Earthquake	Seismometers, accelerometers, tiltmeters, crustal magnetometers, electrical resistimeters for crustal currents, radon meters for groundwater, springwater discharge gauges, turbidity meters for standing water (wells, lakes)
Flood	Stream gauges and water sensors, rain gauges, meteorological forecasting equipment (radar, satellites, etc.), monitoring of groundwater levels by piezometers
Hurricane	Meteorological forecasting equipment (radar, weather stations, satellite images), satellite-based tracking methods, digital prediction models
Landslide and subsidence	Rain gauges, piezometers, extensiometers, tiltmeters, creep meters, geodetic levelling and triangulation, triaxial testing of soil samples
Tornado	Meteorological forecasting equipment (especially Doppler and Nexrad radar), satellite imagery (especially AVHRR – advanced very high-resolution radiometry)
Tsunami	Seismometers, tide gauges, bottom-pressure transducers, taut-wire buoys, satellite-based data transmission in real time
Volcanic eruption	Seismometers, tiltmeters, infrared sensing of crustal heat fluxes (ground-, air- and satellite-based), gas-emissions monitoring, air-quality monitoring, stability monitoring (landslides, crater collapses, lahars, etc), satellite-based atmospheric emissions monitoring (especially TOMS – total ozone monitoring spectrometer)

authorities, who are responsible for transforming scientific predictions into warnings. Predictions and warnings are very different things: the former is an expression of what is expected to occur, while the latter is a recommendation for action. Hence, a warning should explain what is likely to take place (on the basis of a prediction) and what to do to avoid the worst consequences of the impact. This requires that information furnished by scientists be clear, unambiguous and sufficiently simple to be fully intelligible to people who are not experts in the field. Valid predictions, moreover, are those that leave little doubt about what is likely to happen: low probabilities are not helpful to decision makers, however justified they may be scientifically.

The need for timely prediction is self-evident. Besides the time requirements of evacuation and other preparations, rapid mobilization can reduce the short-fall between demand for emergency services and their supply (Fig. 5.2).

It is useful to arrange with the scientific community the information to be furnished, both in the case of impending disaster and during post-disaster surveillance (i.e. monitoring of continued threat or hazard). Formulae can be agreed beforehand for the messages that will be emitted when hazards or impacts reach particular levels, to help reduce any misunderstanding and incomprehension in the heat of the moment. At the same time, civil authorities can be instructed in the meaning and implications of certain scientific events, and the physical channels of communication between scientists and civil-protection authorities (telephone, fax, Internet, etc.) can be secured (Fig. 5.3).

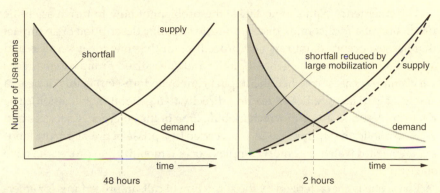

Figure 5.2   Temporal curves of supply and demand for emergency assistance in a disaster or major incident. Prompt, efficient large-scale mobilization of emergency resources and personnel can reduce the initial shortfall between supply and demand.

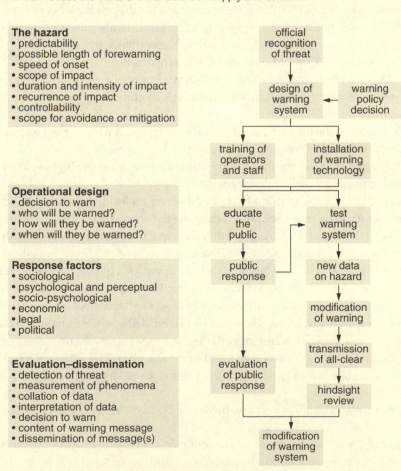

The hazard
- predictability
- possible length of forewarning
- speed of onset
- scope of impact
- duration and intensity of impact
- recurrence of impact
- controllability
- scope for avoidance or mitigation

Operational design
- decision to warn
- who will be warned?
- how will they be warned?
- when will they be warned?

Response factors
- sociological
- psychological and perceptual
- socio-psychological
- economic
- legal
- political

Evaluation–dissemination
- detection of threat
- measurement of phenomena
- collation of data
- interpretation of data
- decision to warn
- content of warning message
- dissemination of message(s)

official recognition of threat

design of warning system ← warning policy decision

training of operators and staff

installation of warning technology

educate the public

test warning system

public response

new data on hazard

modification of warning

transmission of all-clear

evaluation of public response

hindsight review

modification of warning system

Figure 5.3   The warning process. Partly after Foster (1980).

The emergency plan should tackle the problem of how to turn a scientific prediction of a forthcoming disaster, or the scientific description of an impact that has just occurred, into an alert procedure for the emergency services and a warning process for the general public and specialized groups. Only in the very simplest cases (or when the impact cannot be foreseen) is this done in a single stage (i.e. a prediction leading directly to a general mobilization). The next simplest case is one in which mobilization of the emergency services and general public, the warning stage, is preceded by a phase of pre-alarm or stand-by, the **hazard watch**. The most complex general mobilization process has five stages:

- Phase one is characterized by the activation of call-up procedures for emergency workers. During this period they either report for duty or are made aware that they may soon have to report.
- Phase two is one of stand-by, indicating that conditions have become more threatening, but not to the extent that it is worthwhile putting emergency workers in the field or shutting down non-essential activities.
- In phase three, if conditions continue to worsen, normal activities are curtailed and emergency workers are deployed in readiness. The emergency-operations centre is fully functional at this point and call-up procedures have been completed.
- Phase four is the result of a further worsened threat and is marked by public evacuation and other preventive measures.
- Phase five involves the onset of the disaster, or at least the highest level of threat, leading to full-scale emergency action.

It should be borne in mind that not all emergencies mature into full-scale crises. A hurricane may veer out to sea at the last moment, a tornado dwindle, or heavy rain cease to fall. As a result, any sequence of emergency states may end prematurely with stand-down (regression to previous levels) or all-clear, the announced end of the crisis. Even a fully developed impact will eventually end with an all-clear, at least as far as the hazard that has caused it is concerned.

We have dealt with the temporal phases of an alert, but there are also spatial (i.e. geographical) divisions. These can be established by a combination of hazard or risk microzonation and immediate assessment of emergency conditions. Different schemes have been used in different places, and for different hazards (Table 5.2). Here is a suggested generic scheme for emergency zonation:

- white: safe areas.
- green: areas that are potentially threatened but are not yet subject to special measures.
- yellow: areas subject to intense surveillance; evacuation of special groups is needed; other occupants to be placed on stand-by; or, total evacuation with limited re-entry by members of the public (e.g. daytime only).

144

Table 5.2   Staged alert model for natural hazards.

Stage 0 – White or green

Physical situation   Area subject to significant hazard.
Forecast   No event expected in the near future.
Scientific action   Define areas subject to particular hazards and levels of hazard. Calculate return periods of events of a given size. Estimate the maximum probable event. Microzone the responses of soils and topography. Identify possible secondary hazards. Monitor precursory phenomena.
Civil-protection action   Long-term mitigation action and planning. Campaigns to increase public awareness; field simulation exercises.

Stage 1 – Yellow – Hazard watch stage 1

Physical situation   Early precursory signs have been detected. At least one form of mildly enhanced activity or detectable anomaly.
Forecast   Could lead eventually to an extreme event.
Scientific action   Intensify monitoring efforts and re-establish contact with civil authorities.
Civil-protection action   Open priority communications channels to scientists engaged in monitoring activities. Check readiness of equipment and personnel and functionality of emergency plan.

Stage 2 – Orange – Hazard watch stage 2

Physical situation   Several premonitory signs observed and they continue to grow in importance. Accelerating pace of change.
Forecast   Extreme event probable within a relatively short time window, such as a few weeks.
Scientific action   Intensive round-the-clock monitoring activity. Constant contact and close collaboration with civil authorities. Issue periodic information bulletins.
Civil-protection action   Emergency personnel informed and on stand-by. General public given preliminary information about the situation. Contacts resumed with mass media. Transfer hazardous substances to safe places. Begin preparations for emergency action. Shut down some hazardous and sensitive processes. At a late stage, evacuation of vulnerable groups and those who require more time to evacuate. Preliminary warning of the population.

Stage 3 – Red – Hazard warning

Physical situation   Intense premonitory activity observed, with increasing rates. Short-term precursors occur. Minor destructive events may already have started to occur.
Forecast   Major destructive event expected within the near future (usually 1–72 h).
Scientific action   Intensive round-the-clock monitoring activity of all sources of data. Real-time analysis of the results. Constant contact and close collaboration with civil authorities. Frequent information bulletins issued. Advice on civil-protection measures such as evacuation.
Civil-protection action   Frequent issue of specific warnings. General evacuation of threatened populations. Careful monitoring of public action to ensure that it is not inappropriate. Final pre-impact preparations. All emergency services either in action or ready to go into action. Regular transportation in the expected impact area ceases. Interdiction of main risk areas. All sensitive and hazardous processes shut down.

Stage 4 – Purple – Emergency

Physical situation   Major impact has begun in at least one area.
Forecast   Forecasts will be concerned with the duration of impacts, the probability of secondary or repeated impacts, and any variations in the intensity of impacts.
Scientific action   Essentially the same as for stage 3.
Civil-protection action   Evacuation, if it has been undertaken, should cease. Emergency services should be working. Management of evacuees. Prevention of entry into worst-affected areas.

- red: areas subject to total evacuation and interdiction on public access; only very limited access by emergency personnel, and always under controlled conditions and surveillance.

The imposition of such a scheme will require consistent definitions of the zones, wide and successful publicity for the resulting map, round-the-clock monitoring of hazards, and constant control of access at the principal points of entry to the restricted zones. Definitions *must* be clear, widely disseminated and firmly imposed. Many of the 62 people who died in the interdiction zone around Mount St Helens (Washington state, USA) when it erupted on 18 May 1980 had no authority to be there and had slipped past the road blocks.

5.2.1 Warning processes

One of the principal purposes of an emergency plan is, obviously, to save and safeguard human lives. Two essential elements of this are warning and evacuation procedures.

The warning process involves a combination of technical and social measures. It requires adequate knowledge of impending threats, a means of disseminating instructions, and reasonable certainty that these will be understood and acted upon. Once a scientific prediction of impending hazard impact has been received, the principal disaster manager will have to decide whether it merits the dissemination of a warning. Failure to warn when there is a high likelihood of impact may be regarded as culpable negligence; on the other hand, false alarms will reduce public confidence in the warning process. Moreover, whereas most scientific information will convey uncertainty if such exists, perhaps by assigning a probability to the likelihood of impact or by using confidence limits for a prediction, this cannot easily be relayed to the public, who are unlikely fully to appreciate the meaning of probability. Messages should say no more than that impact is "highly likely" and it is better that they express absolute certainty whenever this can legitimately be done. Hence, the disaster plan should designate a person whose role is to receive and interpret prediction messages, and to formulate warning messages. This person should have at least some familiarity with both scientific method and public perception – the more the better.

A successful warning will give the recipients enough time to react, but will not allow so much time to pass that precautions or credibility lapse. It will express a reasonable certitude that specified events will occur in a defined zone during a time window that is not too large (i.e. hours or days, not weeks or months). It will explain clearly what action should be taken and who the warning is aimed at. A simple example might run as follows:

The National Meteorological and Hydrographic Service has informed the County of Alpha civil-protection office that the River Beta is expected to overflow its banks within the next 12 hours. Flooding in Delta and Gamma districts is expected to reach depths of up to 2m. Residents of these zones are required to evacuate their homes and must do so by 15.00 hours (3 p.m. today). Do not delay. Reception centres for evacuees have been opened at Lamda and Omega. Police will direct traffic, and buses will be waiting to evacuate residents who do not have their own transport.

Note, however, that warning should be a repetitive process. The first message should be followed by others that detail changes in the hazard, the impact and the required response. The last message should be an all-clear that informs people that the danger has passed and may tell evacuees that they can return to their homes or workplaces. The emergency plan should therefore designate a spokesperson to emit warning messages, channels of communication to carry them, and a mechanism to monitor the efficacy of responses to the message. Regarding the general public, radio and television are usually the preferred media for carrying the messages. Fax, telephone and Internet can be used for institutions. In restricted areas, police cars with loud-hailers may be used, especially as these vehicles immediately identify a well known source of authority. Sirens are sometimes used, but their signal may be ambiguous and unclear to people who do not know why they are being sounded (the pros and cons of different methods of disseminating warnings are listed in Table 5.3). However, no channel of communication is likely directly to reach all people who need to receive the warning message; one must rely on word of mouth to inform the residual population and on imitative behaviour to ensure that those who initially resist the instructions later follow suit. The emergency plan can provide for feedback from emergency-service workers to ascertain whether the warning messages have taken effect; if not, they can be reissued in modified form. Finally, the plan may need to provide different messages to different groups, perhaps in different languages. In short, flexibility and a marriage of technical and social expertise are the key to the success of the warning process.

Public reaction to impending disaster will only be as good as the information that people are given. It is therefore important to exercise quality control with respect to the warning system, once this has been designed. As such a system is, in effect, embedded in the broader context of scientific prediction of hazard impacts and management of the ensuing emergency, it is useful to evaluate it in context. Here are some of the questions that can be asked as part of a broad evaluation process.

Legislative and organizational framework:
• Does the issuing authority have a coherent set of policies on warning?

Table 5.3 Pros and cons of public warning methods.

Siren

PRO A simple means of sounding the alarm with the benefit of immediacy. Can be used during the night to wake people up and get them to respond.

CON Sirens are easily ignored or misinterpreted by people who do not understand why they are being sounded. Some people may not even hear them or know what action to take if they do hear them.

Police car with loudspeaker

PRO Can broadcast simple verbal instructions about what to do (e.g. evacuate). A police car is a visible, easily interpretable sign of authority.

CON It is a slow process to cruise around an area that needs to be warned. Not all people in the area may see the police car and hear the message.

Radio message

PRO It is easy to broadcast, and re-broadcast, a simple set of verbal instructions about what to do to avoid the disaster impact.

CON A large proportion of the public will probably not be listening to the radio, especially at night. Even some of those people who are listening may not be attentive. The message will need to be broadcast on *all* channels that are received locally.

Television message

PRO At any point in the daytime or evening, television audiences are likely to be relatively large. A visual and an aural message can be broadcast, if necessary repeatedly. The impact is therefore greater than that of radio.

CON As with radio, many people may not be watching television. All channels that can be received locally must be utilized. Very few viewers after midnight and before 0800 h, restricted numbers in the morning and mid-afternoon.

Newspaper announcement

PRO Can be combined with maps, human-interest stories and interviews to increase its impact. Can be given a considerable amount of detail.

CON Not useful for short-term warnings (less than 24 h). Only a certain, relatively small, proportion of the public buys a daily paper.

General publicity campaign

PRO Can utilize all available media, often in creative combinations, to get the message across.

CON Despite this, the message may be misinterpreted by some people who receive it. General publicity campaigns against hazards and risks are not really appropriate to immediate warning processes; rather, they serve the needs of intermediate and long-term warning.

- Will the recipients consider the warning legitimate and authoritative?
- Does it have adequate foundations in law and support from the legal framework?
- Is a procedure in place for monitoring the effectiveness of warning plans and strategies?

The context of forecasting and prediction:

- Are forecasts or predictions detailed, precise and reliable enough to enable warning?
- Do forecasts or predictions tie in properly with warning?

The logistics of warning:

- What area and time window will warnings cover?
- Will special arrangements be needed for night-time warnings?

148

- Are communications networks sufficiently robust to carry warnings under all conceivable circumstances?
- Will warnings be carried by more than one means of communication and does the system of dissemination have sufficient redundancy?
- Have the broadcast media understood their role in the warning process and are they properly connected up to the authority that will issue the warnings?
- Are arrangements adequate to warn minority groups, people with special needs and people in remote areas?
  Pubic attitudes:
- Has the content and level of detail in warning messages been properly determined in relation to the objectives of warning (e.g. to stimulate timely evacuation)?
- Have the probable recipients of warnings been consulted?
- Is basic information on relevant hazards freely available to the probable future recipients of warnings?
- Have attempts been made to raise awareness and educate the public about hazards and emergency situations?
- Does the warning system allow recipients to confirm what they hear?

Positive answers to these questions will go a long way to assuring the functionality of the warning process when it is most needed.

5.2.2 Evacuation

One of the most effective means of reducing danger to human lives is evacuation (Table 5.4). In this section we will concentrate on precautionary evacuation that is carried out shortly before predicted disaster impact.

Table 5.4 Two classifications of evacuation in disaster situations.

(a) Forms of evacuation		
Timing	Short term	Long term
Before impact	Precautionary	Preventive
During impact	Self-preservation	—
After impact	Rescue	To allow reconstruction

(b) Phases of preventive evacuation		
Phase	Time period	Definition
1	Decision	Time between detection of threat and decision by authorities to order evacuation
2	Notification	Time required to warn evacuees
3	Preparation	Time before beginning evacuation
4	Response	Time between departure and arrival
5	Verification	Time needed to ensure success of evacuation
6	Return	After danger has passed

Although the details of the evacuation exercise can be worked out only on the basis of emergency conditions, the broad strategy needs to be incorporated into the plan. It involves processes of decision making for both the authorities, who have to decide whether to order evacuation, and residents, who must decide whether to evacuate (Fig. 5.4). On the basis of hazard and risk maps it is possible to ascertain which areas (residences, factories, etc.) may need to be evacuated under given circumstances. Routes and evacuation centres may have to be designated. The former should conduct evacuees progressively farther away from the source of danger; it should not lead them into new dangers. The latter, which may be set up in schools or other public buildings, should be safe, stable constructions with space to store provisions and set up beds and communications equipment. They may also require generators in case electricity supplies fail. Schools, in particular, may have dining rooms, kitchens and other facilities that make them particularly suitable as temporary evacuation centres.

In view of the costs and problems associated with having them idle and unoccupied most of the time, fully dedicated evacuation centres are seldom constructed. However, it is often expedient to design new public buildings – such as community centres – with inbuilt provision for alternative use as evacuation centres. The number of users of a centre should not be calculated on the basis of the expected total number of evacuees, as many of these will find their own accommodation.

One possible solution to both precautionary and preventive evacuation is to requisition hotels, guest houses and second homes. The plan can provide for a census of these, but it cannot determine the vacancy levels, or possibly even the willingness of proprietors to comply with requisition orders. However, it can provide for compensation in the event that requisition takes place.

The evacuation process need to be planned so that it includes a high degree of supervision. Evacuees required time to receive their orders, organize their departure, leave and travel to their destinations. Bottlenecks, road congestion, vehicle breakdowns and resistance to compulsory evacuation orders all need to be combated. Evacuation needs to be directed, monitored and verified,

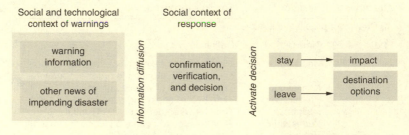

Figure 5.4 Evacuation decision-making processes.

perhaps by the police or civil-protection marshals. Hospital patients, senior citizens, handicapped people and prisoners all require special facilities and treatment. They also need to longer lead times. Finally, registers should be kept of isolated individuals with special evacuation needs.

Evacuation therefore requires the considerable application of resources, principally manpower, communications and vehicles. Buses, vans, breakdown trucks, police cars and ambulances may be needed, needs that must be foreseen in the plan. Considerable thought needs to be devoted to the avoidance of road congestion, especially as clear roads may be needed for emergency vehicles.

Where single buildings or particular institutions are at risk, the plan should designate nearby open spaces where occupants can re-group in safety. Bells or sirens (or more apocalyptically earth tremors, mudflows or smoke and flames) may trigger the evacuation, and designated marshals (monitors) should then assemble the evacuees and check that no-one is missing. Designated open spaces can also be used so that people can leave their homes during earthquakes and re-assemble in safety as they wait for help.

At indoor evacuation centres, the following items will need to be either stockpiled or supplied rapidly when the need arises:
- beds and bedding
- clothing and footwear
- food and drinking water
- medicines, disinfectants and sanitary supplies for bathrooms
- generators and fuel
- emergency lighting and heating
- emergency communications equipment.

Certain types of impact may require evacuation to protected environments that are either within the building in which the evacuee finds himself or herself, or are very close by. Tornadoes are a good example of this. Shelters can be created in basements or other low-lying reinforced structures. In the short term, people can be evacuated from floodable areas to upper floors, but generally intramural evacuation is an under-utilized strategy. It is useful only in certain types of disaster (floods especially, and perhaps hurricanes) and may require specialized planning to ensure that evacuees are not isolated when in need of assistance, for example, for delivery of medical supplies.

Precautionary evacuation is usually safe only if it is carried out before an impact. If there is no time to complete it before disaster strikes, people are generally better off seeking shelter at home or at work, without attempting to evacuate. Moreover, evacuation will require strong justification and enforcement if it has to be prolonged while waiting for an impact that fails to materialize. However, note that the legal authority to compel people to move varies widely from one country to another. Even compulsory evacuation orders are hard to

Box 5.2 Hurricane Floyd evacuation

Hurricane Floyd formed on 8 September 1999 in the western Atlantic Ocean near the Leeward Islands. Over the following nine days this 400 km-wide storm system moved slowly across the Bahamas and up the eastern seaboard of the USA. The forward tracking speed averaged 21 km per hour and sustained rotational winds reached 225 km per hour, making Floyd a relatively rare category 4 storm (on the Saffir–Simpson hurricane magnitude scale). It reached landfall at Cape Fear, North Carolina, on 16 September 1999, causing 500 mm of rainfall that resulted in substantial flooding in coastal river basins. On the east coast of the USA, storm surge heights reached a maximum of nearly 5 m and damage occurred from Florida to New England. Floyd was promptly followed by another category 4 storm, Hurricane Gert, but this made landfall in greatly weakened condition in the Canadian maritime provinces much farther north.

The human impact of Hurricane Floyd was quite exceptional, even considering the US$40 billion of damage caused by Hurricane Andrew to settlements in Florida and Georgia in 1992. Floyd killed 70 people, mostly by drowning them in the floodwaters of the storm surge, which resulted from the friction of wind on sea water and the low atmospheric pressure at the centre of the storm. To escape such a fate, some 2 350 000 people left the coast in the largest peace-time evacuation in US history. Four counties in Florida and six in Georgia were subjected to compulsory evacuation orders and several major interstate highways were converted to one-way traffic flows. In the meantime, major installations were closed down throughout the risk zone. These included Disney World and the Cape Canaveral space centre in Florida, and the naval base at Norfolk, Virginia. Upon their return, some 1.9 million evacuees were temporarily deprived of electricity supplies, some for weeks on end. The cost of damage exceeded $6 billion.

In many respects the Hurricane Floyd evacuation exercise represented a triumph of US hurricane forecasting and emergency planning. The high death tolls of the early 1900s (e.g. 6000 at Galveston, Texas, in 1900) have not been repeated for decades, even though population has grown relentlessly all along the eastern seaboard.

Despite considerable advances in tracking equipment, software and procedures, predictions of the time and location of hurricane landfall still carry a substantial margin of error. This necessitates preparations, including evacuation, that may end up being precautionary rather than demonstrably necessary. They are nonetheless expensive: during Floyd, evacuation alone cost about $600 000 per kilometre of coast. Because the hurricane tracked northwards along the east coast of the USA, it proved difficult to forecast exactly where it would make landfall and what level of storm surge (and damage) would be sustained by coastal communities along its route. This meant that evacuation needed to be carried out over very wide areas.

In Jacksonville, Florida, and Savannah, Georgia, obligatory evacuation put hundreds of thousands of people on the roads at the same time. West of Jacksonville, Interstate Highway 10 accumulated a 300 km tailback of vehicles. The price of hotel rooms went up by 500 per cent, and gasoline supplies dwindled almost to nothing at a time of maximum demand. The exercise ended in a welter of criticism and recrimination, leading possibly to weakening of public confidence in hurricane warnings.

Obligatory evacuation cannot be enforced, and neither can anti-profiteering laws. In the Savannah area (Chatham County, Georgia) residents were informed that they must evacuate and if they did not there would be no rescue or emergency services. About two thirds of them left, only to find themselves stuck for many hours in traffic jams. In the end, they were subject to considerable expense and inconvenience, only to find their homes and businesses undamaged when they returned.

> **Box 5.2 (continued) Hurricane Floyd evacuation**
>
> The first lesson of this event is that evacuation times needed to be staggered to avoid mass congestion on the roads. In such a huge logistical exercise it was not sufficient merely to superintend traffic movements and increase the westward flow of vehicles. Moreover, those drivers who sought to buy fuel at the last moment found that supplies quickly ran out or filling-station proprietors soon shut their businesses and left the area. Radios, bottled water, flashlights, packaged food and other useful or essential items soon disappeared from those shops that remained open. Although some culprits were arrested and fined, it did not stop unscrupulous retailers and hoteliers from profiteering.
>
> In previous years, the US Federal Emergency Management Agency had spent considerable sums of money on free distribution of video cassettes giving instructions on how to prepare for disaster. Although these undoubtedly helped, they did not have much impact on those people who had no intention of behaving prudently and who could have avoided many subsequent problems by making simple personal preparations for the hurricane, such as boarding up windows and accumulating fuel and water when they were freely available.
>
> Although the hurricane did not strike many of the areas where people were forced to evacuate, it could have done. This underlines the importance of ensuring that evacuation is completed in time. If hurricane-force winds and high storm surges had struck long lines of traffic, there could have been many more casualties than the 70 who died.
>
> In conclusion, the Hurricane Floyd evacuation exercise illustrates the need to persist with both public and personal preparations for disaster. Concerning the former, detailed scenarios need to be constructed for the logistics of complex evacuation programmes, and problems need to be anticipated as far as possible using flexible solutions. Regarding the latter, public education programmes need much fine tuning and their effectiveness needs to be monitored.
>
> Sources: L. A. Rossbacher, 1999, "The hazards of hazard prediction", *Geotimes* (December), 44; and reports by Associated Press, United Press International, Agence France Presse and Reuters.

enforce, as people are unlikely to be arrested because they refuse to evacuate.

On a more practical level, the following exemplifies the procedure for designing a plan to evacuate the general public from a town with a population of 10 000–50 000 when the area is threatened by an imminent impact. It is appropriate for places where the hazard is well known and where impacts can be scientifically forecast at least eight hours in advance and preferably rather more than that. It may also be appropriate for post-earthquake evacuation.

First, the planner should divide the urban area into sectors, perhaps of a few hectares each. This can be done in relation to topography, accessibility, social groups (e.g. neighbourhoods), or density of buildings, especially residences. With respect to each sector, one or two public open spaces (e.g. a park or sports field) are designated as **assembly areas** for evacuees. This is especially important when earthquakes are involved, as most people will want to remain out of doors during the period of major aftershocks. The assembly areas should have permanent signs to indicate this use and should be signposted at key road junctions. Publicity material should be distributed to residents to inform them of procedures when an evacuation is required, including the location of assembly points. Where evacuation is bound to be required sooner or later, it may be appropriate to require residents to state on a questionnaire form whether they intend to use their own transport or rely on provided means such as buses. This will help estimate the probable demand for public transport for evacuees. Arrangements then need to be made to co-opt buses, taxis or other vehicles and their drivers when needed.

The map of sectors (shown with different coloured shading) and assembly points (shown green) should then have added to it the routes to be used for evacuation. If possible these should be separated from the routes of incoming emergency traffic. There are obvious problems in ensuring that routes are used as planned during an evacuation. Probably it will be impossible to enforce the desired traffic flow, but the plan should at least mark the position of road blocks and traffic control points, where the police will ensure that vehicles go in appropriate directions. The last part of the evacuation plan concerns the distance to, location and means of reaching evacuation reception centres.

Additionally, it may be necessary to estimate the number of transient visitors to the area. In seasonal terms these may be tourists of migrant workers, and diurnally they may be commuters. They may need special attention to ensure that they too are evacuated when necessary, or to keep them out of the area during a developing emergency.

Special forms of planning include the possibility of evacuating people vertically (i.e. to upper floors of buildings) when flooding threatens. In this case one needs to be sure first that they will remain safe, which may not be the case if hurricane-force winds are blowing, and secondly that they are equipped to endure a period of isolation before help arrives or the flood abates. Other limited forms of evacuation include removing people only short distances, perhaps to local public buildings. This may be appropriate when unexploded munitions or terrorist bombs are discovered and must be made safe.

Having made the plan and communicated it to officials and the public, it would be a good idea to test it. This can be done with a control group or the entire population of a single sector. I can recall one case in which all the over-60s of a mountain village were evacuated for several hours in an exercise designed to test the efficiency of volunteer rescue workers. The settlement was vulnerable to earthquakes and floods, and with this in mind the pensioners submitted quite willingly and cheerfully to the exercise. It is important, however, that the results of the evacuation be monitored carefully so that the plan can be adjusted to iron out glitches.

It is important to recognize that no evacuation plan is likely to be 100 per cent successful. In some cases one may be lucky to succeed in evacuating two thirds of the people at risk. But emergency planners should not be discouraged by this, for evacuation is still the single most effective way of saving lives in many cases of impending disaster. In addition, it may be possible to combine evacuation planning with general urban planning. To begin with, the need for assembly points may help preserve and conserve public open space. Moreover, the needs of evacuation are a powerful argument to have wider, less cluttered main roads and simpler, more easily controllable road junctions.

One final point about evacuation is that there may be more problems in getting people back into homes and businesses than getting them out. They may be returning to buildings that are no longer habitable. However, this is the subject of Chapter 7.

5.3 Search and rescue

In may disasters there is considerable potential to rescue living survivors, but for those who are trapped, injured or exposed to the elements, survival rates fall exponentially within 24 h after impact (Fig. 5.5). The process of search and rescue can be divided into urban (USAR), rural (or regional), underground and marine components, and into the following phases:
- systematic search
- location of victims
- their extraction or recovery from the location where they were found or from the source of continuing danger

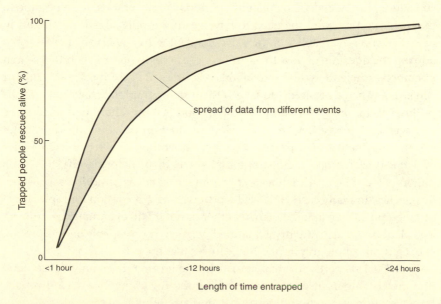

Figure 5.5 Numbers of trapped people rescued alive from various earthquakes as a cumulative percentage over time.

155

• stabilization of their condition pending transportation to medical facilities. Responsibility for organizing this is split between the disaster-plan coordinator, the SAR operational units, and disaster managers at the scene. This section will examine the process of search and rescue in the light of what the disaster plan can contribute to its prior organization. It will not address the problem of single individuals, small groups, or craft that have gone missing, which is a separate issue in SAR, although obviously of more common incidence than disaster work.

Search processes can be divided into aerial and ground-based survey (plus underground survey in the case of tunnels, caves or deeply piled rubble). Marine surveys are conducted with air- and sea craft. Aerial survey is usually carried out on box-pattern, spiral, zig-zag or clover-leaf sortie patterns by helicopters or slow-flying turbo-prop aircraft (Fig. 5.6). It is rapid and efficient when identifying large objects that are easily visible from above, but is difficult or impossible and risky in darkness, fog, storms or very low cloud cover. It can also be difficult to carry out in zones of congested air space, for example, the approach paths or holding areas of airports. Aircraft and their crews must be on stand-by when mobilization is required, and it must be possible to scramble crews quickly, including, where necessary, specialist rescuers and on-board medical personnel. These are points that the disaster plan must address.

Helicopters, in particular, require prolonged and expensive maintenance, which reduces their instantaneous availability. The plan should therefore consider not only general availability periods but also ratios of flying time to maintenance time, and the need to have readily usable substitute aircraft in case of the non-availability of the usual helicopter. It may also be necessary to address the question of how to partition large areas in order to conduct search operations efficiently with the available resources. If visibility is good, aircraft can be equipped – or hired – to take vertical stereographic aerial photographs of the disaster area, from altitudes of around 3000 m. There will need to be arrangements for getting these developed and printed very quickly (digital systems may help reduce the time factor considerably). When printed at scales of 1:15 000 to 1:20 000, stereographic positives may help identify the gross pattern of damage or delineate phenomena such as mudflows or snow avalanches. In many cases, it is well worth writing a provision for rapid aerial photograph survey into the plan, whether or not it will eventually be utilized. The plan should also identify a qualified person who is capable of interpreting stereographic photographs and the features that they show.

Table 5.5 lists pertinent characteristics of some aircraft (mainly helicopters) that are commonly used in emergency work, including search and rescue, transportation of goods and personnel, and air-ambulance services.

In all but the very largest and most apocalyptic disasters, damage is likely

Simple box pattern   **Zig-zag**

Clover leaf   **Spiral**

Figure 5.6   Four sortie patterns for search and rescue or aerial reconnaissance.

to be concentrated in certain easily identifiable locations, such as the sites of major landslides, large building collapses, or urbanized sea fronts or valley bottoms. Ground-based searches can then be concentrated at the site, using trained dogs, sonar, heat sensors, tunnels, probes, microphones, fibre-optic cameras, and so on. The emergency planner should ascertain the level of equipment and expertise that available groups offer and aim to reinforce it if possible. Equipment for on-site search and rescue includes the following:

- sensing devices, as listed above, and dog teams
- flexible probes for use in snow or mud where survivors may be buried
- baulks of wood for supporting tunnels into rubble
- props, jacks, hoists, and hydraulic or pneumatic lifting gear
- arc lights and generators for night-time use

157

Table 5.5 Characteristics of some aircraft commonly used in emergency-operations.

Model	Range (km)	Maximum speed (km/h)	Capacities Lying	Sitting
Helicopters				
Agusta 109-K2	440	260	1 or 2	4
AB206 Jet Ranger	650	220	1	4
AB212/412	450	180	6	14
Sikorski SH-3D	650	260	15	25
Chinook CH-47	200	180	24	44
Alouette SA 319	600	200	1	6
Eurocopter BK-117 C1	540	245	2	4
Fixed-wing aircraft				
G222	2200	360	36	44
Piaggio P-180 air ambulance	2950	480	2	3
C-130 Hercules	4000	600	93	120
Boeing 747	10000	870	160	500

Sources: Page 141 in *Manuale della protezione civile*, Associazione Italiana Medicina delle Catastrofi/Croce Rossa Francese (Casale Monferrato, Italy: Piemme,1990); *N&A mensile italiano del soccorso* (Pistoia, Italy, various issues, 2000–2001).

- all-terrain stretchers for removing injured victims
- basic medical equipment: bandages, splints, neck braces, pain killers, oxygen supplies, etc.

If specialist USAR units are not available, the emergency planner may seek to form them. A full-scale **unit** is a large multidisciplinary group that may consist of the following complement:

- leader (2)
- search team: manager (2), dog handler (4), technical search specialist (2)
- rescue team: manager (2), rescue-squad officer (4), rescuers (20)
- medical team: doctors (2), paramedics or nurses (4)
- technical team: manager (2), structural engineers (2), hazardous-materials specialists (2), heavy-equipment and rigging specialists (2), communications specialists (2), logistics specialists (2), documentation specialists (2).

The suggested complement of at least two of each specialism is designed to ensure as far as possible that at least one of these is available at any point in time. The 56-person team is of a size that is manageable and includes all the main specialities that are required. Doubling competencies allows personnel to be used in relays, and thus to reduce problems of tiredness and concomitant deterioration of performance. It is helpful to clothe team members in identical reflective overalls with badges that distinguish the unit and perhaps also wording across the shoulders that distinguishes each person's specific competence.

In summary, after disaster has struck, search and rescue may need to be carried out in rubble, wreckage, debris (e.g. mud and rocks), snow and ice, fog

Box 5.3 Cable-car incident

The incident

A cable car connects the coastal town of Rapallo with a religious sanctuary on a nearby mountain top at 600 m above sea level. The incident takes place on a Saturday in April with clear skies and light winds. At about 15.10 h the two cabins grind to a halt between pylons. The ascending cabin contains ten people, including one child, and is suspended 40 m above ground level; the descending cabin (which has stopped lower down the mountain) has 23 aboard, including five children, and is dangling 30 m in the air. Both cabins contain attendants who operate the doors and start the cabins moving.

The traction cables have become crossed and are stuck. Both of the cabins sway in the breeze, which causes panic among their occupants, especially the children. The cabin attendants raise the alarm by radio link with the control centre at the upper end of the cable-car track. In a few minutes operators at the centre have found out what is wrong and ascertained that the cars cannot be moved. They immediately call the emergency numbers of the fire brigades (115) and ambulance services (118).

At 15.30 h a helicopter of the Genoa Fire Brigades rescue service is scrambled. It is a relatively large machine (an Agusta AB-412) and takes off with two pilots, a navigator, a doctor, a nurse, and three firemen who are trained rescuers. On the radio from the emergency control centre, the team is made aware that the cable cars are stuck in mid-air, and informed of how many people (and children) are inside each cabin.

Meanwhile, a primary advance medical post is set up on a soccer field at Rapallo. Local Red Cross and public-assistance organizations send six ambulances to this site. A vehicle holding area is then set up in a field on the side of the mountain down slope from the lowest of the suspended cabins. A local family puts a nearby house at the disposal of the rescuers, and a triage station is established there. The nearest general hospital is put on general alert and communicates the availability of beds to the emergency medical control centre (the "dial 118" telephone operators). Local emergency-service commanders and members of the municipal council gather at the advance medical post.

Using its winch, the helicopter lowers two of the rescuers into the upper cabin, which has the most occupants, and one into the lower cabin. As no-one in either car requires immediate medical assistance, the doctor and nurse are not winched down. As the airflow from its rotors makes the cabins sway alarmingly, the helicopter hovers over the cars for the minimum time necessary to lower the rescuers.

Each cabin contains a winch with a harness. At 16.00 hours the process of lowering each occupant down to the ground begins. It is accomplished successfully. Five assorted vehicles transport rescued people from underneath the cable cars to the triage point. Constant communication is maintained between all units in the field and with the fire brigades and medical services emergency control centres.

All the rescued people have their blood pressures and cardiac rhythms checked. Two of them, who have heart conditions, are transported to hospital for more intensive check-ups but are released later in the day. Almost half of the 33 people rescued show signs of pathologies connected with severe stress, although none is too serious to be treated on site. The incident ends four hours after it stated with a meeting for all participants at the town hall in Rapallo.

Evaluation

Fortunately, this accident ended happily and was dealt with in exemplary manner by all the emergency services. Coordination was excellent, tasks were handled smoothly, professionally and without duplication. Rescuers worked courageously and efficiently in close cooperation with each other. No significant hitches occurred in the rescue operation.

Box 5.3 (continued) Cable-car incident

In this event we see the conjunction of two technologies with rather particular inherent risks: cable cars and helicopters. Cable-car incidents are unfortunately not uncommon, despite highly evolved technologies for preventing them or limiting their consequences. In other instances, cabins have fallen to the ground as a result of breaking cables, machinery fouling up, support pylons failing, or aircraft crashing into the cable-car equipment. As noted elsewhere in this book, helicopters are particularly at risk of flying into overhead wires or other obstructions that are difficult or impossible to see from aloft. The operation at Rapallo depended on, as much as anything, the skill of the pilots.

Despite these reservations, it is difficult to see how the emergency could have been dealt with as effectively without the use of a helicopter. It underlines the importance of having an appropriate aircraft with sufficient manpower constantly available. The Rapallo operation was a form of mountain rescue and as such the helicopter vitally needed its winch. Moreover, the presence of winches in the cable cars was very important. Such equipment needs to be constantly maintained and tested to ensure functionality at all times.

The incident underlines the importance of fully functional, dedicated communications channels. Although it occurred in a fairly restricted geographical space, it nevertheless involved firemen, helicopter rescuers, ambulance services, civil authorities, medical personnel, and civil-protection volunteers, and these worked from three base sites (the advance medical post, vehicle holding area, and triage station), from the air, and, of course, from the slopes beneath the two cabins. Coordination is the key to good emergency management in such circumstances, and it depends on the effectiveness of spoken communication.

Although the incident ended with no injuries, it nevertheless involved significant danger. It therefore called upon rescuers to put themselves at risk: even in the best-organized and most meticulously planned emergency operations, one can still end up relying on personal heroism to save the day. Moreover, at Rapallo the emergency services were justified in making substantial preparations for a mass-casualty situation, as fatal cable-car accidents do occur.

Lastly, the Rapallo incident highlights the importance of spatial organization in tackling emergencies. Besides the problem of winching down passengers from the cable cars, the terrain was rugged and posed problems for vehicular access. Crowd control also had to be practised in difficult conditions, as there were many onlookers. Separate triage, holding and advance medical post areas were necessary and all of them had to be organized quickly, with an eye to maximum efficiency.

Source of information: Cremonesi, Paolo 2000. Rapallo: intrappolati sulla funicolare. *N&A mensile italiano del soccorso* anno 9, vol.103, pp. 18–21.

and low cloud, rugged terrain, tunnels, caves, stormy seas, or floodwaters. It is a complex, dangerous, time-consuming, tiring and labour-intensive process to locate, rescue, stabilize and treat living victims (see Tables 5.6 and 5.7 for further details of procedures and dangers). Special skills and equipment are absolutely necessary. Above all, the disaster plan should ensure that accredited specialists in SAR are fully trained: in the aftermath of the 1985 Mexican earthquake more than 200 rescuers were killed in Mexico City and the majority of them were not trained to operate in hazardous conditions. Thus, the disaster plan should address the question of liability for injuries received on the job.

One final point needs to be made. The vast majority of search and rescue in

Table 5.6 Tactical programme for technical rescue.

Technical rescue officers begin to size up the situation.

Other personnel begin to establish control of the scene
• working area cordoned off
• non-essential people removed from cordoned area
• hazards to rescue workers (e.g. electricity and gas lines) identified and mitigated
• establish unimpeded traffic flow to and from site

Technical rescue officers identify a plan of action

Rapidly brief the team
• identify procedures to be used and equipment needed for operation

Appoint command officers and issue identification vests
• rescue leader, rescue-equipment manager, rescue safety officer, rescue personnel manager

Sort out casualty removal procedures
• assign role of patient removal officer
• plan procedures for removing casualties
• identify and assemble necessary equipment

Set up equipment and vehicle staging area close to working area
• assemble and check equipment
• track use and location of equipment throughout the rescue

Set up personnel staging post and use it as place to assign roles to personnel who arrive at the site
• withdraw exhausted personnel to the staging post and substitute with new workers as necessary during the incident
• personnel officer ensure that unnecessary personnel report back to the staging post and limits personnel at work to those who are strictly necessary

Procedures for specific types of rescue work
• *structural collapse*: conduct full reconnaissance for location of victims; if appropriate request building inspector
• *confined space*: lay out 100 m of air hose per worker; bring in lighting unit if needed; assemble back-up rescue team
• *trenches*: no one to enter unshored trenches deeper than 1.5 m; place ground pads at trench edges
• *rope operations*: double-check rope, winch and pulley systems prior to use; helmets to be worn by rescuers; rope system movement commands to be called out only by team leader; rescue frame needed?; hand winch needed?

Source: Derived from written procedures used by Fairfax County Fire and Rescue Department, Virginia, USA.

major disasters is carried out by untrained local people using rudimentary equipment or none at all, because specialized teams generally do not arrive in time to play leading roles. A wise disaster planner will seek out sources of SAR expertise and create procedures to reduce the time to arrival on site. In the case of large disasters, the plan should allow for fast authorization of both foreign SAR teams and those from other parts of the same country. On the one hand, one does not wish to become liable for inexpert aid furnished by inadequately trained units that put themselves excessively at risk, but, on the other hand, one does not wish to prevent bona fide teams from taking part in the rescue effort.

Table 5.7 Common risks and hazards of urban heavy rescue.

Situation	Associated hazard and risk situations
Cliffs, gorges, ravines and mountains	Falling debris, high-angle rescue by climbing or being lowered to site, hypothermia or hyperthermia, improper rigging, rope or rope-system failure
Coastal sea waters	Drowning, fast currents, high waves, hypothermia
Earthquakes, hurricanes, tornadoes and major landslides	Building collapse, tunnelling into rubble and debris, other risks associated with disaster response
Farms and other agricultural facilities	Confined spaces, dust explosions, entrapment in machinery, fertilizers, hazardous materials
High-rise buildings	High-angle rescue, elevator rescue, fire
Industrial plants and facilities	Confined spaces, explosions, exposed utilities, falls, hazardous materials, high-noise environments, machinery entrapment, stored-energy release, toxic gas emissions
New construction	Entrapment in building, entrapment in machinery, exposed utilities, high-noise environments, structural collapse, trench rescue
Old buildings	Entrapment in building, exposed utilities, fire, hypothermia or hyperthermia, oxygen-deficient atmospheres, structural collapse
Rivers, flood-relief tunnels, floodable areas, lakes	Boats, drowning, electrocution, entrapment in submerged hazards, floating hazards such as logs, hypothermia, ice rescue, surface and underwater rescue, swift water rescue, toxic water conditions
Sewers, tanks and cesspools	Confined spaces, oxygen deficiency, toxic gases
Transportation facilities	Confined spaces, derailments, hazardous materials, machinery entrapment, toxic gas emissions, unstable heavy equipment
Trenches	Cave-in and burial, confined spaces, exposed utilities, falls, hazardous material releases, oxygen-deficient atmospheres
Wells and caves	Confined spaces, drowning, hazardous environments

Source: Figures 3.1 and 9.4 in *Technical rescue program development manual*, US Fire Administration (Washington DC: US Federal Emergency Management Agency, 1995).

With this overview in mind, the next section will consider the specific demands of rescue in typical sudden-impact disaster situations. These demand what has been termed "urban heavy rescue".

5.3.1 Urban heavy rescue

Urban heavy rescue is the principal subset of search and rescue associated with disaster work. It is important to recognize at the outset that in sudden-impact incidents and disasters the vast majority of living victims are rescued (a) within a few hours of the event and (b) by *local* rescuers, not specialists from far afield.

In areas periodically affected by disaster it is therefore essential to have trained rescue units on call. These may be part of fire services or they may be volunteer groups, but in either case they must receive specialized instruction in rescue techniques and must conduct regular exercises to keep the training current.

Rescues occur in four settings that are not mutually exclusive: confined spaces, structural-collapse situations, high-angle situations necessitating the use of ropes, and trenches. Even though a major earthquake, for example, will produce a complex and widespread pattern of damage, it can nevertheless be broken down into a series of sites where rescue is needed (the principal effects of most earthquakes are concentrated at specific sites of spectacular collapse of occupied buildings). The standardized procedure is as follows.

Upon arrival at the site of operations, the incident commander or team leader must size up the situation. Information required includes establishing who gave the alarm, and when. Witnesses should be interviewed to determine what happened and where trapped victims are located. It will be helpful if maps and plans of the site can be procured quickly.

Site conditions to determine include the ruggedness, elevation and orientation of terrain, the type and scope of collapses, the occupancy of collapsed structures and the likelihood of entrapment. Site instabilities to be assessed may include groundwater excesses, tension cracks, overhangs, subsidence, bulging, spoil heaps, or superimposed loads. Officers should determine whether utilities such as gas and electricity lines are damaged or exposed.

Meanwhile, the scene needs to be brought under control (see Fig. 5.7). The working area must be cordoned off and usually the perimeter should be manned to exclude interlopers. Non-essential people should be removed from the cordoned area. Hazards to rescue workers should be identified and as far as possible mitigated. Communications systems need to be set up for relaying messages on site and between the site and command centres elsewhere. The local communications node is the command post.

On the periphery of the site, and in close proximity to access roads, a primary staging area should be set up with the command post in it. This can either be within the cordoned area or immediately outside it. Here equipment will be stockpiled, vehicles parked, manpower assembled and emergency workers checked for health and safety. Next to it there may need to be a first-aid station, where triage will be performed on victims recovered from the site, and ambulances will load them for transport to medical centres (see §6.1.2). A secondary staging post may be needed for emergency vehicles and personnel that arrive after the start of operations and are held in reserve until they are needed. The objectives here are (a) not to overwhelm the site with too many workers or too much equipment, and (b) to ensure the unimpeded flow of emergency traffic.

The incident commander and his deputy should then establish a plan for

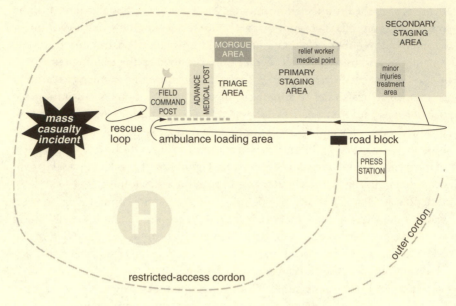

Figure 5.7 Pattern of operations in a mass-casualty incident.

rescue actions. There should be a short briefing of emergency squads (by sector and team), at which they are informed of their tasks and acquainted with health and safety concerns arising from the rescue. The risk of injury on the job increases dramatically with tiredness, and hence rescuers working in difficult and dangerous conditions should be substituted regularly with fresh personnel and allowed to rest and recuperate. At this stage it is essential to appoint a team safety officer who monitors the conditions of both the site and the rescuers. The identity of this person must be made known to all rescuers. It is also helpful to establish a mutual surveillance system in which pairs of emergency workers are responsible for monitoring each other's safety.

At the briefing the commander should appoint leading officers and issue identification vests: the rescue leader, rescue-equipment manager, rescue safety officer and rescue personnel manager. He or she should define procedures to be used and identify the equipment needed, which should be assembled at the staging posts (a primary area for the main equipment and secondary area for back-up equipment). The equipment manager should issue and track the location and use of equipment. The personnel manager should authorize rescue workers to enter the site and conduct specific tasks, and track personnel assignments during the operations. Most importantly, the personnel manager should, in consultation with the commander, withdraw exhausted personnel to the staging post and substitute with new workers as necessary during the incident, and limit the number of workers to the necessary minimum.

One of the first operations at the site should be to identify and mitigate hazards to rescuers, such a broken electricity or gas lines. There may need to be monitoring of ventilation and air quality, water levels, fire hazards and explosion risks. With regard to utilities and machinery, equipment should be secured, supported, blocked, locked out, shut down, tested and checked as necessary to ensure that risks are minimized.

Besides safety, working conditions at the site need to be established so that rescue can proceed efficiently. This may need lighting systems (floodlight batteries attached to motorized electricity generators) and ground pads. Unstable masonry, vehicles and equipment will need to be shored up with timber or steel joists, ropes or hawsers, air-pressure systems, hydraulic jacks or other such equipment. The stability of tunnels and trenches should be checked before rescue workers are allowed to enter them.

The following are some procedures for specific types of rescue work:

- *Structural collapse*   Conduct full reconnaissance for location of victims; if appropriate, request the help of a building inspector; ensure that tunnels built into rubble are shored up sufficiently and are safe enough to enter.
- *Confined space*   Lay out 100 m of air hose per worker; bring in a lighting unit if needed; assemble a back-up rescue team.
- *Trenches*   No-one should be allowed to enter unshored trenches deeper than 1.5 m; ground pads should be placed at the edges of trenches.
- *Rope operations*   Double-check the condition of ropes, winches and pulley systems prior to use; helmets should be worn by rescuers; rope system movements are to be commanded only by the team leader; is a rescue frame needed?; is a hand winch needed?

The principal objectives of the rescue operations will usually be to identify the location of victims, stabilize their conditions, remove them from the site, and abate continuing hazards and risks at the site. Casualty-removal procedures need to be defined at the initial briefing and a patient removal officer may need to be designated. The first task is to establish, as far as possible, the number, location and condition of trapped victims and ascertain the nature of their entrapment. Before they can be rescued they may need oxygen, intubation, cardiac defibrillation, an *in situ* medical examination or equipment such as helmets, splints, or neck braces. In extreme cases, for example of limbs crushed beneath fallen beams that cannot be removed, *in situ* amputation may be needed. This will require a medical doctor to be present among the rescuers and one who is able to operate in difficult and dangerous conditions.

Basic life support is an essential part of rescue in mass-casualty situations. For instance, in earthquakes, crush victims may die very soon after rescue if they are not given appropriate aid, and eventually dialysis to avoid the effects of crush syndrome, in which kidneys are mortally clogged by the products of

Box 5.4 Texas bonfire tragedy

The incident

The tragedy occurred on the night of November 18, 1999 at Texas A&M University, in College Station, Texas.

In order to commemorate sporting rivalry with the University of Texas at Austin, Texas A&M had developed a long-standing tradition of constructing a huge bonfire once a year. The 1999 version consisted of more than 9000 logs, each about 9 m long and up to 20 cm thick, erected around a 20 m centre pole and bound together with wire. At 02.30 h, while about 50 students were putting the finishing touches on the stack, it collapsed into a 9 m-high tangle. Students suspended from ropes around the bonfire were caught up in the wreckage.

Six members of an emergency medical unit were present at the scene when the collapse occurred. By 03.10 h the team had stabilized the condition of 27 victims, put them through triage and sent them to the two hospitals that serve the College Station area, both of which were trauma centres.

When he arrived at the site, the College Station Fire Department battalion chief set up an incident-command system and began to request reinforcements from an increasingly wide area. At the same time, College Station activated the municipal emergency-operations centre, which liaised with state agencies, the National Guard and governor's office.

The incident commander organized a 120 m perimeter around the site and had it manned by police to keep out interlopers. Inside, facilities included a triage station, morgue, rescue base, rehabilitation area for rescuers, canteen, debriefing area and command post. A post for the mass media was established outside the limited access area.

Most search-and-rescue units were based far away in major urban areas more than 150 km away, but 20 SAR specialists were immediately called up from a local firemen's training school. Others arrived from increasingly farther afield during the emergency. Eventually the rescue operation utilized 335 emergency workers and a support staff of 3200. It lacked a designated safety officer, but fortunately no workers at the site were injured, despite the dangers involved in working with a huge, unstable pile of logs and the fact that operations differed considerably from what is usual in urban search and rescue.

Twelve students were killed in the collapse. Two others were trapped alive and were successfully rescued respectively 3 h and 6.5 h after time zero, despite the difficulties posed by the precarious, intertwined state of the logs and the impossibility of shoring up the stack. While trapped they were given oxygen and first aid through gaps in the lattice of logs. Once the living victims were rescued, cranes with grappler arms tore quantities of logs from the stack without needing to cut through the wires that bound them together. The process was carefully monitored and was suspended when bodies were spotted, so that they could be recovered by more delicate means. The last body was brought out almost 24 h after the collapse.

Problems experienced with emergency operations

In synthesis, emergency operations were successful, as one would expect in an area that is fairly well organized to cope with disaster, and given that the incident occupied a small site and generated only 39 significant casualties. However, problems were experienced with emergency communications, superabundance of resources relative to needs, and logistical and medical aspects of operations.

Box 5.4 (continued) Texas bonfire tragedy

Communication among emergency-service units was hampered by incompatibility between the radio systems of ambulance crews and fire-department vehicles, and by the unfamiliarity of private ambulance crews with the fire-service radios installed in their vehicles. As a massive convergence reaction brought thousands of people to the site, cellular and landline telephone systems became largely inoperative as a result of overload. This was increased by the fact that the university authorities encouraged students to call home. Failure to assign radio frequencies reduced the ability to communicate among emergency-relief forces, hospitals and command centres. In addition, calls from journalists, who arrived *en masse*, were not routed to a representative of the emergency services.

Both staging and accredited access were poorly organized within the controlled access zone. Visual identification of key personnel could have been improved, for example, by using reflective vests with "Incident Commander" or other appropriate wording written on them. As requests for clothing, lumber, cranes, wire cutters and other needs were honoured to a level of superabundance by local suppliers, logistical problems gradually mounted up during the emergency phase.

In hindsight, triage was poorly handled. Tags proved too small to contain the information accumulated on each patient. Patients were not evenly distributed among local medical centres, some of which were not even included in the emergency medical plan. Likewise, medical helicopters were not utilized but could have helped spread the casualty load more evenly. Finally, mortuary facilities at local hospitals proved inadequate but alternative arrangements were slow to be instituted. Blunt trauma made identification of bodies difficult, especially as few of the dead were carrying identity documents.

These problems should not detract from the overall impression of an emergency operation well handled. Above all, rescuers showed great courage and professionalism during the exceptionally difficult, dangerous and delicate operation to dismantle the stack of tree trunks and retrieve the casualties. Participants in the emergency exercise were also well catered for in terms of critical-incident stress counselling and support services.

An event such as the Texas bonfire tragedy underlines the importance of learning from the negative aspects of urban heavy rescue. For example, no doubt future incidents in this part of Texas will be handled with more attention to coordination of radio communications. Mass-casualty events will doubtless benefit from improved arrangements for distributing the injured and storing and identifying the dead. But one problem that this example illustrates which is not so easy to rectify is how, in the early stages of an emergency, to avoid over-reacting by committing too many resources (which can simply clog up a rescue operation, or under-reacting by committing too few. Incident commanders therefore need to give serious thought to how they will assess developing situations, and how they can try to develop a feel for needs and anticipate them, especially in unfamiliar sorts of emergency. This can be a tall order.

Source: D. White (2000), "Bonfire rescue at Texas A&M", *Firehouse*, November 2000, 46–53.

ruptured tissue cells. It is therefore important that medical aid be given *during* the rescue, and that the rescue team includes medical personnel.

It is also important that triage at the first-aid post be conducted rationally (see §6.1.2). Triage should be conducted by very few medical officers (perhaps one or two) and should take place contemporaneously with first aid. Transportation to hospital should not begin until the principal victims have been dealt with by **triage officers** and full priority established. Care should be taken to note appropriate details of a patient's condition on triage tags and to record where patients were taken by ambulances or other transport. Record keeping is a vital part of triage.

Lastly, a site commander should know when to declare rescue operations over. Rescuers easily become obsessed by their work, which demands tremendous concentration. They need to be told when to stop.

In conclusion, the process of urban heavy rescue involves assessing the situation, determining the needs it has created, assessing and abating risks to emergency workers, setting objectives, assigning tasks, and, finally, evaluating the success of operations.

5.4 Communications

Information is one of the most vital commodities, and potentially one of the most scarce resources, in disaster. The key to efficient emergency management is to ensure an adequate flow of information about what has occurred, what is happening and what needs to be done. This requires that normal channels of communication be maintained and emergency channels put into action.

Information becomes a premium commodity at precisely the time when it is least available. Hurricanes and icestorms can put telephone lines, antenna and radio masts out of action: earthquakes and landslides can break underground cables and interrupt electricity supplies; floods can inundate telephone exchanges. Hence, a well planned disaster communication system needs to be resistant to structural damage caused by the hazard. There must be provision for rapid repair when damage occurs. In addition, considerable redundancy is needed, with at least one back-up system and multiple paths for information flow so that, when a message must pass repeatedly from person to person, the chain is not broken by failures of transmission or reception.

If communications in the absence of disaster represent a "normal" model, they can be considered to be highly abnormal during the impact, and progressively normalizing in the subsequent aftermath. In this, conditions will remain unusual with respect to the four main constituents of the communication process: the technological hardware used for the composition and transfer of messages; the procedures, formats and conventions used with the technology; human factors of perception and operability; and the organizational context of communication, which imposes a framework of rules, procedures and cultures. All four of these aspects need attention from the emergency planner, who should specifically address the following questions:

- In an emergency, who will need to communicate what to whom? Scenario modelling of this is needed (see §3.1.1) in advance of the next impact.

- Are available channels sufficient to cover these needs? Are all available modes (AM, FM, SSD, digital, etc.) and frequencies (HF, VHF, UHF, etc.) usable?
- Do the tools and spare parts exist to repair broken communications systems and will the necessary expertise be available promptly when it is needed?
- Will there be problems of congestion and slowness, or will bottlenecks occur when large volumes of information must be transmitted rapidly?
- Can systems be shared between groups of users to even out the load?
- Do different communications systems connect adequately with each other?
- Will the systems be robust enough to survive a disaster of the kind and magnitude that are expected in the area?
- Do operators and maintenance people have adequate training to be able to cope with disaster-related communications needs?
- Can field communications systems be moved in and set up quickly and easily? Are they sufficiently autonomous to work well in the field?

Problems experienced with communications during disaster will vary from the simple to the highly complex and technical. One of the more common of the former is the fact that telephone systems rapidly become overloaded by non-emergency traffic, as victims and survivors are contacted by relatives and friends. Therefore, one cannot expect even functioning normal systems adequately to handle the increased traffic generated by civil-protection communications in disaster. Dedicated channels are required.

The emergency planner, or his or her delegate technician, should conduct a comprehensive review of the technology and technical problems of communication in the local area. This should include an evaluation of the probable strength and adequacy of communications between the disaster area and the outside world, with special emphasis on communicating with the main seats of government. On this basis, the plan should identify the principal channels and their operators, outline the protocols that will govern emergency usage and, where necessary, mandate improvements. For example, planning measures might include emergency distribution of cellular telephones to fieldworkers, installation of temporary cellular repeater masts, or renting a mobile satellite-based telephone connection system with 50 lines. Close liaison with telecommunications companies is needed when designing the plan.

The other side of the emergency communications coin relates to the content of messages. Although this will vary considerably with needs, and with the kinds of tasks carried out by agencies sending and receiving messages, some standards need to be adopted. These may include codes for rapid, unambiguous identification of the interlocutors, conventions and protocols for the presentation of different kinds of information (including time, date and site of transmission), rules to ensure that messages are simple, and procedures for archiving messages so that they can be reviewed as needed at a later stage.

Table 5.8 A simple form for recording the details of emergency messages.

Page _____ of:_____	
Date (dd/mm/yy): _____/_____/_____	Day:
Message sent by:	Unit & location:
Means of transmission: Telephone ❑ Cellular ❑ Fax ❑ VHF radio ❑ UHF radio ❑ Satellite phone ❑ Other _____	
Priority: **EXTREMELY URGENT** ❑ **Urgent** ❑ Normal ❑	
Received by: Time _____:_____ h	Unit & location: Received successfully? Yes ❑ No ❑
Message sent to: Time _____:_____ h	Unit & location: Delivered successfully? Yes ❑ No ❑
Other recipient(s): Time _____:_____ h	Unit(s) & location(s): Delivered successfully? Yes ❑ No ❑
Other action required (specify):	
Message:	
Signature of operator:	Signature of recipient:

There may also need to be procedures for verifying that messages have been received, understood and acted upon. Most of these procedures can be worked out by consultation with the appropriate groups and taught to users in training sessions. A simple pro forma for messages is given in Table 5.8.

The following is a résumé of the types of information that will circulate during an emergency:

- general situation reports ("sitreps")
- specific reports on damage to and operability of critical systems
- predictions and forecasts
- other urgent information about hazards
- warning messages
- requests for assistance or resources
- casualty reports
- epidemiological surveillance data
- interagency coordination information
- public-service information

Box 5.5 Nevado del Ruíz volcanic eruption, November 1985

The case of Nevado del Ruíz, 1985, should serve as a terrible warning to the emergency planning community throughout the world. A cataclysmic eruption, coupled with inadequate monitoring, warning and civil-protection activities, led to what in world historical terms was the fourth worst death toll ever experienced in a volcanic emergency.

The disaster

The town of Armero (population 35 000) in Colombia's Tolima province is situated at the foot of the snow-capped Ruíz–Tolima volcanic complex, which rises to 5389m above sea level. The settlement was built on a site that had experienced lahars (volcanic mudflows) in 1595, 1845, 1935 and 1950. However, being of recent construction, Armero had never been destroyed by this sort of phenomenon, although 1000 people were killed in the vicinity by the 1845 lahars.

The average return period of very large eruptions at Ruíz is 1000 years, whereas moderate eruptions recur in the range 160–400 years. In late 1984 the volcano began to show signs of renewed activity. Initially, this meant increases in gas and heat emissions, and then earthquakes and hydrovolcanic explosions linked with the melting of ice on the volcano's summit. This stimulated the interest of the scientific community and the volcano began to be studied and monitored by Colombian, US, Italian and Swiss volcanologists. A consensus was reached that the activity was unusual, but initial interpretations suggested that seismicity posed a greater risk than eruption. However, by March 1985, the Columbian Institute of Geology and Mines (INGEOMINAS), together with scientists working for the United Nations, suggested that a major eruption was possible. Several commissions were established to monitor the situation.

By May 1985 scientists were campaigning for a program of seismographic monitoring on the volcano and a parallel initiative to increase public awareness among the communities located around its base. Following substantial bureaucratic delays, three geophones and four portable seismometers were installed. However, the seismometers could not send data by radio to a monitoring station.

On the evening of 11 September 1985 a hydrovolcanic eruption sent a lahar 27 km down the Rio Azufredo stream at speeds of up to 30 km per hour. On curves the mud climbed as high as 20 m up the canyon walls. The inhabitants of the Azufredo valley and other areas were alerted but not evacuated. The probability of a larger eruption was estimated at 20 per cent. This was low enough for one newspaper, *La Patria*, to claim in a banner headline on 13 September that the volcanic activity at Ruíz was innocuous.

Nevertheless, within a few days, the Colombian civil-defence authorities declared a state of alert. Nevado del Ruíz falls within the provinces of Caldas, Tolima, Risaralda and Quindío, each of which agreed to produce an emergency plan. On 18 September the president of the Red Cross chapter of Tolima province drew attention to the danger that Armero could be completely overwhelmed by a lahar. Civil-protection authorities then started a volcano awareness program in local schools. At the same time, accusations of alarmism began to fly, and that this was leading to the devaluation of property prices and the spread of what detractors called "false information". The local chamber of commerce criticized the press for irresponsible reporting "liable to cause economic losses in the region".

In late September, 4000 cows were evacuated from the region affected by ashfalls from recent eruptions. Volcanic ash may cause fluorosis in cattle and can accumulate in cows' stomachs, poisoning them. Meanwhile, intermittent hydrovolcanic activity continued to deposit ash up to 20 km from the summit of the volcano. However, this abated by early October.

Box 5.5 (continued) Nevado del Ruíz volcanic eruption, November 1985

At a press conference on 7 October, INGEOMINAS presented the first hazard map of the volcano. Risk analysis suggested that, if warnings were disseminated, residents of Armero would have two hours to evacuate once the lahars had begun. INGEOMINAS distributed only ten copies of the map, mostly to scientific institutions, but agreed to have a new version ready in a month's time. Meanwhile, seismicity and gas emissions from the volcano began to increase again. Scientists insisted that a telemetering seismometer be installed and data be processed more rapidly. Gradually some of their demands were satisfied, and by early November it became apparent that intensive monitoring would offer a clue to the volcano's short-term eruptive potential.

However, there were no distinct premonitory signs when the main eruption occurred on 13 November. Ash emissions began at 15.06 h and reached Armero at 17.00 h. At 19.00 h, the Red Cross of Tolima province demanded that Armero be evacuated immediately, but the order was not communicated systematically to the population. The paroxysmal phase of the eruption began at 21.08 h and pyroclastic surges (flows of superheated gas and ash) occurred, melting 10 per cent of the volcano's ice cap. At 23.35 h a wave of mud and debris 40 m high disgorged from the canyon upstream of Armero and flooded through the town at 30 km per hour, subsiding to depths of 2–5 m and burying about 80 per cent of the buildings. It was estimated that 21 015 people were killed, of whom 1000–2000 died slowly while trapped in the mud, from which only 65 were rescued alive. In addition, other lahars, for example on the Rio Chinchiná, caused a further 2000 fatalities. Some 4000 people were treated for injuries during the first week of the disaster, but an unacceptably large number of them died as a result of the disorganized nature of medical assistance.

The revised volcanic hazard map, almost identical to its prototype, was published on 14 November. It proved to have been substantially accurate in predicting threats to the populations in the area.

The causes of disaster

The Nevado del Ruíz catastrophe was one of those events that show the terrible power of hindsight. In retrospect, qualified by knowledge of the tragic losses that occurred on 13 November 1985, many decisions and strategies were wrong. But would they have appeared as mistaken during the circumstances that prevailed before the eruption? Volcanic emergencies are notoriously difficult to cope with when long timescales, uncertain precursors and incomplete predictions are involved. However, one fixed point is the Ruíz hazard map, which was available before the event, was substantially correct, and could have had more impact on preparations than it did.

From the human rather than the volcanic point of view, the disaster was effectively caused by weak leadership, bureaucratic procrastination, and lack of decisive action. The highest levels of civil authority, perhaps insulated from the risks, failed to heed the scientists' remonstrations, to appreciate the significance of their findings, and to visualize the whole panoply of evidence that pointed to the dangers of a cataclysmic eruption. Thus, in the settlements around Ruíz, civil-protection decisions were neither taken nor implemented rapidly enough and foresight was lacking. Moreover, the separate development of emergency plans in the four provinces involved meant that cooperation was inadequate and plans varied in their degree of success.

In retrospect, the whole area around the volcano required precautionary evacuation of its populations. This in turn required decisive action from central government to induce the four provinces involved to implement the measures. Politically, it is usually very difficult to justify evacuation in the face of a threat that, however real, does not point to certain and immediate losses if it is not carried out. Experiences from the Caribbean (for example, Montserrat and Soufriére islands), and even from the 1980 eruption of Mount St Helens in Washington State, USA, underline the strength of opposition to evacuation in the face of an uncertain threat of volcanic eruption.

Box 5.5 (continued) Nevado del Ruíz volcanic eruption, November 1985

It was widely believed that Armero would have two hours to evacuate before dangerous lahars struck. Yet the hazard map made it clear that no safe place existed within 1 km of the town. In fact, experience showed that, despite the shorter time estimates mentioned above, evacuation would not even *begin* until 4–7 h after the danger had first been recognized. In the end, local civil-protection officials were alerted by late afternoon on the day of the main eruption, but they undervalued the abnormal volcanic conditions, which reduced the impetus for action. Hence, if evacuation was ever ordered, it came too late, and attempts to warn Armero by civil-defence radio after the eruption started were unsuccessful.

As usual, prior mitigation would have cost much less than what was spent on post-disaster relief. Had only part of the huge sums spent on helicopter operations after the event been spent on volcanological monitoring before it, evacuation would probably have been made more feasible. This conclusion is highlighted by the danger of helicopter missions during volcanic disturbances of the atmosphere: at Ruiz five people died and three helicopters were destroyed in crashes.

Besides the failings of the civil authorities, the scientific effort also had its deficiencies, including inadequate instrumentation, slow processing of seismic data (which for a while had to be mailed to Bogotá) and rivalry, rather than collaboration, between scientific groups. No responsible agency reacted sufficiently to calls by volcanologists to install more instruments and increase the degree of automated data processing.

The Nevado del Ruíz case offers the following lessons for emergency planning:
- plans should be fully integrated among neighbouring jurisdictions and in terms of the administrative hierarchy (e.g. national, provincial and municipal governments)
- plans should be based on careful evaluation of hazard and vulnerability, which means collecting substantial data on both of these and creating detailed risk scenarios
- in volcanic hazard areas, emergency plans must be able to guarantee public safety in the face of substantial degrees of uncertainty about future and impending threats
- areas of major volcanic hazards require strong civil-protection authorities armed with robust communication systems that have a high degree of redundancy
- emergency communication requires that attention be given to both the technical and the human aspects of the process
- public awareness is the key to any emergency plan that involves rapid mass evacuation.

In the years after the disaster, Armero and the other communities around the base of Ruíz acquired the status of monuments to natural disaster, as well as an up-to-date volcanological observatory. One hopes that the event will be remembered for what it teaches us about planning and organization.

Sources: M. L. Hall (1990), "Chronology of the principal scientific and government actions leading up to the catastrophic eruption and subsequent debris flows of November 13, 1985", *Journal of Volcanology and Geothermal Research* **42**(1/2), 101–115; B. Voight (1990), "The 1985 Nevado del Ruiz volcano catastrophe: anatomy and retrospection", *Journal of Volcanology and Geothermal Research* **42**(1/2), 151–88.

- news reports
- communications network information.

Each type of message may follow specific conventions and protocols and require a specific set of communications methods to disseminate it. Also, the disaster plan should include a means of monitoring the impact of information. It is easy for rumours to circulate in a disaster area, which can distort both the information and the emergency-operations situations (e.g. groups of rescuers can be sent rushing to areas where there are supposed to be trapped victims who in reality do not exist). The best way to tackle rumour is to disseminate authoritative factual information so that people no longer give credence to the rumour. This requires that the disaster plan include a spokesperson, that he or she be well supplied with reliable information on what is really happening, and that he or she be well trained in how to impart information. In sum, although all sources of information cannot be kept under control in disaster, uncertain reports can be verified and corrective statements issued if the plan makes prior provision for such activities.

This section has referred almost exclusively to communication between emergency workers. Managing the mass media is dealt with in §6.9.

5.5 Transportation

Emergency transportation can be divided into the following aspects: vehicles, traffic control, drivers (or pilots), maintenance, and fuel supply.

The disaster plan should designate a logistics officer with responsibility for overseeing the transportation sector and ensuring coordination between agencies and between different modes of transport. Aspects that need to be checked and coordinated include the following:

- The number, size, power, equipment and design of vehicles in relation to the transportation tasks that must be performed. Road, rail, all-terrain, air and water transportation are involved. Inventories need to be maintained of all vehicle fleets.
- The safety and adequacy of fuel stocks, and means of replenishing them.
- The levels of staffing for drivers, mariners and pilots, and the adequacy of their training with special respect to how they are likely to perform under emergency conditions; alternative personnel will be needed in order to ensure a consistently adequate level of crewing.
- The adequacy of routeways, mooring points, landing places, vehicle parks and staging posts, and the means of ensuring their freedom from congestion at critical moments.

- The nature of maintenance schedules and cycles, and their predicted impact on the availability of vehicles at any point in time. Obviously, disasters cannot be relied upon to occur during normal working hours, and vehicle availability must reflect this fact.
- The adequacy of transhipment points (airports, ports, docks, stations, loading and unloading areas), warehouses and equipment for trans-shipping goods (e.g. cranes and fork-lift trucks) and casualties (e.g. wheeled beds).
- The procedures for dispatching and tracking vehicles in order to know where they are located at any point in time.
- The ability of air-traffic control systems, port management, waterway pilot systems, road-traffic control and railway management to cope with the exceptional traffic generated by disaster-relief efforts, especially as they may have to work under difficult meteorological and environmental conditions.

During a disaster, the principal logistics officer will have to evaluate the ability of the relief system to move supplies and personnel to where they are needed. This will involve locating bottlenecks and identifying routeway blockages and other barriers to transportation, and devising appropriate remedies. This in turn will require close liaison with communications officers, disaster managers and the leaders of field teams. In complex systems, computerized disaster-response mapping will be a useful aid to this (e.g. with GIS, see §2.3.1).

Transportation during disaster aftermaths involves some special risks. For instance, in the wake of the December 1988 Armenian earthquake, more than 80 relief workers died in air crashes. Air-traffic control at Yerevan airport was not adequate for the large number of relief flights, and fog and snow made taking off, flying and landing hazardous. Similar risks may exist with respect to rail transport or shipping and ports, although road transport is more likely simply to encounter traffic jams, unless floods, landslides or avalanches make the roads hazardous. If traffic-control systems cannot be upgraded so that they are capable of dealing with increased movements during emergencies, then procedures should be in place to limit such movements when hazardous congestion might occur. In addition, it is especially important to ensure that a sound link exists between meteorological, oceanographic or geological forecasting services and traffic control units. If storms, landslides or tsunamis are likely to make transportation hazardous, robust procedures are needed to forecast the events, clearly advise the traffic controllers, and put into practice well designed procedures to limit traffic by closing airports, suspending flights, trips or voyages, or delaying entry into the hazardous area. The disaster plan may not need to provide for these matters, as most transportation facilities are accustomed to dealing with hazardous conditions, but the disaster planner should at least verify that arrangements are adequate for the very exceptional conditions that disasters produce.

5.6 Engineering

Engineering for disasters embraces the design and construction of temporary protection works, the survey and stabilization of damaged structures, and the rectification of problems with utilities. Hence, several branches of the profession are likely to be involved: civil, structural, hydraulic and sanitary engineering. The disaster plan should include a register of qualified engineers: indeed, at the national and international levels, such lists already exist. The processes of emergency engineering may require surveying equipment, water-quality testing apparatus, and heavy plant with extended fuel stocks. These should be planned for.

The first task of the civil or structural engineer may be to reinforce defences against hazards. Levees and barriers may need to be constructed against floodwaters, lava flows, mudflows, or snow and ice avalanches; fords or temporary bypass roads may be needed. In such cases, the work may need to be carried out very quickly and in difficult or dangerous conditions. Geotechnical engineers may need to verify slope stability or assess the risk of subsidence, rockfall or soil swelling. Hydraulic engineers may be required to calculate the capacity of channels to convey floodwaters or assess the timing of flood peaks. Each time, calculations may need to be made with minimal data, or variables whose values cannot accurately be measured, and the results may be somewhat unreliable. Hence, the engineers should be encouraged to bring questions of reliability – in both diagnosis and calculation – into the open.

Once disaster has struck and damage has been done, experienced structural engineers may be needed to prescribe buttressing and bracing for buildings whose stability has been compromised, and to construct temporary bridges and buildings. Sanitary engineers may be called upon to construct latrines or sewers for temporary camps occupied by displaced populations. Units of army sappers are often very skilled at assembling Bailey bridges and can do so with great rapidity. In addition to defining the roster of qualified engineers, the disaster plan should provide for the stockpiling of materials for shoring up buildings, such as baulks of wood, steel girders and scaffolding.

Sudden-impact disasters that do considerable damage to buildings can lead to instantaneous homelessness on a massive scale. In a few seconds, a major earthquake in a populous area may leave as many as half a million people homeless. This leads to an unprecedented demand for structural survey. In general, surveys will be conducted in two phases, although these can sometimes be combined. The first is necessary in order to determine which homes and other buildings can be occupied or utilized without undue risk to public safety, and which must have evacuation orders applied to them. Typical survey categories for vernacular buildings affected by disaster are:

- green: undamaged
- yellow: lightly damaged and certified for occupation pending repair
- red: heavily damaged; evacuate pending reconstruction
- black or purple: very heavily damaged; evacuate permanently and demolish. With regard to the last of these categories, emergency demolition should be limited to the minimum compatible with ensuring public safety. This will both prevent unnecessary damage to repairable buildings and avoid the loss of recyclable building materials.

Table 5.9 gives some categories and magnitudes to assess for damage census work. The table is compiled specifically for earthquakes, but the categories can easily be modified so that they fit other sorts of impact, such as that of land-slides or explosions. Tables 5.10, 5.11 and 5.12 refer to some of the problems of operationalizing the survey of structural damage. Again, the information given has been developed on the basis of experience in earthquakes, but is easily adaptable to other sudden disasters that reduce the impact area's building stock to a precarious state.

The first stage of post-disaster structural survey needs to be completed within, at the most, a few days after impact, as this is needed to establish a firm figure for the number of homeless people who will need accommodation. The second stage needs to be completed within a few weeks. As a whole, structural survey may be a gigantic task requiring scores of qualified surveyors in a large disaster area. There are likely to be problems of superficiality: time will be short and resources scarce. A large public building may take a team of structural

Table 5.9 Typical post-earthquake damage census for use in areas of small vernacular buildings (not for use on large public buildings).

Property type (100 percent census)
- Dwelling
- Commercial property
- Artisan or professional use
- Store-room or cellar
- Other
Σ Total (number and percentage in each category)

Damage level
0 No damage
1 Irrelevant damage: building is habitable and usable; repair is not urgent
2 Light damage: habitable or usable; to repair
3 Moderate damage: to evacuate partially; repairable pending reoccupation
4 Severe damage: to evacuate fully; repairable at high cost
5 Very severe damage: to evacuate and demolish
6 Partially collapsed: to repair and demolish
7 Totally collapsed: site must be cleared of rubble.
Σ Total number of buildings by category, overall total

Table 5.9 (continued)   Post-earthquake damage census for small vernacular buildings.

Type of horizontal structure (walls)
- Masonry
  - ashlar (dressed stone blocks)
  - random rubble with mortar
  - brick (note *type* of brick)
- Reinforced concrete (is it a well integrated structure?)
- Steel frame (is it an integrated structure?)
- Wood frame and laths
- Adobe, cobb (stones and mud), pisé (rammed earth) or mudbrick (Is it rendered – i.e. covered with cement, stucco or other surface?)

Date of construction
- Before 1900 (often too difficult to assess more precisely)
- 1901–1945 (early twentieth century)
- 1946–1965 (early post-war period)
- After 1965 (late post-war period)

Miscellaneous
- Urban or rural building
- Number of floors, rooms and residences
- Does the building have stalls, garage, or a communal frontage with the buildings next to it on the street?
- Total number of residential and non-residential properties
- Total number of rooms, occupants and evacuees
- Survey of emergency accommodation, requisitioning of buildings

---

Table 5.10   Guidelines for emergency entry into damaged buildings.

Posting	Placard colour	Condition	Entry allowed*
None(not yet inspected)	–	Serious structural damage	Only for search and rescue, and at own risk
INSPECTED	Green	Minor structural damage	Yes
RESTRICTED USE	Yellow	Some structural damage, generally of limited severity	Yes, but according to restrictions; entry into the restricted area only with permission of the local building department
UNSAFE	Red	Structure has serious structural damage, but is *stable*	Yes, according to guidelines
UNSAFE	Red	Structure has serious structural damage and is *unstable*	No. Entry only with written permission of the local building department
UNSAFE	Red	Posting due to *other* than structural damage	No. Entry only with written permission of the local building department

* During the first 24 h, entry into seriously damaged buildings should be avoided in case the damaging earthquake is a foreshock and a main shock follows.

Source: Earthquake aftershocks: entering damaged buildings, R. P. Gallagher, P. A. Reasenberg, C. D. Poland (ATC TechBrief 2, Applied Technology Council, Redwood City, California, 1999)

Table 5.11   Guidelines for classifying damaged buildings as unstable.

Unsafe buildings that have at least one of the following characteristics should be classified as unstable:

• May collapse or partially collapse under its own weight.
• Likely to collapse in a strong aftershock, from additional damage.
• Ongoing (progressive) lean.
• Ongoing creep or structural deterioration.
• So heavily damaged that its stability cannot readily be determined.

Source: Earthquake aftershocks: entering damaged buildings, R. P. Gallagher, P. A. Reasenberg, C. D. Poland (ATC TechBrief 2, Applied Technology Council, Redwood City, California, 1999).

Table 5.12   Recommended days to wait after an earthquake main shock before emergency entry to buildings posted unsafe, but stable (see below for posting conditions).

Mainshock magnitude	Enter for 2 hours	Enter for 8 hours	Enter for 24 hours*
M≥6.5	1 day	3 days	8 days
6.0≤M>6.5	1 day	2 days	4 days
M<6.0	1 day	1 day	2 days

* For continuous emergency access only; full-time occupancy is permitted only when approved by the local building department.

Source: Earthquake aftershocks: entering damaged buildings, R. P. Gallagher, P. A. Reasenberg, C. D. Poland (ATC TechBrief 2, Applied Technology Council, Redwood City, California, 1999). Source: ATC 1999. Procedures for Post-Earthquake Safety Evaluation of Buildings. ATC-20 Report, Applied Technology Council, Redwood City, California.

ATC-20 Posting Classifications (Applied Technology Council)

Inspected (green placard)
The building has been inspected by the local jurisdiction. It may or may not have been damaged. Observed damage, if any, does not pose a significant safety hazard. No limit is placed on the use or occupancy of the building.

Restricted use (yellow placard)
The building has been inspected and found to have damage or some other condition (e.g. a falling hazard) that precludes unrestricted occupancy. It can be entered and used, but some restrictions have been placed on its use. Certain areas may be restricted.

Unsafe (red placard)
The building has been inspected and found to be seriously damaged or have a serious hazard (e.g. a toxic spill). Generally, buildings posted unsafe have serious structural damage. Many are at risk of partial or complete collapse. Some unsafe postings are made when normal occupancy is inadvisable, such as when a wooden house has become detached from its foundation. Generally, entry into a building posted unsafe is not permitted without the approval of the local jurisdiction.

Source: *Procedures for post-earthquake safety evaluation of buildings*. ATC-20 report, Applied Technology Council, Redwood City, California (1989).

surveyors an entire day to survey thoroughly. Moreover, there may be problems of access. Not all damaged buildings will be open, and occupants with keys may be hard to find, especially during the chaotic early aftermath of disaster. Special problems are represented by multiple-occupancy buildings in which the stability of one apartment may affect that of others. If a building cannot be examined from inside, serious structural faults may go unnoticed. Furthermore, surveyors have no automatic rights of entry, although these can in effect be created by making survey a precondition of granting repair funds.

The disaster planner should consider these aspects and include in the plan a procedure for instituting structural survey as efficiently as possible. There should also be a procedure to ensure that architects, engineers and surveyors co-opted for this sort of work have the necessary qualifications and experience. If this is not the case, then training courses may be needed.

During the restoration phase that follows the initial post-disaster emergency, engineers and technicians will be required in order to repair electricity, gas, water and sewerage mains. Water supplies may need to be hyperchlorinated (i.e. treated with large doses of chlorine disinfectants) until their freedom from contamination can be guaranteed. The survey and repair of water distribution networks is a complex and time-consuming process that may take weeks or even months. Further details of repair works are given in §7.2.

5.7 Shelter

Expert opinion inclines to the view that shelter after disaster is a dynamic rather than a static phenomenon, an evolving process rather than a fixed goal. Needs evolve over time (Fig. 5.8), on occasion with bewildering rapidity, and hence strategies designed to satisfy them must emerge with corresponding rapidity. When a major hurricane or earthquake occurs, housing may be destroyed on such a large scale that alternative arrangements cannot absorb all the homeless victims. At the same time, families and individuals cannot manage out of doors for more than a very few days, especially if weather conditions are harsh. In the same way, business, commerce, manufacturing and employment may suffer if alternative accommodation cannot be provided rapidly.

When disaster has seriously damaged housing, the need for alternative shelter is immediate and pressing. In societies where vernacular architecture is very simple and is built of cheap, easily available materials, reconstruction may be almost immediate. This is the case in some of the villages of tropical countries, where the principal building materials are bamboo, banana leaf, small wooden beams and laths, and sheets of corrugated galvanized steel. In areas where rebuilding is complex, time-consuming and expensive, interim solutions are needed. The following discussion will concentrate on the latter case, which is typical of industrialized countries.

After sudden-impact disaster has created a demand for shelter, the solutions adopted tend to be progressively more sophisticated and expensive. They also

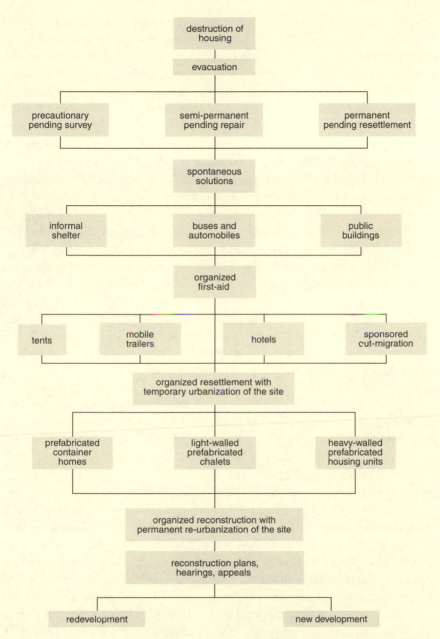

Figure 5.8 The sequence of options in accommodating people who have lost their homes in disaster or have been evacuated on a precautionary basis pending survey and repair of damage.

tend to interact with the problem of structural survey, in as much as the certification of damaged buildings for occupation, repair or demolition has a direct effect upon the number of homeless people. Unless prior hazard, vulnerability and risk studies have been carried out to extraordinarily detailed levels, it may be very difficult to forecast the post-disaster demand for shelter. Nevertheless, the disaster plan must make provision for this aspect.

Much depends on climatic conditions in the local area. Together with actual weather conditions at the time of the disaster, these will determine the pattern of solutions to the shelter problem. Some general considerations are as follows. Temporary shelter (i.e. that occupied between the destruction of buildings and their permanent reconstruction) tends to conform to the following hierarchical sequence (which is also given in increasing order of cost):

- *Plastic sheeting* This is a simple, cheap, easily transported material that can offer rudimentary protection until something better can be organized.
- *Tents* Lightweight vacation-style tents are not appropriate in disaster. Heavy canvas army tents are usually used. They can be erected in less than an hour but require a site that is well drained, reasonably level and free from mud and puddles if there is rain. Sports fields are often the preferred sites for tent encampments after disaster, as they are level, well drained and usually free from encumbrances. Supplies of sand and gravel can be located or stockpiled for spreading across sites that threaten to become muddy.
- *Mobile trailers (caravans, roulottes)* These lightweight vehicles can be towed easily behind ordinary passenger cars. Like tents, they require level, well drained, mud-free sites. Central facilities for washing and communal latrines or portable toilets are required for encampments of tents or trailers. The trailers can be stabilized with jacks and connected to bottled gas or temporary electrical lines.
- *Container housing* These monoblock facilities have evolved out of building-site offices. They are designed to be loaded by crane onto an articulated truck and offloaded onto a level, well drained site, where they will sit on jacks. Purpose-designed **container homes**, of the exact dimensions of the steel boxes used for multiple-mode transportation of goods, include a bedroom at either end, a compartment with a shower unit and washbasin, and a small central living area with a stove and utility cooker. They are constructed of steel frames, aluminium siding and glass-fibre panels, and they have built-in flexible connections to electricity, water and sewerage facilities. They may require access roads, although these can be unmetalled if necessary. However, container parks need electricity, water and sewerage connections to be in position before the containers are emplaced.
- *Light-walled prefabricated housing* These units tend to require full-scale urbanization of the site before they can be erected. Roads, retaining walls,

concrete base plates, drainage pipes and utility distribution systems have to be constructed first. The prefabs consist of factory-assembled panels, which are usually made of wood and glass-fibre insulating material. They are erected at the site, made impermeable and bolted to their base plates. Some ideal characteristics of this type of shelter are listed in Table 5.13.

- *Heavy-walled prefabricated buildings* These consist of concrete panels assembled in factories, and they are more expensive than light-walled prefabs. They are also more difficult to transport to a site and erect there. As a result, heavy-walled prefabs are not widely used, except as a cheap, alternative form of permanent reconstruction. However, there may be problems of decay and degradation before units reach the end of their design lives.
- Undamaged buildings can be requisitioned in order to provide shelter.

Costs vary considerably from perhaps US\$150 for a tent to \$10000 for a monoblock container, to as much as \$20000 for a prefab. Design lives also vary: tents may survive months of constant usage, trailers up to two or three years (although usually less), containers up to ten years, and light-walled prefabs up to fifteen years, depending on the level of wear and tear and the harshness of the local climate. Monoblock containers, in particular, accumulate heat in summertime and may require double roofs and air conditioners, at extra expense. Humidity may attack light-walled prefabs and cause rising damp and salt or mould efflorescence. However, units can be used not merely for homes but also for other types of building. Prefabs, in particular, can be used as churches, canteens, washrooms, public conveniences, municipal offices, shops, artisans' premises and clinics. Hence, units of different size may be needed, especially if different kinds of building are put out of use by damage.

Table 5.13 Ideal characteristics of a light-walled prefabricated dwelling for disaster survivors.

Capacity	1 family unit of 4–6 people
Floor area	40–45 m^2
Weight of all components	less than 10 tonnes, and preferably < 6 t
Maximum weight of single components	preferably not more than 50 kg
Principal materials	wood (solid, laminated, ply), glass fibre
Secondary materials	limited amounts of plastics, resins, glass, steel
Time necessary for 4 trained workers to erect one unit	no more than 12 hours
Tools and equipment requirements	power screwdrivers, spirit level, small-scale mobile scaffolding
Environmental requirements	concrete base pad, utilities, drains, access road
Minimum resistance to weight of snow on roof	2.75 kN/m;
Minimum resistance to loading by sustained winds	28–30 m/sec
Minimum resistance to seismic acceleration	0.35 g (35% of 9.81 m/sec^2)
Minimum expected duration (design life)	10 years

The disaster planner should try to forecast the post-disaster demand for shelter. Difficult though this may be, it should be done for each category of building with the aim of maintaining housing and public services at or above a minimum level after catastrophe strikes. Next, it is helpful to identify sources of each kind of shelter and plan for the acquisition and deployment of units. Where necessary, tents can be stockpiled.

The disaster plan should also tackle the problem of locations for shelter. Each urban area should designate one or more sites, according to estimates of the number of shelter units needed. These should be relatively level (or well terraced) areas that are free from hazards (such as flooding, landslides or snow avalanches), where it is not too difficult to provide utility services, basic urbanization and vehicular access. Such sites should be free of encumbrances so that they do not need to be cleared before they are used, and they should be available on a permanent basis. This may mean that they need to be purchased or officially expropriated before the next major disaster strikes.

The question of shelter after disaster is intimately linked to morale and socio-economic conditions. It needs to be in the right place, so that people can travel to work or till their fields. Thus, it needs to take account of economic activities so that people can use their living space to carry on productive activities that they would otherwise have conducted at home. In this respect, in areas that are dominated by agriculture, it may be as important to house animals or farm machinery as people.

Solutions to the shelter problem must not infringe local cultural traditions, or they will run the risk of being spontaneously abandoned by occupants. The same is true if social groupings are ignored when units are assigned. Therefore, the disaster planner should study social and economic patterns before configuring the plan for shelter provision. This should include measures to reinforce the patterns, rather than inadvertently sowing dissent and demoralization by ignoring them. It is especially important to try to preserve neighbourhood groupings when many shelters are assigned to families; assignments should not be made on a random basis. Note also that the use of supplied shelter will diminish over time and initially involve less than 100 per cent of evacuees, some of whom will find alternative sources of accommodation or leave the area. In fact, many homeless survivors will lodge with friends and relatives, although over time this may become problematical and they may eventually seek official assistance to find alternative accommodation. In the longer term, it is to be hoped that permanent reconstruction will reduce the use of temporary shelter.

Finally, a well planned program of shelter provision may eventually become the victim of its own success. Survivors of disaster should not be encouraged to depend on free assistance, which is unlikely to continue indefinitely. They should not be made so comfortable in temporary accommodation that they begin to lose the incentive to reconstruct, or allow the momentum of the reconstruction effort to lapse. In all cases, carefully phased planning is the key to the successful provision of shelter.

Full-scale reconstruction, which is not technically a part of emergency planning, is dealt with in Chapter 5. For the time being it is worth noting that adequate reconstruction requires enough time to plan, execute and administer it. Interim shelter is therefore important and one should beware of reconstruction that is too rapid, as this is not a sign of efficiency.

5.8 Emergency food programmes

In the aftermath of a large disaster there may be many displaced persons and relief workers in the area. They will need to be fed, although relief units should be encouraged to be self-sufficient wherever possible. Not all sources of food or food storage and cooking facilities will be destroyed in the disaster; hence, estimates of how many people will need or want to participate in mass-feeding programs should be based on residual, not maximum, figures in order to avoid waste and overcapacity. The main planning and organizational questions relating to food programs involve storage of comestibles, preparation, hygiene, perishability, and the role of donated foodstuffs.

It is easy for mass outbreaks of gastroenteritis to occur in the communal canteens that are set up after disaster. On occasion, salmonella or hepatitis can be transmitted. However, good planning and strict observance of a set of rules, especially those concerning hygiene, can reduce the risk of infection to very low levels. Not all aspects of the problem can be foreseen and planned for, but it is nevertheless worth reviewing each aspect, as there is a chain of interlocking processes, from production, through transportation, storage, preparation and serving, to waste disposal. Problems can occur at any stage and affect subsequent links in the chain.

Food that is brought into the disaster area needs to be evaluated in several ways. The first question concerns the nature of the victuals and whether they are wholesome, appropriate and acceptable to people who will consume them. If foodstuffs have been donated, rather than requisitioned or ordered, are they what is needed? If not, can donations be blocked and can the donated foods be hygienically disposed of so that they do not take up valuable storage space? If acceptable, when were they manufactured, how have they been transported to the area and what climatic conditions have they endured during transportation? Are "sell by" or "consume by" dates stamped on packaging and, if so,

185

have they been passed? Such information will not always be forthcoming, but when it can be reconstructed (at least partially), an idea can be gained as to the foodstuff's resistance to perishability.

Food may also be obtained from sources within the disaster area and, again, its history and safety should be checked. Where possible, a qualified food inspector should be part of the emergency personnel assembled under the auspices of the disaster plan. A related problem is the quality of water used for drinking, in food preparation and for washing dishes and food-preparation equipment. Water whose safety cannot be guaranteed should be sterilized by boiling, hyperchlorination or other methods, or should be substituted by clean water brought in by tanker. Again, the planned complement of personnel should include a water engineer or other competent tester.

Food storage is the next link in the chain. Storage conditions should be cool, dry and free from rodent and insect pests. Spoiled food should be disposed of by incineration, burial or removal to a landfill site. Perishable food will require a cold store, and all food will need regular supervision of its storage conditions. These are aspects that the emergency planner should address. If cold storage facilities are not available, then freezers and refrigerators could be trucked into the area, with diesel-power electrical generators if necessary. If this is impracticable, then fresh food needs to be delivered to the area on a daily basis. Pest control with insecticides (taking care to avoid contamination of the food), traps and bait may be necessary to avoid infestation. Appropriate materials can sometimes be assembled beforehand and stockpiled. Food should be distributed only by authorized handlers, with adequate supervision of its condition, and with satisfactory accounting procedures with receipts for disbursements.

Nowadays most food is protected from contamination by the packaging in which it is shipped. Problems, then, are more likely to arise when packages are opened and food is prepared, served and consumed, at which points it may come into contact with contaminants unless strict rules of hygiene are observed. Conditions after disaster are likely to be chaotic and both victims and relief workers, assailed by stress, exhaustion and perhaps injuries, may be vulnerable to infection. But bacterial contamination of food is not inevitable. Food preparation, serving and eating areas need to be well supervised and cleaned frequently and thoroughly. There should be an adequate supply of brooms, mops, cloths, brushes, scouring pads, detergents and disinfectants, and the utensils should be kept clean. Supervision of cleanliness needs to be particularly intense in the early stages of the disaster when food is being prepared around the clock. Cooking utensils should be cleaned, and surfaces, including cutting boards, disinfected. Animals, as well as people who are not involved in cooking operations, should be kept out of the kitchen. Kitchen staff should wear clean overalls and caps. Refuse should be deposited in bags, which should be taken away and properly disposed of. It should not be allowed to accumulate. Bins should be cleaned regularly. There should be adequate and clean toilet and hand-washing facilities for both the preparers and consumers of food. It is helpful to keep facilities separate for the two groups, where possible, to prevent cross contamination. Risks of infection can also be reduced by using (but *not* re-using) disposable cups and plates. Finally, methods of food preparation should be under the supervision of a trained professional, especially so as to ensure that food is properly cooked before being served.

In the matter of hygiene, much depends on the kitchen workforce, who may be volunteers, armed-forces recruits, or other emergency responders. People who are obviously sick, have diarrhoea, nausea, exposed areas of infected skin, or have recently been ill, should not be involved in food preparation, although it has to be admitted that in disaster situations health checks are difficult to conduct. Cleanliness should be enforced rigorously, especially hand washing. Food should be prepared and served by adequate numbers of people without skimping on hygiene. Cooks and cleaners should be separate groups, and the latter group should not be allowed to touch food.

Careful attention to these problems, both at the emergency-planning stage and during disaster management, can reduce them to a minimum. An army marches on its stomach, and likewise, good regular meals can do a great deal for morale and energy levels during a complex and demanding relief effort. Well organized canteens or camp kitchens can supply up to several thousand meals an hour and thus resolve one of the most fundamental problems of sustaining relief workers and offering primary care to survivors. In addition, meals in canteens offer a degree of conviviality that enables relief workers and survivors to discuss their problems in a more relaxed atmosphere than that of the front line of disaster work.

Planning for emergencies and the training of disaster managers should include instruction in how to cope with excesses of unwanted relief supplies, including victuals that are not needed or have perished. If not disposed of, refused or returned to senders, these will reduce the efficiency of relief efforts by taking up storage space and absorbing the efforts of supplies managers. Decomposed food that has not been disposed of can attract pests and give off noxious odours. In the end it may pose a health hazard in its own right and could be a source of pollution, for example, for water supplies.

5.9 The care of vulnerable and secure groups

Special problems are posed by the care of vulnerable and secure groups – for example, young children and infants, the residents of retirement or nursing homes, the physically handicapped, mentally ill people, hospital patients and prisoners – and specialized solutions may be required. Singly or in very small groups, such people do not usually pose a problem of specialized planning, but special measures are needed for the occupants and inmates of institutions.

The first task is to survey all institutions of relevance in the planning area and develop a profile of their sizes, locations and special emergency needs. Evacuation and shelter are likely to dominate the needs. There may be a demand for extra personnel, appropriate vehicles (e.g. secure vans for prisoners or wheelchair-accessible vehicles for the handicapped), and equipment, such as handcuffs for prisoners and blood-pressure monitoring devices for the sick. It is important to assure the continuity of essential activities, which involves providing for custodial services where appropriate, maintaining the treatment of ill patients or the care of the physically frail, and ensuring the continued availability of medicines and treatments. In addition, it is necessary to work out where to send evacuees: prisoners to other secure institutions, the sick to other medical centres, the aged to relatives or other old people's homes.

Having developed profiles of institutions and their needs, the former must be integrated into the logistical part of the disaster plan and the latter must be compared to the resources that can be devoted to the problem. Where necessary, reports on shortfalls can be written and submitted to the appropriate authorities with requests for additional resources.

The question of specialized needs ushers in a consideration of the separation of plans with respect to level in administrative hierarchy and the types of environment to which they relate. Hitherto, we have dealt with general emergency planning in an all-embracing sense and a local context. The next chapter develops some of the requisites of specialized plans for particular organizations (e.g. industrial concerns), facilities (e.g. airports), and sites (e.g. school buildings).

CHAPTER 6

Specialized planning

Particular organizations and situations will require specialized solutions to the emergency planning problem. This is the case where general municipal or regional plans are unlikely to offer the degree of response capability that is needed. In some instances, as is the case when archives, libraries, works of art or historical patrimony are seriously at risk, the answer may be to create a specialized module in the general plan and to assign personnel, equipment and facilities explicitly to it. In other cases, such as airports, factories and hospital systems, separate plans may be required, and these should be integrated with the main emergency plan. In addition, special attention should be given to the management of mass-media resources, as these are the key to information dissemination among the general public.

6.1 Emergency medical planning

The provision of health and medical services under emergency conditions involves a plexus of problems that include the following: the provision of first-aid and triage at the site of the disaster; transportation for the sick and injured; the management of hospitals and clinics so that they can cope with mass-casualty situations; disease surveillance and the prevention of epidemics; and physical and psychiatric health maintenance and rehabilitation (Fig. 6.1). The field also overlaps with sanitation, public hygiene and the provision of shelter, as these affect a displaced population's state of health and ability to recover from injuries sustained during disaster.

Disasters are likely to give rise to a sudden and perhaps massive demand for medical facilities. Normal procedures of queuing people or allocating services will not work and must be suspended during the emergency. New protocols and methods are required to allocate resources and services to maximize the number of lives saved, induce the best and most rapid aggregate improvement

189

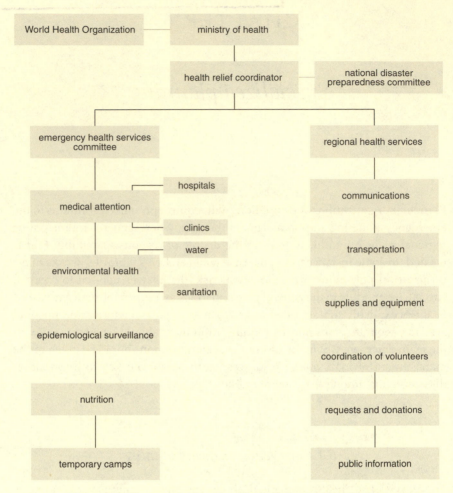

Figure 6.1 The organization of medical relief in disaster situations.

in patients' prognoses, and ensure the most equitable chances of maintaining good general health levels. This will involve intensifying and concentrating resources where they are most needed and making decisions rapidly, accurately and fairly to determine who has first access to treatment.

6.1.1 Emergency medical services (EMS)

Saving human lives is the primary objective of disaster prevention and response. It therefore follows that the organization of emergency medical activities is one of the primary components of disaster management. It should either figure prominently in the general emergency plan or be the subject of separate emergency planning, to be integrated fully with the main plan. When the number of people killed or injured is expected to be very small (say, fewer than 25–30), normal medical facilities and operations will be able to cope with them. In contrast, larger mass-casualty incidents require special organization and extra resources. Moreover, the number of lives saved will be proportional to the rapidity and efficiency with which **emergency medical services** are provided.

There are three main phases in the medical emergencies that accompany disasters. The first is that of impact, when medical facilities are damaged and some medical personnel may be lost, either because they become casualties themselves or because their main objective is to save and look after their immediate family members. Communications and transportation facilities may be damaged, and so may routeways, which will further reduce the efficiency of the medical response to the disaster. The second phase is that of emergency and isolation, in which initial medical relief is administered solely by available local resources and manpower. The duration of impact is a function of the type of hazard, the severity of its effects and the size of the area affected, while the duration of the emergency isolation period is related to the speed with which aid can be furnished from outside the immediate area. Generally, however scarce the initial medical aid, medium-term relief tends to be adequate; indeed, healthcare may for a time be better than it was before the disaster, such is the attention that is likely to be devoted to health matters once national or international relief efforts really get under way. However, it is as well to recognize that successful efforts to save lives will be heavily concentrated in the first 24 hours after impact. Hence, **field hospitals**, for example, that arrive more than a day after catastrophe has struck are often relegated to secondary tasks, such as the provision of back-up and routine medical care. The more heroic work tends to be carried out by units that were already in place before disaster struck, and thus the efforts of the emergency planner should be focused on these.

The principal requisites of a medical emergency are: the stabilization, recovery to medical facilities and care of injured people; the recovery and disposal of the dead (with autopsies and coroner's inquests, where necessary); the monitoring and control of **communicable diseases**; the care and health maintenance of displaced populations; and special care for infants, the handicapped, the sick and the elderly. Normal healthcare procedures are based on several systems of preference: first come, first served, and ability to pay, for example. In a mass-casualty situation, these are likely to be overwhelmed or invalidated by the number of patients requiring treatment and the impossibility of keeping full and accurate records on all of them. Hence, the procedures must be substituted by a system maximizing aggregate welfare, for example, by saving the largest number of lives possible or restoring the largest number of people to full health.

The first stage of medical work in disasters begins at the scene, where damage has occurred and there may have been casualties (see §5.3 and §6.1.2). Generally, the geographical pattern of the occurrence, or **incidence**, of casualties will be nucleated: although small numbers of people may be killed or injured over a wide area in, for example, an earthquake or a hurricane, in all but the most lethal of such events casualties will tend to be concentrated in a few particular locations (the cardinal points for primary EMS work), such as large, densely occupied buildings that have collapsed.

Table 6.1 gives a summary of the expected nature of casualties in particular

Table 6.1   Sources of death and injury in disaster

Effect of disaster upon buildings
- total collapse of structure (1,2,3,4,5,10)
- progressive collapse of structural elements as load shifts (1,2,3,4,5,10)
- compression of one or more floors (1,2,3,4,5,10)
- collapse of roof (1,2,3,4,5)
- collapse of structural wall, beam, stairs or whole of façade (1,2,3,4,5)
- collapse of angle, part of façade or elevation, part of roof (1,2,3,4,5)
- collapse of ceiling or partition wall (1,2,3,4,5)
- breakage of glass (5,6)
- fall of interior fittings and furniture (1,2,4)

Direct effect of primary or secondary natural hazards
- landslide or avalanche (rock, ice or snow) (1,2,10)
- volcanic blast or ashflow (1,6,8,9)
- sudden inundation, water wave or tsunami (1,7)
- windstorm (1,2,3,4,6)
- lightning strike (9,11)
- wildfire (8,9,10)

Secondary effect of natural hazards
- animal bite or insect sting (8)

Direct effect of primary or secondary technological hazards
- catastrophic failure of machinery (1,2,3,5,11)
- transportation accident (1,2,3,4,5,6,9,11)
- toxic spill (8,9,10)
- explosion (1,4,5,6,9)
- electrocution (9,11)

Direct effect of social hazards
- violence and use of firearms (1,4,5,12)
- explosion (1,4,5,6,9)

Main physiological pathologies
1 general trauma (cranium, thorax, spine, limbs)
2 crush injuries
3 fractured bones
4 contusions and concussion
5 loss of blood, internal bleeding
6 lacerations and cuts
7 drowning
8 toxic shock or toxicity effects
9 burns
10 suffocation
11 wound shock
12 gunshot wounds

(a)

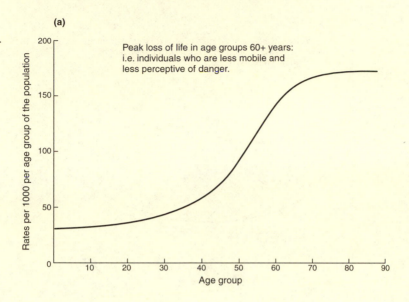

(b)

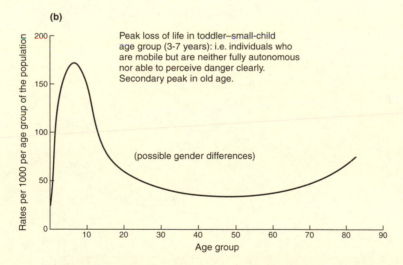

Figure 6.2 Above and overleaf: four hypotheses of the age distribution of mass casualties in disaster.

types of disaster. Figure 6.2 shows four of the possible age distributions among victims of various kinds of natural catastrophes, especially earthquakes; and Figure 6.3 shows the varying medical prognoses of disaster victims. In these, the majority of victims will succumb to the collapse of buildings or damage to their contents. Thus, crush injuries and fractures will dominate. Burns and smoke intoxication may be important categories if fires break out after the

193

(c)

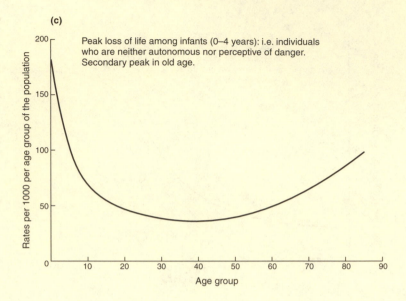

Peak loss of life among infants (0–4 years): i.e. individuals who are neither autonomous nor perceptive of danger. Secondary peak in old age.

(d)

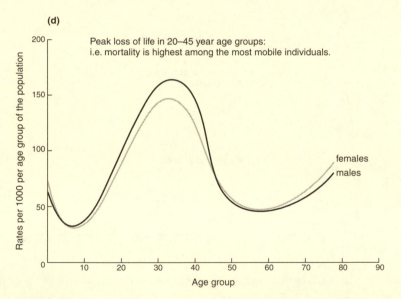

Peak loss of life in 20–45 year age groups: i.e. mortality is highest among the most mobile individuals.

shaking. The fact that patterns of casualties are broadly predictable for given disaster scenarios has two implications for disaster planning. The first is that initial field diagnosis of injuries must take account of what is to be expected (see §6.1.2, which is about triage) and the second is that medical facilities must be equipped to cope with mass casualties of particular types and patients of different ages and infirmities (Fig. 6.3). Where fire is a risk, patients may need to be sent rapidly to burns units or respiratory detoxification facilities. Crush

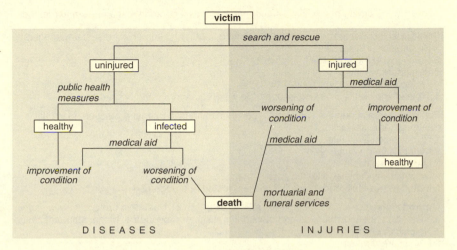

Figure 6.3 Flowchart for post-disaster medical conditions.

injuries in earthquakes can cause serious or fatal risks of kidney failure, and so dialysis equipment is needed. If it is impossible to have enough of such equipment on hand in the local area, there should be arrangements to transfer patients rapidly to medical centres elsewhere, which are expected to have enough expertise and capacity to treat them promptly. In short, capacity to treat given types of injury must be estimated both locally and regionally, or interregionally, and it should be exploited in full under the tenets of the EMS plan. It should also be borne in mind that damage to medical facilities and loss of personnel will reduce capacity to respond locally.

The medical-disaster planner should begin by making an assessment of available personnel, supplies and facilities (Table 6.2). Personnel include qualified doctors, internees, registered nurses, paramedical workers, hospital orderlies, ambulance crews, administrative, secretarial and managerial staffs, cleaning staff and other auxiliary workers. To what extent are they trained to cope with the demands of a mass-casualty disaster? For example, are doctors and nurses versed in emergency medicine? The EMS plan should provide for training and retraining. It should establish a pattern, or chain, of command and a call-up procedure (see §4.1.2), with provision to substitute key personnel who do not or cannot report for work.

Large medical centres should have an internal disaster plan that is well integrated with external plans and that begins to function as soon as the alert is sounded or disaster strikes. It requires an emergency-operations centre at the hospital that is well connected by radio and telephone both to other medical centres and to the main local EOC, which is probably the municipal centre of command. Emergency staff should be organized into operational units for disaster

195

Table 6.2 Estimate of medical disaster preparedness: personnel, materials and methods.

Personnel (sum scores for each unit) – doctors, nurses, paramedics, disaster managers
1 No personnel available
2 Personnel being appointed
3 Personnel available
4 Trained and certified personnel available
5 Trained and certified personnel available with regular drills and upgrading

Materials (sum scores for each unit) – communications, personal protection, medical treatment, ambulances
1 No materials available
2 Materials being purchased
3 Materials available
4 Materials available and tested
5 Materials available and regularly checked, tested and upgraded

Methods (sum scores for each unit) – attack plans, triage, treatment protocols; ambulance assistance, patient distribution and monitoring; disaster procedures, triage, standardization
1 No plan available
2 Plan in preparation
3 Plan available
4 Plan available and tested
5 Plan available and tested with regular drills and upgrading

Methods (EMS)
• General disaster plan
• Specific disaster and evacuation plans
• Ambulance assistance plan
• Victim distribution plan
• Hospital disaster procedure
• Decontamination protocol
• Triage protocol
• Treatment protocols
• Transfer protocol (disaster site–ambulance); (ambulance–hospital); information release
• Reporting procedure and records

Primary EMS services
• Industrial plant medical units
• EMS hospitals
• Ambulance services
• Red Cross
• Public-health services
• Regional search-and-rescue services

Secondary EMS services
• Military, civil and national guard units
• Hospital specialist trauma units
• Main pharmacies
• Department of environmental health
• Fire services
• Police services
• Coastguard and harbour services
• Volunteer units and organizations

Source: Modified from "Criteria for the assessment of disaster preparedness", J. de Boer, *Prehospital and Disaster Medicine* **12**(1), 13–16, 1997; and "Medical disaster preparedness of Rotterdam–Europort: an application of methodology for assessing disaster preparedness", J. de Boer, *Prehospital and Disaster Medicine* **12**(4), 288–92, 1997.

work. These include: general surgery; intensive care for cardiac and cardio-vascular treatment; general medicine; re-animation and anaesthesiology; diagnostic biochemical analysis for laboratory work; radiology; computer-aided tomography, magnetic spin resonance, ultrasound and other sophisticated diagnostic scanning techniques; and haematology and transfusion.

Procedures need to be designed to register emergency patients rapidly and to create and update records of their treatment and medical condition. Where possible, measures are needed to requisition copies of patients' medical records quickly, especially as some patients may be allergic or resistant to particular treatments or in need of medications that they were taking before the disaster. The EMS plan should also make provision to conserve and safeguard medical records if the hospital itself is struck by disaster: data on patients should not be allowed to disappear under rubble or be saturated by floodwaters or consumed by fire. In the event, hospital diagnosis will need to be streamlined for major emergency situations, as the next section will show. Lastly, healthcare eligibility status may need to be established, including means of eventual payment for treatment received, or insurance documentation. This may require streamlining of hospital bureaucracy. As a matter of principle, however, when disaster strikes, hospitals should not refuse to treat seriously injured patients on grounds of inability to reimburse costs.

Widespread disasters may give rise to a need to establish the functionality of medical facilities. Are hospital buildings damaged to the point that they are unusable and must be evacuated? Has equipment been lost? Are emergency electrical generators disaster resistant and fully functional? Are hospital records unretrievable? How successfully has the roll call of personnel been carried out and has the call-up procedure succeeded in establishing a full complement of disaster workers? Hospital disaster managers must be able to answer such questions very quickly. Even in very large disasters, such as major earthquakes, the peak admissions figure for patients is likely to occur within 24 hours of impact, and further admissions will decline to background levels within six days (Fig. 6.4). If the number of casualties greatly exceeds the capacity of locally available facilities, vehicles, routeways and plans must be able to cope with a potentially massive need for the interfacility transfer of patients, which will arise within 72 hours of the disaster. In addition, EMS plans must aim to maintain routine activities where they cannot be suspended for the duration of the emergency. In a disaster aftermath, it may still be necessary to have ambulances on call for rescue activities unconnected with the event.

Experience shows that the number of hospital beds is not an ideal variable on which to base EMS plans. It is not even ideal when viewed in terms of occupation ratios (i.e. the probable number of unoccupied beds when disaster strikes). To a certain extent, hospital beds can be improvised and rotated by

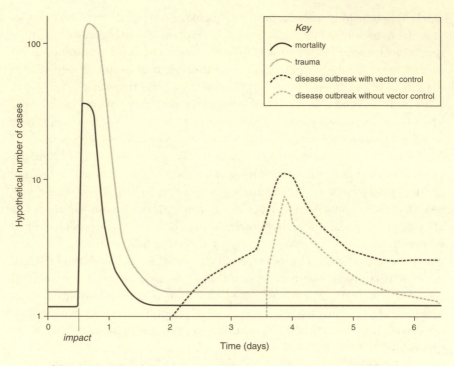

Figure 6.4   Emergency medical case load after a mass-casualty disaster.

careful management and intensive nursing. Much more significant is the capacity and quality of treatment facilities. Thus, the number of *general* beds is much less significant than the number of *intensive-care* beds or of operating theatres. In addition, medical treatments can be abruptly halted by lack of supplies. Fortunately, it is relatively easy to identify the items that are most likely to run out. These include supplies of blood, splints, bandages, antiseptics, analgesics, X-ray film, sterilization preparations and vaccines (e.g. against tetanus). Emergency supplies of such materials must be stockpiled and kept current (i.e. not allowed to pass "use by" dates). The EMS plan must include arrangements to distribute them rapidly and efficiently on the basis of immediate need.

EMS plans also need to tackle pharmaceutical problems. The range of drugs needed will not be limited to the main ones associated with emergency treatment. The care of survivors will also involve ensuring continuity of treatments that were begun before the disaster struck (thus the continued availability of general medicines). This may mean setting up emergency pharmacies, co-opting local pharmacists into the emergency-planning process, and arranging for emergency distribution of general and specific medicines in the difficult conditions of reduced accessibility that the disaster may have created. In this respect, it is particularly important to plan in advance. In disasters, the history of medical-supply donation is very checkered. It is not uncommon for crates to arrive in the disaster area full of drugs that are out of date, damaged by water infiltration or mould, or simply identified in languages or by brand names that give little clue to handlers as to what the drugs actually are. The EMS plan needs to assign a qualified pharmacist to the receiving station to identify and classify drugs by type and usage. It must provide for adequate communication with donors, such that unwanted or undecipherable drugs are not sent to the disaster area. In the end, procedures are needed for the safe disposal of unwanted drugs. These are not easy problems to solve, either by advanced planning or by emergency management, but they need to be anticipated and addressed.

The acid test of a good EMS plan lies in two measures: response time and resource capability. It must minimize the former and maximize the latter, under the constraints of available resources. Are procedures designed such that seriously injured people can be given medical care to stabilize their condition and then ameliorate it, and will their prognosis change for the better before it is too late and the injuries become fatal? Inefficiency in EMS provision during disasters all too easily translates into the loss of lives that could have been saved. Efficiency means allocating resources so as to achieve the greatest aggregate benefit – the most rescues and cures, the fastest improvements in patients' conditions – without concentrating the resources into time-consuming and costly medical-rescue efforts for questionable benefits. These are not easy goals to achieve, as they depend on assessments of the likely pattern of casualties in time and space. At least, assessments of ability to respond can be made, in terms of timing and applied expertise, on the basis of field exercises (see §3.6).

One aspect of the EMS plan that should not be ignored is the security and safety of medical establishments themselves. Besides provision for rapid assessment of damage after a hospital or clinic has been struck by disaster, an evacuation plan is necessary. In some instances, damage or hazard may be limited to one part of the facility, such as a single wing (perhaps flooded or compromised by landsliding or earthquake shaking) and hence occupants can be evacuated to other parts of the building. In other cases, hopefully rare, the entire facility may have to be evacuated. In this instance, and with respect to many other issues, EMS should be conceptualized in relation to the *system* of the hospitals and clinics, not merely the individual facilities, for the occupants of one may have to be evacuated to another. There are, of course, many other benefits to be derived from sharing response capabilities, not least of which are the avoidance of duplication in the medical relief effort and the ability to plan journeys to hospital on the basis of the shortest trips or the most viable routes.

Ideally, evacuation also requires alarm procedures and sirens, as well as protected exits, assembly points, ward monitors and directors of emergency movements. In organizing the process it is opportune to review other security measures. Is the level of emergency training among staff adequate? Are fire-fighting systems such as sprinklers, hoses and hydrants adequate and fully functional? In the case of something like an earthquake threat there are other sources of risk. For example, is laboratory glass likely to break during the shaking? Is there a potential for a toxic or radioactive release? How well will operating theatres survive the impact of the disaster and with what effect upon their functionality? Is vital equipment secure and are the patients' records and other archives protected against damage? The answers can be found only by a comprehensive review of arrangements in the light of expected disaster scenarios.

Although the nature of injuries to individual victims of disaster varies considerably, aggregate patterns of casualty are broadly predictable. But mass-casualty events may produce so many injured survivors that special techniques are required to deal with them. The basis of these is **triage**.

6.1.2 Triage and casualty management

Triage classifies injuries in terms of what benefit a patient can be expected to receive from immediate or short-term treatment, not of the severity of the injuries. It is a form of rationing of healthcare that is utilized when demand overwhelms supply. The purpose of the present section is not to describe triage procedures but to consider how they relate to the EMS planning process.

In many large disasters the injured victims will be many people with relatively minor ailments that are not life threatening, a smaller number with significant but relatively simple injuries, even fewer with serious and complicated injuries, and a group whose injuries have either proved fatal or are likely to do so. Other events may produce nothing but deaths (some aircraft crashes, for instance) or predominantly serious injuries (windstorms or debris avalanches, perhaps, although death may be more likely than survival in such events). In any case, the classification of injuries should give priority to patients who require immediate attention in order to ensure their survival and recovery.

Triage is carried out in two stages: at the scene of the disaster to determine who receives priority medical attention in the field and is first to be transported to hospital, and at the medical centre's receiving bay to determine who is first to be treated (Figs 6.5, 6.6). The process involves noting certain details about the patient. Apart from names and addresses (where known), approximate age, sex, pulse rate and other such basic information, assessments are usually based on a combination of variables. These may include simple investigation of the cardiovascular, respiratory and neurological functions of the body. In the simplest case, the determination is purely visual and intuitive (i.e. it is based on the perception and experience of the triage officer), whereas in other cases, indices

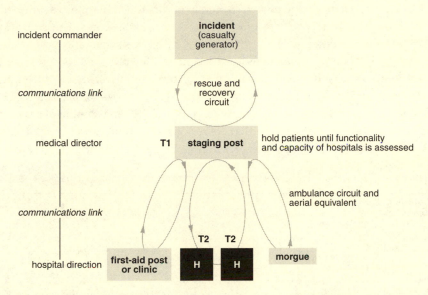

Figure 6.5 Pattern of triage and emergency medical operations at a major mass-casualty incident or disaster. T1: primary triage, which determines who receives priority transport *to* hospital; T2: secondary triage, which determines who receives priority treatment *at* hospital. Hospitals: percentage functionality of hospitals is assessed on the basis of structural damage, personnel loss, and the number of beds available.

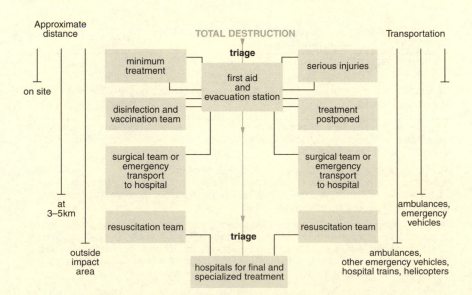

Figure 6.6 Triage and transportation pattern in major disasters.

or scores are assigned to up to seven different bodily functions and are summed (see Table 6.3). Thus, we have injury-severity or trauma scores. These lead to a determination based on four categories. The most serious and urgent are cases in which the vital functions are affected and the patient's condition will deteriorate without immediate medical aid. Internal injuries, profuse bleeding and cranial trauma are examples of problems that give rise to this state. Such patients are usually tagged red and given top priority (an example of a triage tag is shown in Fig. 6.7). Injuries that are serious, or at least significant, but can wait for up to 6–12 hours for treatment may be tagged yellow and given second

Table 6.3a   Some triage categories and injury severity scores: Glasgow Coma Scale (GCS).

Opening eyes
Spontaneously 4
Upon verbal stimulation 3
Upon stimulation by pain.. 2
No response... 1

Best verbal response
Normal conversation 5
Confused conversation 4
Inappropriate words 3
Incomprehensible sounds 2
No response... 1

Best motor response
Obeys command... 6
Localizes (pain).. 5
Withdraws (pain) 4
Abnormal flexion 3
Abnormal extension 2
No response... 1

Maximum GCS score 15, minimum score 3

Table 6.3b   Some triage categories and injury severity scores: Glasgow–Pittsburgh cerebral performance categories.

Good cerebral performance. Alert, able to work and lead a normal life. May have minor psychological or neurological deficits (mild dysphagia, non-incapacitating hemiparesis, or minor cranial nerve abnormalities).

Moderate cerebral disability. Conscious. Sufficient cerebral function for part-time work in a sheltered environment or independent activities of daily life (e.g. dressing, travelling by public transportation, and preparing food). May have hemiplegia, seizures, ataxia, dysarthria, dysphagia, or permanent memory or mental changes.

Severe cerebral disability. Conscious. Dependent on others for daily support because of impaired brain function (in an institution or at home with exceptional family effort). At least limited cognition. Includes a wide range of cerebral abnormalities, from ambulatory with severe memory disturbance, or dementia precluding independent existence, to paralytic and able to communicate only with eyes, as in the "locked-in" syndrome.

Coma, vegetative state. No conscious. Unaware of surroundings, no cognition. No verbal or psychological interactions with the environment.

Source: "A study of out-of-hospital cardiac arrests in northern Minnesota", J. W. Bachman, G. S. McDonald, P. C. O'Brien et al. *Journal of the American Medical Association* **256**, 477–83, 1986.

Table 6.3c Some triage categories and injury severity scores: trauma assessment indices.

Score	Description	Code
0–1	Ambulatory	A
2–4	Moderately serious	MS
5–7	Critical but likely to recover	CR
≥8	Critical and unlikely to recover	CRU

Source: "Design and validation of a practical index for trauma assessment", N. Guzmán, M. X. Paz, N. R. Moreno, F. Niño. *Disasters* **13**(2), 154–65, 1989.

Table 6.3d Some triage categories and injury severity scores: mangled extremity severity score (MESS).

Skeletal/soft tissue injury (energy level)
Low	stab, simple fracture, civilian gunshot wound	1
Medium	open/multiple fractures, dislocation	2
High	crush injury, shotgun wound, military gunshot wound	3
Very high	as above, plus gross contamination of soft tissue avulsion	4

Limb ischemia
Pulse reduced or absent but normal perfusion	1
Pulseless; paresthesia; reduced capillary refill	2
Cool; paralyzed, insensate, numb	3

(score doubled for ischemia longer than 6 hours)

Shock
Systolic blood pressure maintained above 90 mm Hg	0
Transient hypotension	1
Persistent hypotension	2

Age
Less than 30 years	0
30 to 50 years	1
More than 50 years	2

Source: "Disaster triage: START, then SAVE – a new method of dynamic triage for victims of a catastrophic earthquake", M. Benson, K. Koenig, C. Schultz. *Prehospital and Disaster Medicine* **11**(2), 117–24, 1966.

Table 6.3e Some triage categories and injury severity scores: injury severity classes.

Class	Description
I	Vital functions affected
II	Serious injuries needing operation but able to wait up to 12 hours after receiving first aid
III	Moribund
IV	Slightly injured: with proper supervision can wait for treatment

Source: "The injury severity score: a method for describing patients with multiple injuries and evaluating emergency care", S. Baker, B. O'Neill, W. Haddon et al. *Journal of Trauma* **14**, 187–96, 1974.

Table 6.3f Some triage categories and injury severity scores: Champion's trauma score.

Respiratory rate
Number of respirations in 15 seconds (multiply by four)
10–24 4
25–35 3
>35 2
<10 1

Respiratory effort
Shallow – markedly decreased chest movement or air exchange .... 1 (normal)
Retractive – use of accessory muscles or intercostal retraction ...... 0 (shallow or retractive)

Systolic blood pressure
Systolic cuff pressure – either arm auscultate or palpate .... 4 (≥90)
No carotid pulse 3 (70–89)
.... 2 (50–69)
.... 1 (<50)

Capillary refill
Normal: forehead or lip mucosa refill in 2 seconds 2
Delayed: more than 2 seconds capillary refill 1
None: no capillary refill 0

Glasgow coma scale (GCS)
14–15 5
11–13 4
8–10 3
5–7 2
3–4 1

Trauma Score = A+B+C+D+E (maximum 16, minimum 1)

Source: "Role of trauma score in triage of mass casualties", H. R. Champion & W. J. Sacco. *Journal of the World Association of Emergency Medicine* **1**(supplement 1), 24–9, 1985.

Table 6.3g Some triage categories and injury severity scores: Champion's New Trauma Score.

Pulse rate per minute	or	Blood pressure	Respirations per minute	Best motor response	Score
0		0	0	None	0
1–40		0–49	1–9	Extension/flexion	1
41–60		50–69	>36	Withdraws (pain)	2
≥121		70–89	25–35	Localizes (pain)	3
61–120		≥90	10–24	Obeys command	4

(Maximum 12, minimum 0)

priority. The "walking wounded", whose injuries do not make them priority cases, will be tagged green and transported to secondary treatment centres, perhaps by bus, where they will be examined by nurses or doctors as soon as the more urgent cases have been dealt with. Patients who are fatally injured or dead are tagged black and given lowest priority. They will be removed to hospital mortuaries or to morgues (Fig. 6.8).

With respect to the question of the number of injured people, injured survivors may leave the impact area before triage gets under way. Some may be

Table 6.3h Some triage categories and injury severity scores: survival probabilities (Ps) associated with Champion's Trauma Score and New Trauma Score.

Scores	Trauma Score Ps	New Trauma Score Ps
16	0.99	–
15	0.98	–
14	0.96	–
13	0.94	–
12	0.89	0.99
11	0.83	0.97
10	0.73	0.95
9	0.60	0.90
8	0.46	0.82
7	0.32	0.70
6	0.21	0.54
5	0.13	0.38
4	0.078	0.24
3	0.046	0.14
2	0.026	0.076
1	0.015	0.040
0	–	0.021

Source: "Trauma triage", H. F. Champion. *Journal of the World Association of Emergency Medicine* **3**(2), 1–5, 1987.

Table 6.3i Some triage categories and injury severity scores: Gormican's CRAMS triage score.

Circulation
Normal capillary refill and BP>100...2
Delayed capillary refill or 85<BP<100..1
No capillary refill or BP<85 0

Respiration
Normal 2
Abnormal (laboured or shallow) 1
Absent..0

Abdomen
Abdomen and thorax non-tender 2
Abdomen or thorax tender 1
Abdomen rigid or flail chest (penetrating wounds to the abdomen or thorax)....................0

Motor functions
Normal 2
Responds only to pain (other than decerebrate)..1
No response (or decerebrate)..0

Speech
Normal 2
Confused..1
No intelligible words...0

(Maximum 10, minimum 0; major trauma ≤8, minor trauma ≥9)

Source: "CRAMS scale: field triage of trauma victims", S. P. Gormican. *Annals of Emergency Medicine* **11**(3), 132–5, 1982.

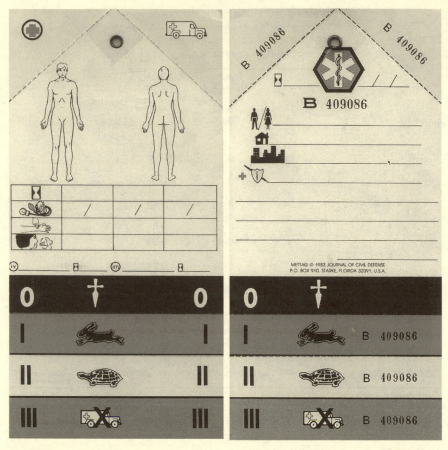

Figure 6.7 Example of a triage tag. The bands in the lower half are, from top to foot, black, red, yellow and green.

transported to hospital by colleagues, friends or relatives; others will seek medical attention on their own. Thus, triage is not as orderly a process as it may seem. Nevertheless, it has its own procedures. Standardized triage tags can be used to record basic information and colour-code patients for transportation and treatment (Fig. 6.7). However, there has been some debate about the need for specialized tags in particular types of disaster. Fires, for example, give rise mainly to burns (to the exposed parts of the skin and possibly to the airways and lungs) and smoke intoxication effects. The triage tags should allow assessors to record more information on these problems than that required for basic determination of priority. Both air crashes and high-rise building disasters have given rise to calls for specialized triage tags to be used only in such events.

The practical process of field triage involves setting up a triage station. It can be positioned, for example, on the periphery of a disaster site or near the main

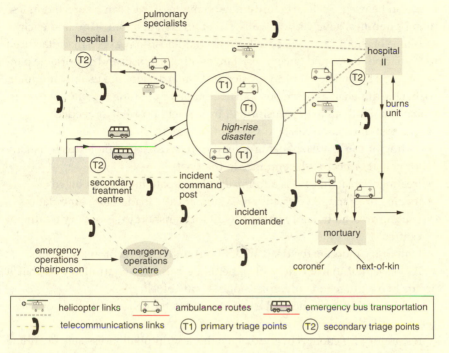

Figure 6.8   An example of the emergency medical circuit for a mass-casualty incident involving a fire in a high-rise building.

exit of a building from which victims are being extracted. It is useful to position the triage post near to the incident command centre, if there is one. Triage should be practised only by trained, qualified personnel who are clearly identified as triage officers. Few of them are needed, perhaps only one per 25 patients with significant injuries. The triage post should have adequate communications with hospitals and the emergency-operations centre (via the incident-command post, if this is located nearby), and it should be set up in such a way that the arrival, loading and departure of ambulances or other emergency vehicles is unimpeded. In this respect it helps to appoint an orderly to direct traffic. It is obvious that adequate supplies of triage tags must be available at the site (they can be stockpiled in ambulances) and their significance and usage should be the subject of training exercises.

Rather than have a separate station for field triage, it can be set in the wider context of an **advance medical post** or **first-aid station**. This can be successfully established in any large room through which the movement of foot traffic and around which that of vehicular traffic can be organized. Structures such as gymnasia, warehouses, schools, hotel lobbies and large restaurants have been adapted to this use. If no existing buildings are available or suitable, inflatable

207

tents can be used. Such structures may be made of PVC, can be erected in less than 20 minutes, have vertical walls, cover an area of about $40\,m^2$ and enclose a volume of about $75\,m^3$. Purpose-built first-aid posts have been manufactured in EV marine compound (a very resistant synthetic fabric) and as portable panels of wood that make a rapidly erected hut. Such designs are usually modular, so that the dimensions of the post can be doubled by joining together two units, which enables them to be transformed into a field hospital, if needed.

The first-aid station can be divided into as many as five zones:

- sorting or triage area: patients' injury statuses are classified (this is usually located at or near to the entrance to the post)
- treatment area: first aid is administered and patients are stabilized
- loading area: triaged patients are systematically loaded into ambulances
- waiting area: ambulatory patients wait for transport to secondary treatment centres
- temporary morgue: the dead are laid out under sheets or in body bags.

It is essential that the flow of patients be carefully controlled and that it occurs in one direction only, so as not to create blockages.

The treatment area should be organized for the following kinds of emergency medical work:

- respiratory-function recovery: oxygen therapy, intubation, assisted breathing
- thoracostomy with drainage tube
- recovery of circulatory functions: haemostasis, volemic re-integration
- analgesia, anaesthetics and dosing with medicines
- first-aid treatment of burns
- emergency amputations
- treatment of injuries.

The principal objective is to render seriously injured patients capable of being transported to hospital where they can receive more intensive treatment in a more sterile environment. The treatment area should be equipped for oxygen and infusion therapies, and should be well supplied with intravenous liquids, pain-control drugs and materials used in re-animation therapies.

Regarding transport to hospital, the emergency medical plan should include a **first-wave protocol** (FWP), which details the number of patients with particular kinds and degrees of injury that can instantaneously be accommodated at each medical facility in the area of the plan's jurisdiction. Once quotas are exhausted, patients are distributed on the basis of the **first-wave ratio** (FWR), which considers the capacities of local hospitals relative to each other. The FWP and FWR are to be used until such time as the first-aid station receives precise up-to-date information on the readiness and actual capacities of hospitals.

It is important to avoid confusion when second-stage triage is conducted at the ambulance receiving bay of a medical centre (which should be maintained in an uncluttered state). One or two doctors should be used as triage officers here and patients should be rapidly removed on wheeled beds, the urgent cases to operating theatres or intensive-care units, and the non-urgent cases to a holding area. Second-stage triage also brings up the matter of tracking a patient's condition. Here, we pass from the triage tag to the medical record. This needs to be streamlined in relation to the normal bureaucratic procedures of charting the patient's progress. An example of a specialized diagnosis and treatment form, in this case designed for flood victims, is given in Figure 6.9. Computerized records can help; some emergency hospitals have experimented with hand-held computers activated by pens rather than keyboards. As a rule, emergency documentation for hospitalized victims of disaster should include the following information:

- personal details of the patient, in as much as these are known or can be ascertained (name, address, next of kin, health-insurance details, etc.)
- details of triage determinations made in the field (priority colour, trauma scores assigned, etc.)
- triage code and any trauma scores assigned at the hospital's internal triage point (usually at or close to the emergency and accident receiving bay)
- details of the patient's clinical condition
- treatments applied
- results of any diagnostic tests
- final treatment to be given to the patient.

Mention of the temporary morgue serves to remind us that the management of dead bodies cannot be overlooked. Prevailing wisdom suggests that corpses left at the scene of a disaster do not constitute a health hazard and hence there is no need to dispose of them hastily. A contrary theory is that a hazard does exist when relief workers inhale skin dust that contains bacteria or touch blood-coated surfaces. Masks and rubber gloves can be used to guard against such hazards. However, there are few indications that living people in the disaster area have been made ill by contact with, or proximity to, dead bodies. Yet if the physical problem can be discounted, the psychological one cannot. The presence of death, and the visible reminder of it, can induce critical-incident stress in relief workers and deep depression among bereaved survivors. The issue of death therefore needs to be handled with sensitivity by emergency planners and disaster managers alike. Bodies should not be wound into sheets and dumped into communal trenches (in wet conditions these may flood before they can be filled in again). Neither should cadavers be sprayed with disinfectant. Instead, there are medical procedures, bureaucratic formalities and religious or familial rites to be observed before bodies are buried or cremated.

In a mass-mortality situation, bodies should be retrieved as soon as possible without compromising the chances of survival of living victims. If they are

Hospital:		Location:	
Date: ___/___/___		Time: ___:___	
Patient's forename (s), FAMILY NAME:			
Home address (if known):			
Age: ____ years		Male ❑ Female ❑	
Medical record no.:		Medical insurance no.:	
Emergency diagnosis			Treatment
Near drowning ❑ Respiratory distress ❑ Intubation ❑			
Neurological: Coma ❑ Focal deficit ❑ Paraplegia ❑ Quadriplegia ❑			
Musculoskeletal Location:			
Fracture: Open ❑ Closed ❑			
Dislocation ❑ Sprain ❑ Strain ❑ Contusion ❑			
Skin damage: Laceration ❑ Puncture ❑; Location:			
CO poisoning: Moderate ❑ Severe ❑			
Burns: 1st degree ❑ 2nd degree ❑ 3rd degree ❑			
% of body surface Location:			
Bite: Snake bite ❑ Animal bite ❑ Location:			
Vision: Impaired ❑ Blindness ❑			
Cardiac: Arrest ❑ Arrhythmia ❑ Other:			
Gastro-intestinal: Haemorrhage ❑ Nausea/vomiting/diarrhoea ❑ Acute/Abdomen ❑			
Genito-urinary: Urinary retention ❑ Kidney failure ❑			
Obstetrics: Spontaneous abortion ❑ Delivery ❑ Other:			
OUTCOME: Discharged ❑ Hospitalized ❑ Fatal ❑			
TREATMENT (enter code (s) in right-hand box): **C** = specialist consultation **IV** = intravenous access **R** = resuscitation **W** = wound care **O** = other (annotate form or attach note to specify)			

Figure 6.9 Specialized form for the assessment of injuries caused by floods.

dismembered, badly burnt or partially decomposed, body bags will be needed. The process of recomposing bodies must be done on site; the task should be assigned to trained and experienced personnel working in relays. In really serious cases of bodily decomposition, breathing apparatus may be needed.

Where bodies are not so seriously damaged, they should be removed on stretchers to a holding post, which should be a dry, well ventilated and reasonably discrete place. Here they should be arranged on the floor in lines, covered with sheets and given numbers. Personal effects should be removed and conserved in bags with numbers that correspond to those of each body. Later the bodies should be removed to morgues. As mortuary capacity may be limited in relation to the number of bodies (mortuaries and crematoria may have been destroyed in the disaster), any available cold store can be requisitioned as a temporary morgue. Death must be certified, with official registration of the cause, and bodies must be formally identified. For the former procedure, a coroner is required and on occasion an autopsy may be needed. Victims who have not been disfigured can be identified visually by next of kin. This is obviously a heart-rending process that needs to be handled with great delicacy and care. It may help to detail psychiatric and religious counsellors to be on hand at mass-identification exercises. Victims who cannot be recognized visually may be identifiable on the basis of skin anomalies (moles, discolorations, scars, etc.), other visible but not ambiguous peculiarities, dental records or DNA. The last two options may take time and require prolonged cold storage of the body. Lastly, coffins and funeral arrangements will be required, followed by burial or cremation, according to custom or rite.

The emergency planner should arrange for the facilities (morgues, hearses, holding posts), supplies (body bags, coffins, etc.) and personnel (fieldworkers, coroners, psychiatrists, priests, dentists, laboratory scientists, etc.) to be available if and when disaster strikes. Procedures for coping with mass mortality need to be worked out in detail in advance of the event that causes it.

Relatively few of these prescriptions can, or should, be included in the disaster plan, but much can be done by the emergency planner to ensure that they are carried out. Training, education, field exercises and logistical preparations should all be used to ensure that triage is widely understood and will work effectively in diverse kinds of catastrophe.

6.1.3 Epidemiological surveillance and health maintenance

Survivors of disaster who have not been injured, or who are making a good recovery, also require medical assistance during the chaotic aftermath situation, in which normal programs of health maintenance will be disrupted, as will normal modes of living. There is little evidence that murder and suicide rates are likely to increase after disaster, and only a tiny percentage of disasters give rise to outbreaks of disease. Hence, it is difficult to gauge the risk of post-disaster epidemics, but it is certain that these can be avoided by taking appropriate measures. Sanitary cordons, which restrict access to the disaster area and require people who enter it to undergo basic cleansing, inoculation or health check-up procedures, tend to be ineffective, as do mass-vaccination and mass-disinfection programmes. These merely hamper the relief effort and cannot nullify possible sources of risk.

Most, perhaps all, living victims of a sudden-impact disaster will be rescued within six days of the event. Risk of injury then relates to normal activities, relief efforts (e.g. in the spontaneous collapse of damaged buildings), recovery work (e.g. electrical shock while repairing cables), diseases and medical conditions that are **endemic** to the disaster area, disruption of habitat or habits of fauna (e.g. snake bite, dog bite or malaria), and the effects on people of disruptions in living conditions. The last of these can give rise to respiratory diseases from the effects of cold and damp, gastro-enteritis or salmonella from food contamination, or giardia, dysentery or gastric upsets from inadequate personal hygiene or contaminated water supplies.

It is important to ensure that hygiene, cleanliness, disinfection and disinfestation are maintained in camps set up for relief workers and displaced survivors. If particular conditions or diseases can be identified as threatening, then particular groups of people can be vaccinated against them. This would be the case if there is, for example, a history of endemic meningococcal meningitis in the area. Children and medical-relief workers may be vaccinated, but any attempt to inoculate all people in the disaster area would fail and be a huge waste of resources, such is the chaotic nature of population movements after disaster. Old people can be vaccinated against influenza, and relief workers can be given drugs to prevent malaria, as appropriate.

The disaster plan should identify potential risks, in accordance with impact scenarios, and arrange to devote appropriate resources to them. This may involve attempting to estimate the future need for vaccines and arranging to have them stockpiled or supplied when needed. Plans to vaccinate appropriate groups of people should be laid in advance, with attention to schedules, especially as some vaccines require a follow-up inoculation some days after the first one. Relief workers should be required to declare their medical history to ensure compatibility with any vaccines that may need to be administered to them.

It is sound practice to take steps to reinforce normal healthcare and preventive medicines in the wake of disaster. Field hospitals and mobile clinics can accomplish this efficiently, provided that they are adequately staffed, given sufficient medical supplies, and connected to adequate diagnostic laboratory facilities. Any apparent rise in the disease rate may be a function of improved diagnosis, as there are likely to be more doctors and nurses in the disaster area than before the disaster, and more patients will be seen more often than usual.

The correct response to the threat of disease epidemics is to establish an epidemiological surveillance system. Its successful implementation requires prior planning. The first stage is to designate epidemiologists for the task. Next, lists of diseases and conditions to be monitored should be drawn up in relation to health risks associated with particular disaster scenarios. These fall into the following categories (see also Table 6.4):

- clinically confirmed cases of certain diseases (cholera, typhoid, botulism, etc.)
- suspected clinical syndromes, requiring laboratory work to confirm the presence of the disease
- groups of symptoms requiring diagnosis (e.g. diarrhoea, rashes, fever, etc.)
- physical trauma resulting from injury (e.g. fractured limbs)
- general medicine (obstetrics, paediatric problems, arthritis, etc.).

When disaster strikes, the epidemiological observation system should move into action within 24 hours of the impact. There are three parts to the process: data collection and interpretation, medical investigation of apparent outbreaks, and prophylaxis of confirmed emergencies. This is far more effective and efficient than any overall measures that indiscriminately try to prevent medical emergencies from happening. Data collection involves contacting all medical facilities in the disaster area at a set time of day (e.g. 12.00 h) to ascertain how many new cases of each condition have occurred during the previous 24 hours. Fax and e-mail tend to be better suited to this than telephone or radio

Table 6.4 Epidemiological surveillance list.

Food shortage not expected	Serious food-shortage and disease
Diagnosed diseases	*Major causes of mortality*
meningococcal meningitis	diarrhoea and cholera
morbillus (measles)	dysentery
pertussis (whooping cough)	malaria
typhoid fever	malnutrition
viral hepatitis	meningococcal meningitis
	morbillus (measles)
Partially classified symptoms	pneumonia
diarrhoea with fever	
diarrhoea without fever	*Epidemic diseases requiring immediate*
fever with cough	*action*
fever without cough or diarrhoea	cholera
	malaria
Partially related to the crisis	meningococcal meningitis
general surgery	morbillus (measles)
trauma	poliomyelitis
	typhus
May be unrelated to the general crisis situation	
freezing or hypothermia	*Special nutritional deficiencies*
general medicine	beriberi
obstetrics	microcytic anaemia (iron deficiency)
psychological disturbance	macrocytic anaemia (folate and B12 deficiency)
	pellagra
	scurvy
	xerophthalmia

- Sum the total number of cases for each category, Σ
- Sum the grand total number of cases, Σ

transmitter, because the process involves filling in a form. Data are collated and related to the developing post-disaster trends and pre-disaster averages (bearing in mind that many diseases and conditions are seasonal in character). A comparison of methods and criteria for collecting epidemiological data is given in Table 6.5. Methods that require more than 24 hours are mainly appropriate to the acquisition of baseline data rather than for direct monitoring.

Table 6.5 Comparison of methods of collecting epidemiological data in disaster settings.

Pre-disaster "background" data (time required: continuing)
- Requires trained staff
- Uses standard diagnoses
- Provides pre-disaster "baseline" data for comparison with post-disaster data
- Well established means of collecting information on patterns of disease and illness
- Not specifically adapted to disaster needs

Surveillance system (time required: continuing)
- Requires some trained staff
- Standardized diagnoses and means of collecting data
- Can detect trends and changes
- Requires continuous application of appropriate resources (see text for description of these)
- Can be expanded, but definitions of data categories must not be changed during the data collection period

Field survey (time required: hours or days)
- Requires experienced statisticians and field epidemiologists
- Can be made very specific to needs
- Slow and labour-intensive
- Generally restricted to sample survey; reliability varies with representativeness of sample

Rapid field survey (time required: up to 3 days)
- Requires few trained staff
- Mainly used to ascertain number of deaths and numbers in broad categories of injury or apparent illness
- Provides data rapidly when need is great
- The survey may be inaccurate and may yield an inaccurate picture of the situation

Rapid health screening system (time required: continuing)
- Requires qualified health workers
- Equipment needed depends on data being collected
- Can be set up and activated quickly
- Data are only collected on people who are present at the place of screening, which may create an unrepresentative sample of the population

Observational tour of the disaster area (time required: hours-days)
- Does not require expert epidemiologists of statisticians
- Places emphasis on direct observation of conditions and interviews with people who are present at the scene of the disaster
- Non-quantitative data are collected, which may be lacking in objectivity and precision
- Rapidly provides an idea of conditions in the disaster area

Remote view from aircraft or via satellite images (time required: minutes–hours)
- Requires aeroplane, helicopter or means of obtaining a useful satellite image
- Can reveal extent of any destruction that is visible from above
- Produces very rapid results
- Does not yield epidemiological data
- Subject to large errors in estimation of the situation

Source: Adapted from table 3.1, pp.40–41 of "Surveillance and epidemiology", S. F. Wetterhall & E. K. Noji, in *The public-health consequences of disasters*, E. K. Noji (ed.) (Oxford: University Press, Oxford 1997).

If increases in daily values are noted, field investigation will be required. It is also warranted in cases of rumours about disease outbreaks. Therefore, the EMS plan should designate field investigation teams, which will primarily be composed of epidemiologists, medical doctors, public-health specialists and experts on hygiene. These should attempt to verify the existence of any apparent increase in epidemiological rates, determine the causes, prescribe remedies and ensure that sufferers are properly cared for.

In the main, rumours will be unfounded and disease outbreaks will be prevented from becoming epidemics. However, adequate health maintenance will require the resources to prevent common conditions: heat-stroke in summer, bronchitis in deepest winter, gastro-enteritis in canteen kitchens, and so on.

In addition, it is important to monitor the social, economic and environmental conditions that could cause public health to worsen. Table 6.6 gives a form

Table 6.6 Toledo area (USA) Disaster Medical Assistance Team (DMAT) outreach form.

Head of household's name: _____
Record no. _____ of _____

Address: _____ GPS location: _____ Map ref.: _____

Date: _____Time: _____

Environmental situation
Electricity..none – intermittent – reliable
Habitable space .... 0 to 1/3 – 1/3 to 2/3 – 2/3 to all
Heating or air conditioning ....none – partial – constant
Transportation..none – unreliable – reliable
Potable water (minimum 4 litres/person/day)........................ <1 day – few days – >1 week
Food..insufficient – few days – >1 week
Suitable clothes...none – insufficient – adequate
Surveyor's overall impression of situation..poor – fair – good

Demographic situation
No. of inhabitants ....>10 – 5 to 10 – 2 to 5 – one
Responsible adults...none – one – more than one
Children...............................infants (no. _____); toddlers (no. _____); children (no. _____)
Old people....... >80 years (no. ____); 65 to 79 years (no. ____); 55 to 65 years (no. ____)
Surveyor's overall impression of situation..poor – fair – good

Medical situation
Pregnancy complications .... major – minor – none
Chronic illnesses...................................... currently unstable – potentially unstable – stable
Acute illnesses .... immediate evacuation – field treatment – minor needs
Continuing care... follow-up in <3 days – no follow-up
Medicine needs...........................<24h supply – 1 to 7 days' supply – 1 to 2 weeks' supply
Surveyor's overall impression of situation..poor – fair – good

Psychiatric situation
Death in the family ....not coping – partially coping – coping
Missing family member ....not coping – partially coping – coping
Dead or missing pet...not coping – partially coping – coping
Surveyor's overall impression of situation..poor – fair – good

N.B.: These assessments can be turned into scores – bad (3), moderate (2), good (1) – and summed as a disaster outreach severity score.

for noting (and where appropriate scoring) the environmental, demographic, medical and psychiatric condition of households that have recently been affected by disaster. Such data need to be updated as necessary and compared each day with epidemiological trends in the disaster area.

6.2 Veterinary plans

Farm animals are an important and valuable resource, the mainstay of certain agricultural economies. Likewise, horses can be very valuable and, like farm animals, may be very significantly at risk in particular kinds of disaster, such as floods, which can drown them, and earthquakes, which can destroy farm or stable buildings and kill the animals as walls and roofs collapse. Areas with significant numbers of cattle, sheep, horses, pigs, poultry, and so on, should also have a plan to reduce the impact in this sector and to manage disaster.

The veterinary emergency plan should include provisions for dealing with catastrophic pollution that poisons or contaminates farm animals. This is perhaps most likely if they become trapped in floodwaters that contain sewage or toxic chemicals, or if animal husbandry is practised in close proximity to hazardous industries. Where such an eventuality is likely, the risk scenario should indicate what treatment, drugs and protective measures will be necessary.

Evacuation of livestock is not necessarily a straightforward process. It may require vehicles to transport the animals: horse boxes, partially enclosed trucks, trailers, and so on. If evacuation of humans is contemplated, it may be necessary to develop a parallel plan to evacuate animals from farms, zoos, pet shops or other places where they are usually kept. This requires that a corresponding place be designated where the evacuated animals can be kept, fed and tended until they can be returned. In addition, wildlife specialists may wish to develop plans to trap and transport wild animals (perhaps endangered species) if these are seriously at risk.

Clearly, veterinarians need to be co-opted into the emergency plan. Moreover, drugs and medicines may be required for normal and emergency use on farm animals, so supplies need to be ensured during the disaster.

Some of the needs likely to be generated by catastrophe can be ascertained in advance by conducting a survey of farms in terms of their vulnerability to damage, the census of animals (e.g. head of cattle), and resources available locally to ensure continuity of husbandry in the event of damage. The general disaster plan may require a section for the integrated management of, for example, dairy farms and horse studs. This will include provisions for rapid survey of damage and arrangements for substitute facilities where needed.

Weather disasters can have a particularly harsh impact upon farms, especially in hilly and mountainous areas. During snowfalls, intense rain and winds of hurricane force, upland farms can become isolated. The disaster plan in such areas should tackle the problem of how, systematically, to contact each farm. If needs cannot be ascertained by telephone or radio transmitter (although ice and snow storms can render both of these inoperative), then they will have to be estimated. Supplies for both farming families and farm animals may need to be airdropped from aeroplanes or helicopters, or moved in by snowmobile or other off-road vehicle. Again, it is helpful to construct a scenario of possible future needs in this respect so as to know what to stockpile in the future.

In floods or snowstorms, particular problems are created by the dispersal of farm animals. Arrangements therefore need to be made to search for, locate and rescue them. Where such equipment is available, thermal sensors can be used for rapid aerial survey, as they are able to detect heat emissions from an animal of moderate to large size. Rescue may be a thornier problem, although helicopters with winches can sometimes resolve it when terrain is too rugged or conditions are too harsh to permit easy access at ground level.

In the event that large animals are killed in a disaster, arrangements may need to be made to cremate them. This should be done, after identifying labels or marks on each animal have been recorded, under qualified veterinary supervision. Veterinary officials should also ensure that no animals have been improperly slaughtered without respecting statutory hygiene conditions and that dead animals have not been secretly butchered for consumption.

Domestic pets can easily be overlooked by disaster-management teams. Although this may seem unimportant in comparison with emergency services for people, and even compared with those for farm animals, it is not entirely so. To begin with, people can be extraordinarily attached to their pets, to the extent that some would not think of evacuating without them. Other people may attempt heroic but inadvisable rescues of pets trapped by raging waters, fallen masonry or crackling flames and billowing smoke. Finally, people who depend on pets for companionship may suffer great losses of morale if the pets are not rescued and cared for. Hence, this is a problem that needs to be tackled under the aegis of disaster planning and emergency management.

For the disaster planner, it may be worth discussing the question of rescuing domestic pets during briefing sessions held with search-and-rescue groups. In areas of high hazard it may be a good idea to involve in the plan veterinarians who deal only with domestic animals, so that their services are more likely to remain available during the aftermath of the next catastrophe. Vets, the proprietors of kennels, and the directors of animal shelters may be able to arrange temporary shelter for disaster-stricken pets when their owners can no longer accommodate them, perhaps through loss of their homes.

Lastly, stray animals – dogs in particular – may need to be rounded up after disaster so that they do not become carriers of pathogens, such as rabies, that can harm people. It is more humane to impound dogs than attempt to shoot them in the street, although it does make continuing demands on care facilities. Obviously, unusual behaviour in both domestic and wild animals should be reported if it occurs after disaster, in case it indicates an outbreak of rabies or constitutes a danger to people in some way. If such behaviour is observed, expert help may be required from animal specialists.

6.3 Emergency planning and schools

In 1933 an earthquake caused severe damage to dozens of school buildings in the Long Beach area of California. Fortunately, the tremors occurred when schools were not in session, but if they had been in use at the time there would have been large-scale loss of life. The resulting outcry led California to adopt to some of the earliest comprehensive anti-seismic building codes. Yet elsewhere the problem persists: a 1995 study of seismic hazards in the Boston (Massachusetts) metropolitan area suggested that school buildings continue to be vulnerable structures, especially as they often contain many people.

In 1966, 116 children were killed and 29 injured in the Welsh mining village of Aberfan. They were in two schools, which were partly destroyed when a mudflow swept through them at 09.15 h on a weekday morning. In 1986, children at school in El Salvador were crushed by seismically generated structural collapse while sheltering under their desks. In 1995 an earthquake occurred in Egypt that caused scores of deaths in schools as a result of chaotic unplanned attempts to flee buildings in severely dangerous conditions. All three tragedies might have been prevented by better planning and emergency management.

Whatever hazards prevail locally, schools need special protection, in terms of both structural measures and planning to protect pupils and staff. Particular attention should be given to plans to evacuate the schools. Evacuation remains the most efficacious short-term non-structural means of ensuring the safety of pupils and staff. In order to ensure that it functions and does not cause problems, it requires careful planning and frequent exercising.

In some respects, the process of developing an emergency plan for a school is a microcosm of disaster planning in general. As far as possible, the resulting plan should be integrated with disaster plans at the level of the whole community, especially as, in an emergency, support will be required from outside the immediate bounds of the school.

The process begins with the assessment of hazards to the school precincts

218

Box 6.1 Emergency planning for higher education: U. of Massachusetts at Amherst

In the USA there is increasing interest in planning for unforeseen emergencies in medium-size and large organizations such as universities. Recently, hurricanes and tornadoes have done considerable damage to campus environments in Florida and Texas, seismically active faults run through several branches of the University of California, and flashfloods have damaged university environments in Colorado. Finally, the threat of terrorism or riots, or simply the impact of power failures on complex technological systems, also require contingency planning.

The following account summarizes some aspects of preparedness for emergencies at my home institution, which can be considered a good example of a large campus university that bears a modest risk of disaster. Some of the problems and procedures used may also be relevant to other large institutions, such as military bases, industrial complexes, science parks and the headquarters of large corporations.

Assessment of vulnerability

The University of Massachusetts at Amherst was founded as an agricultural college in 1869. It is situated amid the forests, farmland and small towns of the Pioneer Valley of New England about 160 km from Boston. This state-run institution of higher learning has 24 400 students and 5300 staff. The campus, which is adjacent to the town of Amherst (population 30 000), contains 337 buildings worth an estimated US$1.6 billion and infrastructure valued at $400 million. This includes more than 25 km of roads, 14 bus routes, 183 km of footpaths, 40 km of heating ducts, 40 km of electricity lines, and 21 600 telephones. Each year the campus uses 115 million kWh of electricity (obtained from external sources) and its heating plant burns 30 000 t of coal. More than 11 000 students live in the 41 residence halls on campus. The library system, dominated by the 28-storey main library building, houses more than 3 million books. But, because of underfunding by state authorities, by the year 2000 a shortfall of $393 million had been accumulated in deferred maintenance of physical plant.

Since 1750 there have been 14 natural disasters in the area, comprising five major floods, three minor earthquakes, three tornadoes, two blizzards and one significant hurricane (in 1938). In 1913 a scarlet-fever epidemic occurred. Since then there have been occasional health scares, as well as water and electricity supply breakdowns, icestorms, lightning strikes, hailstorms, fires, riots and episodes of civil unrest.

Hurricane reference scenario

Scenario modelling for emergency planning requires the selection of one or more reference events. In 1879 a hailstorm and tornado destroyed several buildings at the university, and Massachusetts experienced devastating tornadoes in 1953 and 1995. In 1978 areas farther east in New England were paralysed for several days by a huge snowstorm, and in 1998 an icestorm caused massive damage and prolonged disruption to large areas farther north. Any of these disasters might serve as the primary reference event. However, the 1938 hurricane, an unnamed storm, is probably the most appropriate natural disaster to use. Since then, hurricanes have threatened but not damaged the campus: in September 1985 all students and personnel were evacuated for the passage of Hurricane Gloria, which, however, had lost its force by the time it reached the Pioneer Valley.

In contrast, the 1938 storm arrived with sustained winds of 130 km per hour and gusts of up to 240 km per hour. On campus 140 trees were uprooted and 70 more were seriously damaged. More than 300 mm of rain fell, which caused the nearby Connecticut River to rise 2–3 m above flood level. After the storm, 200 refugees from nearby towns had to be accommodated on campus and electrical power was off for two weeks.

Box 6.1 (continued) Emergency planning for higher education

Since 1938 the University of Massachusetts has grown into a completely different class of institution. Besides its mission to educate 25 000 students in a total of 94 degree programs, it handles $80 million a year in research grants and contracts. It is also a major employer and economic resource in its local region. Major disruption by a hurricane or other catastrophic event could have very serious repercussions for research experiments, the ability to teach students, revenue streams from these activities, and hence the prestige and position of the institution in the competitive modern academic world.

The first class of effects would result from direct impacts. Damage caused by wind and rain to classrooms would lead to suspension of classes, delays in completing courses and probable withdrawal of some students from the university. It would also affect research and teaching laboratories. More complex effects would result from loss of heat and electrical power, with eventual knock-on impacts upon student enrolment levels and research funding. In addition, indirect effects would be far from negligible. For instance, although the campus is not considered floodable, its environs are. Several of the most significant access roads would disappear under water. The local wastewater treatment plant would probably be flooded to a depth of about 1 m, leading to contamination of streets and freshwater wells by sewage.

Winter storm reference scenario
After the 1978 winter storm in eastern New England, the governor of Massachusetts suspended driving on the state's roads for six days, except for holders of special permits. Snow drifts reached 5 m in height and 3500 vehicles remained stranded on one road (Route 128) alone. Directly after the icestorm of 4 January 1998 almost 2 million electricity customers were without supplies, and 15 days later more than 80 000 remained without power in New England alone.

Scenario modelling of a major winter storm impact suggested that the university would respond quite well to the immediate emergency. However, food services, gasoline supplies and emergency generating capacity would probably be inadequate. Arrangements for rescuing stranded people appeared to need improvement and more vehicles would be needed that are capable of operating safely on snow and ice. It appeared that problems with emergency arrangements would become progressively more serious if the emergency lasted any longer than four days without being able to send people home. Special needs would include cordoning off of dangerous areas, for example, the vicinity of tall buildings where ice could accumulate and eventually fall off in heavy sheets.

Evolution of emergency preparedness
In response to disasters at various other American universities, and to the "year 2000 computer alarm", in 1999 the University of Massachusetts at Amherst set up a campus-wide emergency committee and charged it with producing a plan. At the time of writing, this was incomplete and hence will not be discussed in detail here. Prior to any special arrangements for autonomous management, any emergency situation experienced at the university would probably be handled under the auspices of the Hampshire County Multiple Casualty Plan. An incident-command system would be set up and the incident commander would be the highest ranking officer available from the town of Amherst fire department. This, of course, begs the question of whether municipal resources are sufficient to handle an emergency that might involve both the town and the adjacent campus (plus also the campus of Amherst College, a large private higher-education institution situated in downtown Amherst).

Box 6.1 (continued) Emergency planning for higher education

However, at the time of writing, a mass-casualty plan was already in place at the university health centre, which is the source of primary healthcare for almost all students and many employees. Although not a recognized trauma centre, in an incident with casualties it would fulfil part of this role and at minimum be ready to practise triage and to dispatch the injured to local hospitals.

Space does not permit a detailed description of the emergency medical plan here, so the following account will be limited to a few details. The plan provides for setting up one or more emergency-operations rooms in the health centre and for determination of the level of crisis. Tier 1 is a classed as a limited victim incident, while Tier 2 is a full mass-casualty event, which may also require mobilization of mental-health support services. The plan gives detailed information on criteria and conditions for determining the level (e.g. number of patients requiring medical assistance), and it describes call-up protocols, staffing levels and medical procedures. It spells out the requirements for triage and routines to follow in each department of the health centre. It describes equipment needs, job assignments and emergency-room designations. Charts, room plans and checklists are included in this 56-page document. Finally, arrangements for testing and updating the plan are described.

In essence, the emergency medical plan is clear, practical and sensible. As with any such document, there is always room for improvement. For example, the draft I reviewed was somewhat vague about what level of medical-disaster work the health centre could handle and what it would be allowed to do. Designated trauma centres are 20–90 minutes away from the university by ambulance, if roads are normally usable. This invites the question of exactly what criteria and conditions should induce a trauma officer *not* to send a patient on such a journey. Other questions concerned the rapid retrieval of health records, the management of emergency medical communications, and arrangements for handling public and press enquiries. Finally, the draft was a little vague about how it would integrate with external emergency plans (e.g. for the university, municipality and county).

These reflections highlight several common difficulties in developing emergency plans. First, it is not easy to follow a coherent strategy when plans are developed in a piecemeal way. Secondly, the knack of emergency planning is one of being able to visualize in the mind's eye what is likely to happen when "all hell breaks loose". Are plans sufficiently robust to impose order upon the sort of chaos that would ensue? Has any simple requirement been neglected (e.g. keeping radio batteries charged or making large signs to hang on emergency-room doors) that would compromise a whole series of operations?

Other aspects of preparedness

Two aspects of emergency preparedness at the Amherst campus warrant further examination. The first is evacuation and the second is functionality during a disaster aftermath.

In the past, campus evacuations have been successfully implemented at the university, but largely because they have been carried out with adequate lead times. As the university experiences occasional "snow days" when winter storms shut down normal activities, it has a detailed policy and procedure for closures. But what would happen if, when the teaching semester were in progress, a complete evacuation were necessary over a short period of time?

Box 6.1 (continued) Emergency planning for higher education

Scenario modelling suggested that rapid evacuation would be possible, although difficult to achieve. In one situation, some local roads might be closed as a result of flooding or traffic problems. Control centres could be set up in two buildings located respectively at the northern and southern ends of campus, and there could be assembly areas in sports fields and car-parks. University police forces could be deployed to direct a system of one-way traffic flows. The university's 40 buses could be used, and a local bus company could be asked to provide more. However, experience with major events on campus, such as graduation day, indicates that the huge convergence reaction that occurs in such cases could easily lead to gridlock, which would be exacerbated by emergency conditions. Moreover, it would be impossible to keep records on who had been evacuated to where. However, it would be possible to design a plan and publicize it through student networks and the university's Internet facilities, although not to test it fully.

In essence the process of rapid evacuation (perhaps in conditions of imminent fire or tornado risk) would be complex and with no guaranteed level of success. Nevertheless, the nettle must be grasped: at the very least, conditions could occur on campus that might make partial evacuation a necessity, for example, if fire broke out or hazardous materials escaped from a laboratory. The solution in such cases is to develop flexible modular plans and to exercise with them on a limited basis.

Short-term scenarios provide one kind of basis for emergency planning, but it is also important to consider the medium-term emergency situation. What would happen if the campus were without electrical power, water, telephone connections, or heating for a matter of weeks? This is not an entirely improbable situation: after the 1998 icestorms some communities in eastern Canada were without electricity for six weeks; even central Montreal was severely disrupted for 12 days.

In brief, the process of building a scenario for the medium term involves considering all significant eventualities. Students would be sent home as soon as road conditions permitted and, until that could be accomplished, many would have to be cared for on campus. Many employees would temporarily be laid off. Activities would be shut down or cancelled. But various scientific experiments would have to go on if they could not be suspended, and laboratory animals would have to be cared for. In the event of an icestorm, a major problem would be represented by prolonged difficulties of communicating with affected people. Damage to radio masts, cellular repeater towers, telephone cables and other equipment would impose a blackout for some days. A vulnerability survey would reveal the likely extent of such losses, and perhaps suggest ways of avoiding them.

Sources: This example is based on the work of students David Brand, Sharon Brodeur, Jonah Choiniere, William Gontarz, James Laing, Mary McRae, George Roberson, and Bonnie Roy, and professionals Robert F. Laford (University of Massachusetts Health and Safety Executive) and Alan J. Calhoun MD (University of Massachusetts Health Service). Their collaboration is acknowledged with sincere gratitude.

and their surrounds. This needs to be worked up into a scenario for hazard impact and a corresponding scenario for response. If more than one type of hazard threatens the area, then multiple scenarios may be needed, although they will undoubtedly have many elements in common. For instance, it may be appropriate to evacuate the building by one route if an earthquake occurs, but by another if a nearby river threatens to flood. As with general emergency planning, all significant hazards should be addressed in the plan, not merely a single dominant risk. A simplified procedure is listed in points 1–3 of Table 6.7.

Table 6.7   Steps in the design and operation of a school disaster plan.

---

1. Assemble materials:
   - street plan of area, with principal buildings identified
   - internal plan of school, with locations of stairways, exits, etc.
   - describe school: size and characteristics of buildings, number of floors, number of staff and students, etc.

2. Describe setting of school: surrounding area – location of residential, commercial, industrial development, roads and their usage; location of nearest emergency services.

3. Identify local hazards, assess their importance, rank them and work out exactly where they are likely to occur and what effect they will have on the school.

4. Work out response scenarios:
   - evacuation routes, with alternatives
   - assembly points, with considerations as to their safety and ease of reaching them
   - refuges (churches, community centres, theatres, libraries, etc.) and routes to them, with consideration of the safety of routes.

5. Procedures:
   - monitors or teachers assemble students at primary assembly point, call the register and take attendance
   - monitors conduct a sweep of the buildings to ensure no one is left behind in closets, restrooms or elsewhere, but only if it is safe to do so (inform incident commander or other emergency-service chiefs promptly of action taken or not taken)
   - evacuation option chosen and implemented by the building's principal after conferring with police chief, fire chief and emergency-management director (if appropriate)
   - shepherd students to secondary assembly point (e.g. nearby church), taking care that no hazards are encountered along the route.

6. Which emergency services will arrive, in what order, from where, and approximately how soon after the start of the emergency? What sort of mutual aid plans will be utilized when necessary and what impact are they likely to have on the situation?

7. Teachers and principal to know who incident commander is and what emergency forces are present or on their way.

8. Describe in detail procedures used to choose evacuation options on the basis of selecting the safest route in relation to the nature of the incident. Evacuation must lead evacuees out of danger and away from it, not through it, to safety.

9. Design a procedure for conducting evacuation drills and carry it out. Where appropriate, use permanent signs (e.g. coloured arrows) to show routes from classrooms to exits.

10. Institute procedures for conducting periodic surveys of obstacles to be removed in order to maintain exits unobstructed.

11. Study means of avoiding congestion and separating evacuees from incoming emergency services.

12. Avoid gymnasia or other large structures with roofs that could collapse during an impact or as one of its results.

13. Evaluate interior hazards (e.g. falling furniture, exits that might block, roofs that could collapse, glass that might shatter).

14. Design a procedure for posting and disseminating disaster plans, including measures to ensure that all personnel who should be aware of the plans have read, understood and remembered what is necessary to know. Share plans with town officials and directors of emergency services.

15. Are all exits safe for use in all possible scenarios? If not, this needs to be made clear in the plans and drills.

Table 6.7 (continued)   Steps in the design and operation of a school disaster plan.

16. Plan the role of the school's first-aid officer.

17. If the impact scenario predicts that the school can be re-entered after the hazard has passed, some rooms can be designated for first aid and other activities.

18. Work out a procedure for contacting parents and safely returning their offspring to them if the school remains closed after the impact. Arrangements must be made to hold and care for students whose parents cannot be located or are not immediately available.

19. Work out on-site and off-site liability, including custodial liability (*in loco parentis*).

20. Design material to acquaint parents with the school's evacuation plans and to inform them of where their offspring can be picked up under particular evacuation situations.

As far as possible, the hazard scenario should indicate how disaster could affect the school, including the risks of secondary impact, such as post-earthquake fire or delayed structural collapse. The response scenario (point 4 in the table) should consider the issues arising from programmed responses to any expected hazard impact. For example, having decided to evacuate the school and regroup pupils in a green space outside, are there any risks at the site of being electrocuted by broken power lines or crushed by falling trees? Might the evacuation route be strewn with broken glass? Simple precautions in classrooms and along corridors are well worthwhile. Bags, books and clothes should routinely be stored against the wall or in alcoves, and not between desks, where they may hinder rapid evacuation. Coloured (and perhaps luminescent) arrows on the floor can guide children to exits when they may be confused or distracted by fear. Furniture and fittings may need to be secured to walls to ensure that they do not fall on people or block exits.

A scenario is a hypothetical construction of a probable future reality and as such it cannot be verified until the events it depicts actually occur. Given this uncertainty, some serious dilemmas may arise in the construction of the response scenario. For example, on the basis of many past experiences, the American Rescue Team (ARC, www.AmerRescue.org) recommends that people *do not* shelter under desks during earthquakes unless such furniture is exceedingly strong and massive. Worldwide, many children have been crushed as beams and masonry have fallen onto the flimsy desks, under which they were sheltering. But where else is there to hide? The ARC recommends taking cover in the angles between floor and wall, as even complete structural collapse will leave about 15 per cent void spaces and that is where they are likely to be. On the other hand, grouping small children under desks stops them from running about out of control and getting themselves into danger.

It should be borne in mind that evacuating a school may require two refuges if it is not judged safe to re-enter the building after the impact. The first will probably be a green space or car-park where children and staff can re-assemble and take the register. If the wait is likely to be a long one, sanitation facilities may be needed. The second will be a place where children can be held until they are collected by parents. It is a principle of evacuation that it should conduct evacuees to progressively safer places. One must therefore survey the routes to both places for possible hazards and also consider the means of getting there (is it going to be safe to drive children to a given refuge area?). It is also essential to ensure adequate separation between the evacuees and incoming emergency vehicles (points 6 and 11 in Table 6.7).

The custodial problem (*in loco parentis*) can be a particularly difficult one if school buildings cannot be re-entered after primary evacuation. Children may need to be calmed or at least kept in order. They may also need to be safeguarded against continuing hazards, kept occupied and continually reassured until they are collected. The US Federal Emergency Management Agency recommends that evacuated schoolchildren be given identification tags that list their names, addresses and home contact numbers, as well as school, class and teacher details. These would need to be prepared in advance of the disaster and distributed straight after evacuation.

Teachers must take registers with them when leading evacuations and ensure that all pupils are accounted for when evacuation is complete. Moreover, children should not be sent home indiscriminately: it is estimated that in a recent Californian earthquake 6500 children out of 12 000 in a certain school district would have been in severe danger if they had been sent straight home.

The cooperation of parents in emergency planning is essential. In the first place, children should be released only into the custody of registered bona fide family members. Others, such as babysitters or domestic helpers, must have written authorization to collect children. Secondly, full parental participation in the emergency plan is the only way to avoid complete overloading of the school's telephone system and major traffic jams outside school gates. Thus, to avoid compromising emergency-management processes, parents must be informed by letter not to telephone the school and not to drive there.

The emergency plan must include procedures to keep track of when, where and to whom each individual child was released after evacuation. In the event of an impact, perhaps a highly localized one, that is not immediately obvious to parents, there must be plans to contact parents by telephone, even when this cannot be done from school offices. It may therefore be necessary for school secretaries to take a cellular telephone and a list of relevant telephone numbers with them when evacuating. In some cases, cooperative schemes may involve parents who live or work near the school being given specific roles as helpers in the emergency evacuation.

In all but the smallest schools, there will certainly be some division of roles in emergencies. The principal and teachers will have primary responsibility for evacuation, but first-aid officers will also have essential roles that must be fully mapped out. School bus drivers may also be essential personnel.

When disaster strikes, emergency workers will arrive at the school. These may include policemen, fire services, ambulance crews or civil-protection volunteers. Plans should arrange for specific forms of cooperation between school staff and external workers, for example regarding search of school buildings or survey of structural damage. Thus, school evacuation plans need to be registered with civil-protection authorities as part of the local emergency plan.

Above all, successful plans require adequate motivation on the part of teachers. These must be encouraged to take emergency preparedness seriously and to play a constructive leadership role in planning and exercising. School emergency plans should be tested at regular intervals, perhaps once every three months: conditions can change, new pupils and teachers arrive, drills can be forgotten. Sociological studies of evacuation behaviour show that, during disaster impacts, evacuees do what they have been trained to do, as there is no time to think the situation through. It is therefore vital to renew the training regularly and fully. Adequate support from teachers is the key to this process.

Although full emergency planning for schools is a complex process, it not only safeguards precious lives but also has implications far beyond its immediate objectives. In the constant battle to improve public safety against the threat of disasters, children are catalysts. They are often more receptive to new ideas than adults are, and they may be the ideal means of introducing to their families notions of safety and participation in civil-protection activities. Setting a good example at school can thus help make families more accustomed to the idea that ordinary citizens have a role to play in disaster prevention.

6.4 Terrorism and crowd emergencies

In this section we consider some of the requisites of planning civil-protection work in terrorist incidents, mass gatherings and similar events that are either emergencies by definition or have the potential to end in disaster.

6.4.1 Public order and security

Looting, mass panic and social strife are not common consequences of disaster. Indeed, anti-social behaviour is likely to occur in the immediate aftermath of a catastrophe only if it was endemic before the impact. Thus, disasters do not create preconditions for the breakdown of society; indeed, social cohesion and

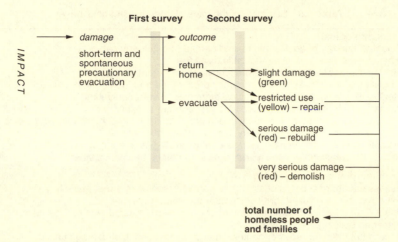

Figure 6.10 Integrating structural survey of damaged buildings with evacuation of their occupants.

solidarity are heightened, as people work together against the common enemy (flooding, earthquake damage, fire, or whatever it is). This is sometimes known as the therapeutic community that disasters produce.

In the context of disaster, public order assumes a new dimension. This involves managing evacuation processes, ensuring that evacuation orders are complied with, and blocking public access to dangerous places. After the impact it means managing the convergence reaction. In this, emergency personnel, volunteer workers, victims, relatives, sightseers, journalists, politicians and VIPs converge on the scene of the disaster. The agents of public order must ensure that emergency vehicles can circulate freely and park where they are needed. Where appropriate, this will also involve keeping landing areas free for helicopters. All this requires traffic to be directed and access to critical areas to be controlled.

When buildings have been damaged, or where areas remain hazardous, public safety must be guaranteed by surveying the buildings or the site in order to make a determination of accessibility potential. Survey procedures therefore need to be integrated with post-disaster evacuation plans (Fig. 6.10). The disaster plan should include a list of accredited specialists in architectural, engineering, geological and geotechnical survey. If necessary, it should also make provision for training them in post-disaster survey work (a list of some of the uses of surveys is given in Table 6.8).

In summary, the main tasks in the field of public security will be the accrediting of emergency workers, ensuring that access restrictions are maintained, managing traffic flows, supervising evacuations, and providing security for VIPs. While providing for these functions by ensuring the presence of an

227

Table 6.8 Summary of the uses of post-disaster field survey.

Epidemiological
- Establish number of deaths and injured people (the latter by category).
- Monitor the progress of communicable diseases.
- Identify disease vectors and the preconditions for potential epidemics.
- Indicate the need for specific emergency procedures and monitor their effectiveness.
- Monitor the nutritional or psychological state of survivor populations.
- Indicate which measures are needed to prevent further injury.
- Indicate which measures are needed to improve the delivery of healthcare and sanitation.
- Investigate rumours of death, injury, disease outbreaks, etc.
- Indicate the most appropriate disease-control strategies to apply.
- Assess the functionality of hospitals, clinics and other medical services.

Architectural and engineering
- Assess danger of collapse of structures and corresponding threat to public safety.
- Assess continuing vulnerability of structures to hazards.
- Assess need for short-term and permanent long-term evacuation of occupants from damaged buildings.
- Assess the need for repair or demolition and reconstruction of damaged buildings.
- Identify routeway blockages and assess functionality of main transportation corridors.
- Assess damage to utility networks and indicate repair needs.

General house-to-house surveys
- Assess post-disaster living conditions among survivors.
- Identify vital needs of survivors.
- Monitor psychological state of survivors.
- Identify "problem" households that require special assistance.

Logistics
- Compile an inventory of available human and material resources.
- Account for relief supplies and indicate where there are specific needs.
- Match relief needs with available supplies.

adequate number of security personnel (e.g. policemen), the plan should not over-emphasize this aspect, as disaster is not likely to destabilize the social fabric unless it was in a precarious condition beforehand. Thus, it should not be assumed, for example, that unfamiliar people are necessarily looters. Heavy-handed policing and authoritarianism are not recommended unless there is a special need for such approaches.

6.4.2 Terrorism and hostage taking

Strictly speaking, terrorist incidents, mass shootings, riots and demonstrations that end in violence are not to be equated with the other kinds of incident, such as earthquakes and air crashes, dealt with in this book. One reason is that some form of human malevolence or misbehaviour is at their root and so one is not dealing with a physical phenomenon that can be regarded as a common enemy, but usually with a conflict of interests between legal and illegal behaviour. Another reason is that the forces of law and order usually have absolute jurisdiction, whether they be ordinary police forces or special commando units.

Nevertheless, civil-protection authorities and volunteers groups become involved in such incidents more often than one might think. Their role is usually to provide back-up assistance, aid to victims, help with stabilizing damaged property, psychological counselling to survivors, and so on. Emergency medical services can be involved, as they would be in other types of incident or disaster (although shootings and bombings involve different kinds of physical injury from those produced by, for example, earthquakes or avalanches).

It is particularly difficult to construct scenarios for attacks, bombings and riots. For example, terrorists usually aim to foil the authorities' preparations, if necessary by changing tactic at the last moment. They may plant a second bomb at the point where people assemble after evacuation or at the exits of a building under threat. Emergency planning then becomes redundant and emergency management is a question of outwitting malevolence.

However, at the very least, contingency plans should detail the forms of collaboration between emergency services and police forces. What forms of alert will be used if a terrorist incident or riot occurs? If barriers, cordons or controlled perimeters are to be set up, where will emergency vehicles be stationed? What signs or procedures will be used to distinguish emergency workers from police or terrorists (in the field and in communications on radios or telephones)? How will emergency workers be integrated into the police or military command structure and exactly what roles will they be expected to play?

A crucial aspect of planning is to ensure the safety of emergency personnel and their vehicles and equipment. Every effort must be made to guarantee that they are not put in a position where they may be shot at, taken hostage or become targets of a rampaging group of people. This requires high levels of collaboration between police and emergency rescue units. It also requires that the chain of command reflect the necessities of the situation: it may need to be more authoritarian and monolithic than in other sorts of disaster. Finally, in some cases it may be preferable that emergency workers are not viewed (by rioters or terrorists) as strict collaborators with the forces of order, but as neutral rescuers, in as much as this can be achieved. This may necessitate keeping them in the background or on low-profile assignments.

A special word is warranted here concerning hostages. Four main types of situation may involve people being taken hostage:

- failed robberies and other acts by professional criminals
- the work of mentally disturbed people
- acts of terrorism
- acts of violence within families.

Negotiations with hostage takers are conducted along broadly similar lines in any of these situations and should be in the hands of trained specialists who must work under carefully managed conditions. Most hostage situations last about 8–12 hours, and it is vital to build up a rapport with the hostage taker during the first 45 minutes of this period. Negotiations should be conducted between one negotiator and the individual or main hostage taker without involvement by third parties. Care must be taken with what the news media broadcast, as ill advised statements can inflame hostage takers who hear them on a radio or television set.

Negotiations should be conducted patiently in an atmosphere of calm, giving the impression that the situation is under control. The negotiator needs to read between the lines to anticipate any developments that might resolve the situation. The aim is obviously to end the siege without violence erupting. As far as is possible, the hostage taker's demands should not be met and ultimata should not be acknowledged. Questions should be non-judgemental and open ended, with no attempt to shame, embarrass or shout down the hostage taker.

6.4.3 Emergency planning for biological and chemical terrorist attack scenarios

Although many scenarios for biological or chemical attacks by terrorists may turn out to be nothing more than alarmism, a worldwide potential exists for acts of gratuitous violence through the use of pathogens, radioactivity or chemical agents. Many terrorist incidents have been limited in scope and intended to force the process of political negotiation; others may be based on a desire to exact reprisals by causing mass casualties or major disruption. This is where biological and chemical terrorism come in. As with bombings and shootings, the forces of law and order may lead the official reaction, but emergency planners, managers and responders are highly likely to become involved. Hence, the subject deserves treatment here – as best it can, given that the terrorist aims to outwit the authorities by creating new and unpredictable scenarios.

Probable nature of the incident
In broad terms, there are four possible kinds of attack:
- viral or bacterial disease pathogens
- chemical toxins
- radioactive substances
- nuclear weapons.
  For biological or chemical toxins, there are five possible means of delivery:
- spraying or jettisoning from aircraft
- explosive diffusion in a bomb
- emplacement and explosive diffusion with a missile
- dispersion by hand
- contamination through a network (e.g. postal sorting and deliveries).

Three sorts of potential events can occur:

- deliberate release of agents that have been refined in ways that improve their effectiveness as weapons
- use of common pathogens or unaltered toxins
- false alarms or hoaxes.

Toxic agents can be dispersed as dust, aerosol sprays, gases or water-borne substances. Most of the potential biological and chemical weapons involve substances that form lethal concentrations in very small or minute doses. Examples include the chemical dioxin, the nerve gases Sarin and Soman, and Legionnaire's disease, a form of bacterial pneumonia spread by aerosol. The following two cases illustrate levels of toxicity. The LC_{50} (lethal concentration producing a 50 per cent mortality among victims) for Sarin is $100\,mg/m^3$ per minute. Thus, if 1 kg is released in an enclosed space measuring $100\times100\times10\,m$ ($100\,000\,m^3$), and dispersal occurs uniformly to an average concentration of $10\,mg/m^3$, the LC_{50} will be reached in 10 minutes. Looked at another way, 80 kg of a chemical with an aerosol toxicity of $0.025\,g/kg$ would be required to cover $100\,km^2$ with a cloud that exposed individuals to LD_{50} (the lethal dose producing a 50 per cent mortality among victims).

A general distinction can be made between common forms of particular substances or pathogens and those that have been "weaponized", that is, refined into a form that makes them more lethal or more effectively dispersed. For example, anthrax spores in numbers large enough to be lethal can be made to attach themselves to very fine powders or disperse themselves in liquids that can be used to make fine aerosols, which will increase their penetration of the victim's lungs. Generally, the world's stocks of biological and chemical weapons involve substances that have been refined to improve their strike capabilities, but terrorists may not have access to the stocks or the laboratory equipment to manufacture their own.

In respect of biological pathogens, a terrorist attack may be intended to produce one of the following results:

- *Epidemics in human populations* These may result from pathogens that are very easily diffused but not contagious, such as anthrax, which comes from the bacterium *Bacillus anthracis*. Alternatively, highly contagious pathogens may be used, for instance, bubonic plague (from the bacterium *Yersinia pestis*), botulism (from the bacterium *Clostridium*), smallpox (from the virus variola) and dengue or ebola haemorrhagic fevers. The effects of biological weapons may not be apparent for days; for example, the incubation period for anthrax infection is 1–6 days, plague 2–3 days, and smallpox 12 days. Smallpox has not threatened public health since 1977, but laboratory stocks of it still exist and could conceivably be stolen and used by terrorists.
- *Epizootics, or animal-disease epidemics, in populations of farm animals* Bovine

231

spongiform encephalopathy (BSE) and foot-and-mouth disease are both virulent and difficult enough to control to be candidates. Mass poisoning could be used where there are very large concentrations of beasts.

- *Epiphytotics, or plant disease epidemics, in standing crops*   These effects can be achieved using any of the following pathogens: Karnal Bunt fungus, rice blast, wheat stem rust (wheat smut) and, according to United Nations sources, about seven other common crop diseases. During the Cold War the superpowers cultivated large amounts of rice blast and wheat rust.

Epizootics and epiphytotics would have consequences for agricultural economics and food supply, although perhaps only BSE has the potential to be passed along the food chain to humans, and this would occur, if at all, over a period of years.

Epidemics would have both direct effects (mortality and **morbidity** through contact, contamination and infection) and a series of indirect spin-off effects. Public anxiety after a major act of biological or chemical terrorism would undoubtedly cause widespread incidence of traumatic stress disorders requiring psychological treatment. There would be hypochondria and somatic disorders (also known as sociogenic illnesses or MIPS (multiple idiopathic physical symptoms)), with associated requests for treatment. Hence, of the 5000 people who sought treatment at hospitals after the Aum Shinrikyo gas attack of 1995 in a Tokyo underground train station, only 1000 were injured and the rest had psychosomatic complaints – they were the "worried well". After the Iraqi Scud missile attacks on Israel in 1991, 54 per cent of patients admitted to hospitals in the Tel Aviv area had symptoms induced either by anxiety or by overdoses of atropine, which they had used as an antidote to non-existent gas attacks. Psychosomatic and stress disorders are likely to require long-term treatment, whereas for genuine victims of attack there may be long-term concerns about reproductive health and disability induced by pathogens or toxins.

Organizing the response

The first concern of any response to a biological or chemical terrorist incident is to recognize the scope of the attack and what substance is involved. Rapid diagnosis and decontamination are essential. Large numbers of patients may need to be cared for, given antidotes or antibiotics, counselled and placed into quarantine. Victims, indoor environments and outdoor areas may need to be decontaminated and the physical spaces cordoned off to prevent entrance by unprotected people.

The safety of emergency responders is obviously an important consideration. Many protective suits and gas masks are not sealed well enough to guarantee protection against micro-bacteria. The only solution is to stockpile specialist equipment in accessible places and to develop protocols for its distribution to and use by emergency responders. New products for the control of biological terrorist incidents are being actively developed and marketed, and so response technology is constantly being upgraded.

Bioterrorism attacks may not be spectacular enough to be immediately detectable. If they are suspected but not proven, intensive epidemiological monitoring is required. If a cluster of cases is detected, or if recognizable symptoms occur, then anti-microbial or anti-toxin therapies can be administered on both a reactive and a preventive basis. This will depend on the availability of stockpiles of antibiotics and antidotes. Generally, prophylaxis should not be administered unless there is a clearly demonstrated need (i.e. not to help calm people's fears). Mass vaccination is, as usual in disasters, a waste of time.

As only very small concentrations of a substance may be needed in an attack, and as it is likely to be contagious or highly toxic, the identification of what exactly has been released involves special problems. The more difficult the problem, the more specialized the laboratory that is needed. In the USA, four levels of laboratory are designated; the very few highest-level institutions are Department of Defense units that have special expertise in chemical and biological weapons. Laboratories must have special analytical equipment (e.g. (electron microscopes) capable of accurately identifying minute traces of key substances. Trained, experienced staff may be needed and special procedures followed to avoid contamination. Results must be obtained rapidly and, finally, special procedures for the transportation of analytical samples may be needed, to ensure both priority and minimization of risks in transit. Although full diagnosis may take up to three days, field-assay kits exist for some common toxins and pathogens. These need to be stockpiled for immediate use.

At the site of the attack, decontamination and sanitary cordons are necessary. An incident-command system (ICS, *q.v.*) is an appropriate means of directing operations, unless the site is so large (e.g. city-wide) that especially Draconian measures must be used. Hazardous materials response teams (HAZ-MAT **teams**) must be on site, and police, paramilitary personnel or soldiers will direct operations. The incident commander will form task forces to tackle each aspect of the response at the site. In many cases, decontamination is best achieved with soap and water (unless this could lead to translocation of toxins into the sewer system). Quarantine may be problematical if the regulations governing it are lax. People who might be contaminated may need to be subjected to prolonged isolation, but the legal instruments to achieve this may be insufficient. If patients are not compliant, they may be difficult to detain.

A sense of perspective
Scenarios for terrorist attacks have included using a light aircraft to disperse large quantities of toxins or pathogens over a major city, causing hundreds of

Box 6.2 Terrorist attacks on the USA, 11 September 2001

The attacks on the United States on the morning of Tuesday 11 September 2001 were unprecedented in scale, daring and degree of coordination. Three of the four targets were hit, with tragic and spectacular results: over 3000 people were killed. At the time of writing, a few months after that fateful morning, it remains to be seen whether these outrages will remain unique or whether they will usher in a new phase of world instability. In either case, they will have profound implications for emergency planning and management during outbreaks of terrorism, as shown in the following analysis.

The events

At 0845 h on 11 September 2001 a Boeing 767 on a routine commercial flight was hijacked and deliberately flown into the upper floors of the north tower of the World Trade Center (WTC) in New York's financial district at the southern end of Manhattan Island. Eighteen minutes later another hijacked 767 crashed into the adjacent south tower. At 1010 h a Boeing 757 was flown into the Pentagon military headquarters in Washington DC, and at the same time another 757 crashed in a field in rural Pennsylvania, apparently missing the target that the hijackers aboard it had intended to fly into.

The Boeing 767 has a fuel capacity of 90 770 litres and the 757 42 680 litres. All had taken off with full tanks only minutes before from airports in the eastern USA. The 266 people aboard the airliners, including all 19 hijackers, died instantly in the crashes. The jet fuel ignited fierce conflagrations in all three buildings. Fireballs were injected into each of the 110-storey WTC towers, igniting floors 95–103 of the north tower and 82–93 of the south tower. The fire at the Pentagon burned for many hours but was contained by the massive structure of the building, which had been designed originally to resist attack.

The WTC was built of steel beams clad in concrete, with a strong central column, containing elevator shafts, stairs and utility wells, from which steel beams radiated outwards to connect with the rest of the load-bearing structure. The fires overwhelmed sprinkler systems and increased in intensity to 800–1100°C. This turned concrete into powder or soot, and the structural steel first buckled, then melted. Sixty-two minutes after impact the south tower collapsed, and the north tower followed suit at 1028 h, 103 minutes after it was hit by the first airliner to crash. The towers were designed to resist the impact of a smaller jet liner, and to retard a fire for two hours (the calculated evacuation time), but it would have been extremely difficult to build them to resist the impact, blast and fireball effects of the deliberate attacks on 11 September 2001.

In total, 189 people died in the Pentagon and about 2890 in the WTC, in the latter case including hundreds of foreigners from a total of 60 countries. The WTC death toll was limited by early efforts to evacuate this complex of seven buildings, which at peak times contained as many as 40 000 people. A few of the victims took their own lives by jumping out of windows to avoid being burned in the fire. Among the dead in New York there were 343 firemen and 78 policemen, who rushed to the scene immediately after the crashes and in many cases went up the stairs of the towers to rescue people or fight the fire.

The collapses could have been much more destructive if the impacts and fires had occurred lower down the two towers, which might have made them fall over rather than subside vertically. As it was, the load shift, which was probably of the order of 100 000 tonnes, led to progressive collapse onto a restricted site, where the 1.2 million tonnes of debris formed a compact heap. The collapse caused tremors equivalent to a magnitude 2.3 earthquake.

Box 6.2 (continued) Terrorist attacks on the USA, 11 September 2001

Analysis
Although precedents for most of the individual aspects of the attacks can be found in previous events elsewhere, in this case they were without parallel with respect to scale and coordination, as well as in the complete absence of forewarning. Hence, it would have been difficult to develop adequate response scenarios through prior planning. However, much can be learned from the event. Here is a brief discussion of some of the main issues.

PROBLEMS IN NEW YORK
- EVACUATION PROCEDURE   This has always been a contentious issue, as such buildings represent large concentrations of vulnerability in the case of fire or structural failure. Photographs of the evacuation of the WTC show that stairwells were narrow, crowded and congested. Instructions and orders given to people on the upper floors of the towers were apparently conflicting, such that not all people went straight to the emergency exits, seeing the debris accumulate in the street below and feeling that they were safer inside. Many of those who did eventually evacuate found darkness, water saturation, smoke and overcrowding. They also had to make way for firemen, with their bulky equipment, who were going up the building. It took people an hour or more to evacuate from the 70th floor and above. It was then difficult to exit safely, because of the smoke and dust wafting around and the debris that rained down around the buildings' perimeter. On a more positive note, eyewitness reports suggest that, as sociologists would have predicted, the incidence of panic was very limited. Most people acted calmly and rationally, even under extreme duress.
- HIGH-RISE BUILDING PLANNING SCENARIOS   Some experts in structural engineering have claimed that once the scope of the fires was evident (i.e. a very few minutes after impact) they knew that collapse of the towers was inevitable. If this is true, then it has profound implications for the sort of scenarios on which emergency plans should be based in cities that have many tall structures. In New York, quite understandably under the exceptional circumstances, the risk of collapse was underestimated and many emergency workers were sent to their deaths, their vehicles crushed by falling debris.

  It is common to base emergency plans on the more probable – and usually less catastrophic – event rather than the unlikely one with profoundly horrifying consequences. Yet emergency planning is about "thinking the unthinkable" and it should furnish the means for decision makers to be able to adapt to highly unfamiliar conditions. With this in mind, the problems of managing major emergencies in very tall buildings evidently need to be more thoroughly investigated and given more detailed consideration in city disaster plans. The safety of rescue workers is a paramount consideration here, as well as the need to speed up evacuation and protect evacuees from harm in the street outside. New buildings over a certain height will require improved safety features.
- EMERGENCY MANAGEMENT   The New York City emergency-operations centre was intended to be housed in the WTC, as the NY–NJ Port Authority had offices on floors 2, 14 and 19 in the north tower. The decision to locate it there was as an act of defiance in the wake of the 1993 bombing of the building and a strategic move, given the central role of the complex in New York's financial district. However, it turned out to be the most unsuitable location possible. Perhaps in environments as complex as Manhattan, several well protected sites need to be set up as alternative EOCs, or subsidiary operations centres need to be capable of being expanded and transformed into the main centre in times of acute need.

Box 6.2 (continued) Terrorist attacks on the USA, 11 September 2001

On a more positive note, once the collapses had occurred, perimeter control of the site proved relatively easy, as the southern tip of Manhattan is surrounded by water on three sides. Casualties were evacuated across the Hudson River to an advance medical post on the Jersey City shore, a procedure which, although time consuming, was much easier than the alternative of treating them among the smoke and confusion of Manhattan. Patients who required hospital treatment were distributed among 83 hospitals in five New York boroughs and three suburban counties. Of the 5284 people treated, 7.9 per cent required hospitalization. Although one or two of the nearest major trauma centres rapidly reached their quota, others waited in vain for a late wave of casualties that never arrived: the main medical emergency (as defined by the need for triage) was effectively over within eight hours of time zero.

SEARCH AND RESCUE   The 450 000 tonnes of rubble of the WTC formed a very compact mass at the site, which was exceedingly difficult to tunnel into, and one, moreover, that remained very unstable and subject to spot fires. Many void spaces were either empty or were filled with pulverized rubble, the same mud that liberally coated the site and all who worked on it. Hence, few survivors were found during the search-and-rescue efforts, which went on for many days. The sheer scale of the collapse necessitated the use of heavy equipment of a kind that is usually inimical to the delicate operations needed to avoid crushing living victims as debris that traps them is removed. Conditions at the site were made exceptionally dangerous by the large amounts of precarious masonry still standing and ready to collapse without warning. This required a constant watch to be kept, alarms sounded and work interrupted at the slightest sign of impending collapse. The fall of the 47-storey WTC Building 7 at 1720 h on the first day of the emergency further complicated search and rescue. The search area was divided into quadrants and 90 000 tonnes of debris were removed in the first week.

PROBLEMS IN WASHINGTON DC

A *Washington Post* editorial published a week after the tragedy concluded that "A review of last Tuesday's events suggests that the District was unprepared for the emergency and therefore unable to react and assist the public in a timely and effective fashion." This is a very serious allegation that demands some investigation.

- EMERGENCY COMMUNICATIONS AND RESPONSES   Despite frequent testing, the local emergency broadcast system was apparently not activated on the morning of the tragedy. Most of the occupants of the city obtained information and instructions on what to do from the mass media, whose interpretations did not necessarily reflect official management policy. In some quarters this appears to have been largely unformulated. For example, the mayor's chief of staff – unable to reach the city administration because of overloaded telephone and cellular systems – used e-mail to order office workers to evacuate, but four minutes later the city administrator countermanded the order, also by e-mail. The first order was a response to erroneous information that three more aircraft were converging on Washington, and the second to a decision to keep the government functioning during the crisis. Besides the obvious deficiencies of Internet-based warnings, overload rapidly affected the celerity of net-based messaging. In the event, inessential workers left for home in large numbers and this led to massive traffic jams, clogging emergency-service routes. They lasted for about three hours, until an improvised emergency-circulation system came into action.

  It was intended that satellite telephones be distributed to key Washington officials and used to overcome the overload of public telephone systems. However, they remained locked in a cupboard throughout the early phases of the crisis. Moreover, the city's health department lacked radios capable of monitoring hospital communications, and thus it suffered reduced ability to participate in logistical decision making and to determine the availability of medical services.

Box 6.2 (continued)   Terrorist attacks on the USA, 11 September 2001

- EMERGENCY PLANS   The District of Columbia Metropolitan Police Department, which has 3800 officers, had no anti-terrorism plan and no guidelines to inform commanders and officers how and where to respond. A plan had to be improvised during the crisis. Moreover, police were not informed that federal workers had been sent home, which led to extra traffic jams, although these were eventually cleared as emergency traffic-circulation systems were organized. Police fell back on millennium "Y2K" plans to identify which road junctions to man. Meanwhile, the US Congress and Senate did not evacuate, which is perhaps just as well, as members were untrained in the procedures, and evacuation route maps were not current.

    On the other hand, the Washington Metro system immediately implemented high-security procedures, although many commuters were not aware that it continued operating throughout the day of the attack. Furthermore, Arlington County implemented its emergency plan within ten minutes of the attack. Yet, according to its administrator, the city's main trauma hospital, the Washington Hospital Center, did not have enough capacity for a major mass-casualty incident, which would have meant that adequate medical response would have depended on the efficient distribution of patients to other trauma centres.

- CONCLUSION   When the Secret Service realized that a third plane was heading for the White House (and then the Pentagon, given that it changed course), there was no procedure in place for shooting it down, and the nearest fighter jets were 200 km away. But apart from such drastic – and hypothetical – measures, it was abundantly clear that the scope of a possible terrorist attack on the capital city of the USA had been underestimated. Scenarios were too modest, especially regarding the level of disruption, and existing plans were neither comprehensive nor sufficiently generic. Moreover, the City of Washington's Emergency Management Agency was too understaffed and underfunded to create and implement an adequate emergency plan. The consensus seems to be that the city and its emergency services needed highly codified but broadly generic plans, such as the US Military's "Threatcon" (threatening conditions) plan, which details the tasks to be executed at a series of alert levels.

Conclusion

Whether the US terrorist outrages of 11 September 2001 remain a unique set of events or usher in a new order of magnitude of such disasters remains to be seen. As the events mainly affected cities that are extremely rich in emergency resources, there was no lack of emergency assistance. Quite the reverse: congestion and the convergence reaction were significant problems. Hence, drastic measures were required to curb the ensuing confusion, and in this respect, terrorism necessarily provokes a more authoritarian response than other forms of disaster, with the sole exception of outright warfare. Emergency planners will have to work around this problem as they seek to integrate civilian forces into the response structure.

    The medical, psychological and economic problems caused by the disaster will go on for years, with deep impacts not only on the families of victims but also on the US emergency-response community. One hopes that there may also be an enduring legacy of improved emergency planning, with more precise "rules of engagement" and more detailed scenarios for such events. We owe it to those who perished to achieve this.

thousands of deaths. Although apocalyptic scenarios should not be completely ruled out, they are unlikely, not least because highly toxic substances are not easy to manufacture, store, transport and disperse. If this is possible, the handler could be incapacitated by contamination before the attack has a chance to occur. Hence, major biological, chemical or radiological incidents would probably need major state sponsorship and highly sophisticated technological support systems.

It is therefore wise to concentrate planning efforts on smaller events. Although the degree of unpredictability is large, protocols, procedures, stock-piles and plans of action can be created as described above. In any event, the planner should assume that a period of alert would involve high levels of disruption to daily life, whether or not an attack materializes. During the Persian Gulf War, 4500 alerts signalled possible chemical or biological attacks, none of which materialized. If main underground railway stations were to have detectors and alarm systems for chemical attacks, false alarms causing major disruption would be inevitable.

Given the relative inefficiency of many forms of chemical or biological attack, some experts argue that the principal effect of such forms of terrorism is to create high levels of public anxiety. Therefore, particular attention needs to be given to mental health and behavioural issues in emergency medical planning and to the communication of risk in management of the public. In the end, plans may never have to be activated, but they certainly should be drawn up.

6.4.4 Emergencies involving crowds

Conflict-based emergencies are one form of social disaster, but the other main type, crowd emergencies, has a different root, the result of aberrations in mass behaviour.[*]

Whether mass gatherings are held in the open air or in stadia or auditoria, they require special planning, in which civil-protection authorities may become involved. Wherever there are crowds, there are risks of mass-casualty incidents caused by crushes, rushes or crashes. Rock concerts and sports fixtures involving many young people may give rise to crowd incidents (crushes or fires in stadia, fighting, crowd surges, mass vandalism, intoxication with drugs or

---

[*] The rest of this section owes a great debt to the following reference, which should be consulted for the wealth of further detail that it offers. It is available on the Emergency Preparedness Canada website www.epc-pcc.ca: *Emergency preparedness guidelines for mass, crowd-intensive events*, J. A. Hanna (Hamilton, Ontario: Scientific/Technical Reports. Emergency Preparedness Canada, 1995; produced within the Canadian Framework for the International Decade for Natural Disaster Reduction).

alcohol, etc.); healing events may leave older people stressed or they may be taken ill; air shows and motor racing may suffer crashes that injure spectators.

Emergency planners and managers can, and perhaps should, become involved in all stages of the event, from its conception to the clear-up process after it has occurred. For some events, it may be possible to select the site with a view to minimizing hazards and ensuring access and freedom of operations for emergency services. Issues to be tackled include logistics, crowd control, public health and the provision of medical aid.

As access roads and site areas to be used by emergency vehicles need to be kept free of traffic and parked cars, it may be necessary to close some roads to general access, perhaps using a temporary local ordinance. This obviously requires analysis of possible overcrowding effects on the rest of the local road system. At the same time, it is prudent to restrict helicopter or light-aircraft flights so that they do not take place over large groups of people. At the Ramstein air show in Germany in 1988, 70 people died and 400 were injured when jet fighters engaged in acrobatic manoeuvres crashed into each other and fell in flames onto the crowd below.

Under less apocalyptic conditions, crowd regulation involves controlling movement and seating arrangements to avoid crushes and entrapment. Estimates of the number of people in a given area may need to be made in advance. Generally, people need an area of at least $0.25\,m^2$ per person to sit in (twice that if they are sitting on the ground rather than in seats); moving crowds need almost $2.5\,m^2$ per person to walk in at normal speed. At $1.0\,m^2$ per person, movement becomes restricted; at $0.5\,m^2$ it is reduced to shuffling; and at $0.2\,m^2$ the press of people becomes dangerous and potentially uncontrollable. It should be the responsibility of the organizers of the event to ensure that people are not allowed to enter any restricted area in too great a number for safety. No more than 25 people per minute should pass along narrow corridors or singly through entrances, and no more than 16 per minute should walk in single file up stairs. Standard escalators can carry no more than 100 people per minute.

Control can be achieved by having marshals limit and superintend access and by using barriers. The latter should be able to resist minor pressures, but neither collapse dangerously (if people push against them or climb on them), nor allow people to be crushed if crowd pressure develops. Barriers can be designed to break in the event of a potentially deadly crush and allow overflow into safe areas. In indoor environments, it goes without saying that exits should be clearly marked and always capable of being opened from inside, and doors or stairwells that appear to be exits but are not should be signposted as such.

Health hazards include the potential that overstressed equipment or temporary fixtures may collapse onto people. Checks should also be made for other hazards, such as the presence of snakes or rodents, and for certain public-health hazards. On this subject, hygiene can be assured only by having an adequate number of toilets, perhaps at least one per 100 people. The ratio of facilities should be something like three for females and to every two for males, or perhaps two to one (no fixed rule seems to exist). Other public-health issues involve the storage, preparation, sale and disposal of food and drink on site. Some of the rules outlined in §5.9 apply here.

In general, no more than 1.5 per cent of the public at rock festivals should be expected to require medical assistance. However, at a large event this would still necessitate a proper first-aid station (a simple example is given in Fig. 6.11). Access to crowded areas for medical-assistance teams needs to be assured by keeping pathways or aisles clear between sections of the crowd. Open-air events may require all-terrain emergency vehicles (perhaps converted four-wheel-drive trucks), especially if conditions become muddy enough for ambulances to become bogged down. Communications with at least one local hospital must constantly be maintained, and access to the site of the event assured for ambulances. Other issues of medical aid include signposting the first-aid station and ensuring that it has adequate capacity, staff and drug supplies.

Finally, there is scope for contingency planning that assumes the worst. At the most basic level, a site plan should be available for use in an emergency. If a grandstand collapses, or a crowd crush occurs, or if fire breaks out in a crowded auditorium, one needs to be sure that essential manpower and equipment can be brought in quickly and efficiently. For this, medical staff may need to be called up in large numbers, and contingents of police quickly drafted in.

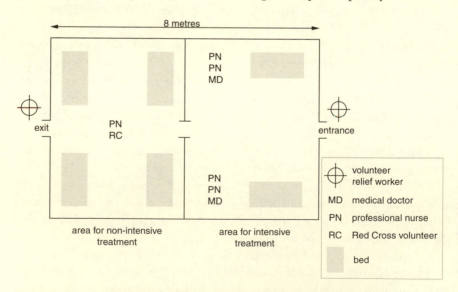

Figure 6.11 A suggested layout for a first-aid station (advance medical post) in a modular tent with vertical walls. Triage may be practised in the area around the entrance.

In short, as with planning for other types of disaster, a little pessimism can be healthy, and a modicum of scenario development will undoubtedly help prepare for any contingencies.

6.5 Emergency planning for industries

Although employment in manufacturing has tended to decline relatively in the Western world, there has been a corresponding increase in automation and in the sophistication of industrial processes. Concurrently, there has been a veritable chemical revolution, which has led to a huge increase in the number and range of liquid, solid and gaseous substances used in manufacturing processes. Many of these are unstable or toxic enough to cause serious injury (see Tables 6.9 and 6.10, and Fig. 6.12). In most countries these changes have been accompanied by a substantial development of norms and regulations designed to ensure industrial safety, although it has often been the case that the development of legislation has lagged behind the increase in risk. The general question of industrial safety is large and complex enough to have filled many books and hence will not be tackled here. Instead, the present account will be limited to some observations on how to tackle risks and hazards in factories and industrial plant in the specific context of disaster.

Broadly speaking, industrial risk breaks down into the following sectors:
- transportation and trans-shipment (of hazardous materials)
- storage or raw materials, components or products
- dangers associated with manufacturing processes
- explosions from the rupture of pressure vessels, the occurrence of exothermic reactions or the sudden generation of steam or gases
- toxic releases as a result of transportation crashes, errors during transshipment, damage to storage facilities, leakage from inadequate containment, human errors from valves left open, and explosions.

Toxic materials include acids, poisons, radio-isotopes and bacterial contaminants; release can occur in solid, liquid or gaseous form, with consequent risk to soil, water, atmosphere or surfaces. The impact of disaster can also be felt in terms of damage to inventory (raw materials, semi-finished components and stockpiled products) and to buildings, equipment and machinery. Sudden-impact disasters that affect dangerous industrial processes can lead to very serious risks of injury (and perhaps entrapment) among workers.

Factories and industrial plants may be required by law to develop emergency-response plans, which are bound to be a necessary precaution. To some extent, these may be microcosms of general risk situations. The analysis of

241

Table 6.9 Some chemical and radionuclides that have given rise to large-scale contamination events.

---

acrolein
ammonia
cesium [137]Cs
chlorine
dioxin (2,3,7,8-tetrachlorodibenzo-*p*-dioxin)
ethylene oxide
hydrogen cyanide gas
methyl isocyanate gas (MIC)
nitric acid
pentaerythritol
perichlorinated ammonia
phosphorus oxychloride
phosphorus trichloride
plutonium [239]Pu
polychlorinated biphenyls
strontium [90]St
sulphuric acid
TNT (2,4,6 trinitrotoluene)
uranium [235/238]Ur
vinyl chloride
white phosphorus

cardiotoxins – attack the cardiovascular system
dermatoxins – attack the skin
neurotoxins – attack the nervous system
genotoxins – alter the genes

Specific antidotes

Aniline	Methylene blue
Arsenic	Penicillamine
Cyanide	Sodium thiosulphate
Hydrogen fluoride	Calcium gluconate
Mercury	Penicillamine
Organophosphates	Atropine
Toxin	Antidote

Source: Table 17.3 on p.362 of "Industrial disasters", S. R. Lillibridge, in *The public-health consequences of disasters.* E. K. Noji (ed.), 354–72 (Oxford: Oxford University Press, 1997).

Table 6.10 Examples of the medical consequences of chemical accidents.

Category	Example	Agent
Carcinogenic	Primary liver cancer	Polychlorinated biphenyls
Dermatological	Sicca syndrome	Toxic oils
Hepatic	Porphyria cutanea tarda	Hexachlorobenzene
Immunological	Abnormal lymphocyte function	Polybromated byphenyls
Neurological	Distal motor neuropathy	Tri-o-cresyl phosphate
Pulmonary	Parenchymal damage	Methyl isocyanate
Teratogenic	Cerebral palsy syndrome	Organic mercury

Source: "Review of major chemical incidents and their medical management", P.J. Baxter, in *Major chemical disasters: medical aspects of management*, V. Murray (ed.), 7–20 (London: Royal Society of Medicine, 1990).

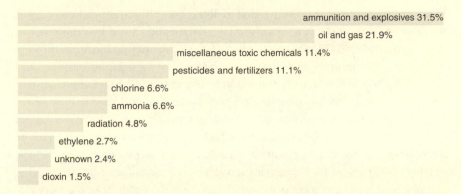

ammunition and explosives 31.5%

oil and gas 21.9%

miscellaneous toxic chemicals 11.4%

pesticides and fertilizers 11.1%

chlorine 6.6%

ammonia 6.6%

radiation 4.8%

ethylene 2.7%

unknown 2.4%

dioxin 1.5%

Figure 6.12 Known sources of hazardous material incidents worldwide, 1900–1990.

vulnerability at the plant level will take several forms. Assessments must be made of the susceptibility of buildings, equipment and storage facilities to damage, the effects of spontaneous damage on functioning equipment and its operators, the immediate safety of workers, the potential for toxic releases, and the behaviour of materials (especially volatile and toxic chemicals) used in the plant under all circumstances, usual and exceptional. Can measures be taken to strengthen storage facilities such that items are not broken during the impact of a disaster? Can potential toxic releases or explosions be prevented or contained? Can potentially disastrous pollution be anticipated and arrested before it contaminates the soil, air or water of the surrounding area? Are factory buildings and fixtures anti-seismic, if the risk is of an earthquake, or **flood-proof**, if waters may rise?

Alarm systems and evacuation procedures will probably be needed. Response plans must be drawn up, displayed prominently in the plant, and carefully taught to employees. Although the process can be expensive, it is essential to make emergency procedures – and indeed ordinary industrial processes – the subject of **failsafe design**. This means that if they go wrong the consequences are not catastrophic: damage will be arrested by other mechanisms, back-up procedures will activate themselves, or dangerous processes will automatically shut down. Anticipating the risks involves "thinking the unthinkable" and examining eventualities that may seem to be far fetched in the absence of disaster. In fact, some of the worst catastrophes have resulted from highly improbable combinations of circumstances, but the fact that they were improbable does not mean that they could not have been foreseen.

One aspect of continuous industrial processes is how to maintain them or successfully shut them down when disaster strikes. Attention needs to be given to the maintenance of power supplies by using alternative sources, such as auxiliary generators that automatically start up when the main power supply fails.

Monitoring equipment may be needed to detect exceptional conditions, such as the start of an exothermic reaction, and automatically shut down the process. Arrangements in advance may be necessary to pump out flooded areas rapidly.

Emergency monitors and disaster managers will be required on site and these will need to be properly trained in both industrial-hazard management and general civil-protection methods. Special attention will need to be given to firefighting capabilities on site, as methods of tackling fires will vary with risk of explosion or toxic volatilization (in which case breathing apparatus will be necessary). Industrial plant is often complex and dangerous enough that the risk of entrapment of firefighters is high. This can be combated by site-specific training and a system in which firefighters monitor each other for personal safety (known in the USA as the buddy system).

Registers of substances and materials kept at the plant will have to be compiled and kept in a safe but constantly accessible place. Table 6.11 lists the international classification of hazardous substances and Table 6.12 gives an example data sheet for a hazardous material, gasoline (petrol). These should give details of the toxicity and release-potential of substances, including those that are inert until they are volatilized or hydrated and thus made mobile. Plans should be drawn up to deal with the risks associated with each material. These will also involve informing local authorities and possibly calling in specialized hazardous-materials response teams (HAZMAT squads) from outside the plant or area. In this context it is essential to supply information about the risks that is as accurate and detailed as is necessary in order to deal with them. This can be done only when records have been kept satisfactorily.

At the same time, assessments need to be made of the possible effect upon the surrounding area of emergencies that happen at the plant. These should include both impacts generated by faults or catastrophes that occur on site and the possible effect of more general disasters (e.g. regional floods or earthquakes) on the factory. Will local populations need to be evacuated? If so, there should be a plan to get them out of the area, in accordance with the principles outlined in §5.2. Are aquifers or rivers likely to be contaminated by the release of toxic substances? Could there be a toxic cloud or radiation plume? Knowledge of prevailing winds is important to evacuation planning here. Might workers become trapped in dangerous conditions on site? Such questions can be answered by developing hazard scenarios, training workers and emergency responders, and creating an emergency-response plan.

Commonly, industrial risk analysis involves the use of **event trees** and **fault trees**. Many industrial processes can give rise to small errors or failures that in their own right are fairly innocuous. But if they combine, or one small failure leads to others, the result can be a major disaster. For instance, a minor gas leak may cause an explosion that sets off other leaks and explosions, or chance

Table 6.11 International hazardous materials classification (United Nations).

20 inert gas

    22 refrigerated gas
        223 inflammable refrigerated gas
        225 comburent (combustive) refrigerated gas
        228 corrosive refrigerated gas

    23 inflammable gas
        236 toxic inflammable gas
        239 inflammable gas that may spontaneously cause a violent chemical reaction

    25 combustible gas

    26 toxic gas
        265 toxic combustible gas
        266 very toxic gas
        268 corrosive toxic gas
        286 toxic corrosive gas

30 inflammable liquid (catches fire at 21–100°C)

    33 very inflammable liquid (catches fire at <21°C)
      X333 spontaneously inflammable liquid that reacts dangerously with water
        336 very inflammable and toxic liquid
        338 very inflammable and corrosive liquid
      X338 very inflammable and corrosive liquid that reacts dangerously with water
        339 very inflammable liquid that may spontaneously cause a violent chemical reaction

    39 inflammable liquid that may spontaneously cause a violent chemical reaction

40 inflammable solid

    41 inflammable solid
      X423 inflammable solid that reacts dangerously with water producing an inflammable gas

    44 inflammable solid that becomes liquid at high temperatures
        446 inflammable solid that remains solid at high temperatures

    46 toxic inflammable solid

50 combustible material
        539 inflammable organic peroxide
        558 highly corrosive combustible material
        559 combustible material that may spontaneously cause a violent chemical reaction
        589 corrosive combustible material that may spontaneously cause a violent chemical reaction

60 toxic material

    63 inflammable toxic material (catches fire at 21–55EC)
        638 corrosive inflammable toxic material

    66 very toxic material
        663 very toxic and inflammable material

    68 corrosive toxic material

    69 toxic material that may spontaneously cause a violent chemical reaction

80 corrosive material

X80 corrosive material that reacts dangerously with water

    83 inflammable corrosive material
        839 inflammable corrosive material that may spontaneously cause a violent chemical reaction

    85 combustible corrosive material
        856 toxic combustible corrosive material

    86 toxic corrosive material

    88 very corrosive material

X88 very corrosive material that reacts dangerously with water
        883 very corrosive and inflammable material
        885 very corrosive and combustible material
        886 very corrosive and toxic material
      X886 very corrosive and toxic material that reacts dangerously with water

    89 corrosive material that may spontaneously cause a violent chemical reaction

Table 6.12   Example data on a hazardous substance.

GASOLINE (PETROL)

UN Code[1] 1203
CAS[2] nos 68606-11-1, 93572-29-3, 94114-03-1
EU nos 649-270-00-7, 649-312-00-4, 649-389-00-4

Physico-chemical properties
Red-coloured liquid; insoluble in water; vapour is heavier than air; odour usually comes from colourings and anti-knock additives.

Root of formula .... from C_5H_{12} to C_9H_{20}
Vapour density 1.4
Density of liquid relative to water 0.8
Boiling point .... 40–200°C
Flammability point .... 40°C
Flash point .... 280°C (60 octane) – 450°C (100 octane)
Lower and upper explosive limits .... LEL 1.4%; UEL 7.6%

Transportation requirements
Orange rectangle .... 33 (highly inflammable liquid) – 1203
Diamond.. danger label: 3 flame
Labelling   **F+** *flame,* **T** *skull*

Risk notation
R12: highly inflammable **R45:** carcinogenic, category 2 **R65:** can damage lungs if ingested
Safety recommendations
S45: in case of incident or injury consult a physician (if possible show physician the labelling of the agent that caused the injury)
S53: avoid exposure to this substance; read instructions before use

Nature of danger
Danger of fire or explosion: vapours form inflammable mixture in air at ambient temperatures; liquid will float on water during hosing operations, on rivers or in drains; water may be incapable of putting out fires if large quantities of gasoline have been spilt; when it catches fire, gasoline containing lead will form toxic smoke that contains lead oxides. No particular problems of reactivity.

Toxicity in humans
Toxic upon inhalation or ingestion. Irritant to eyes and respiratory tracts; inhalation of vapour may cause damage to blood, bone marrow, central nervous system, skin, eyes, or respiratory system; symptoms of exposure include sickness, tiredness, depression, vertigo, narcosis, nausea, vomiting and dermatitis.
Carcinogenesis .... Category 2 (IARC[3] class 1; ACGIH[4] category A3)
Daily limit value–time weighted average (TLV-TWA).............. ACGIH 300 ppm (890 mg/m;)
Short-term exposure limit (STEL) .... 500 ppm (1480 mg/m;)
Immediate danger to life and health (IDLH) .... 3000 ppm
Lead in petrol can be highly toxic to the environment.

First aid measures
Inhalation .... evacuate the injured person from the contaminated area
Contact with skin .... immediately remove contaminated clothes and
.... wash skin with much water
Contact with eyes... wash with abundant water for 10–15 minutes
Ingestion .... if patient is still conscious, feed him or her salt water or
.... soapy water to induce vomiting

1. United Nations Committee of Experts on the Transport of Dangerous Goods.
2. Chemical Abstracts Service of the American Chemical Society.
3. International Agency for Research on Cancer.
4. American Conference of Governmental Industrial Hygienists (ACGIH).

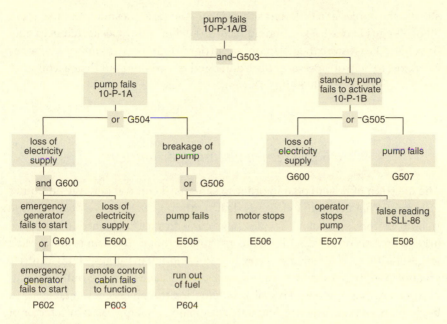

Figure 6.13 Example of a fault tree for a hazardous industrial process. Codes (e.g. G503) refer to a manual emergency operations.

contact between volatile substances that are normally isolated from one another can lead to a runaway exothermic reaction. Event trees follow a simple flowchart logic in which A leads to B or C, B to D or E and so on, so that one can trace the possible progress of incidents towards the so-called **top event**, the maximum expected disaster. In an analogous way, fault trees (Fig. 6.13) use flowchart methods to connect up the consequences of different failures. For example, if a coolant pump fails, a back-up pump should start up. But what are the consequences if the back-up pump fails, perhaps through lack of diesel fuel or as a result of mechanical breakdown? Event trees and fault trees are invaluable aids to scenario planning with respect to industrial processes (and note that the methodology can be adapted to many other forms of disaster where small problems can be magnified by coincidence). However, the one factor that is perhaps most difficult to incorporate into them is human error. Although redundant safety mechanisms, failsafe devices and automatic checks and barriers can be constructed, the sheer irrationality of human mistakes, negligence or wilfulness defy precise analysis.

In factories, archives and records also need to be protected against disaster. This problem is dealt with more generally in §6.7; they should be kept, as far as possible, in protected places (e.g. not liable to flooding or storm damage). There should be a procedure for backing up copies of all documentation and storing the copies at a different site. It is, of course, a question of not keeping all one's eggs in one basket. Should records be damaged, specialized help may be required to restore them. Computer programs and consultancy companies now exist to help businesses restore damaged records and put their accounting procedures back together after disaster.

6.6 Emergency planning for tourism

The question of how to plan in order to safeguard cultural heritage and other attractions that generate tourism is dealt with in §6.8. Here we deal with the risks to visitors, some of whom may have come to a particular area to enjoy the art and architectural treasures that constitute heritage. Tourism is the world's largest industry. Tourist areas – coasts, lakes, mountain ranges, skiing centres, areas of exceptional scenery, parks, historic townscapes, famous buildings and amusement parks – pose special problems for the emergency manager. To begin with, populations will fluctuate seasonally, in many cases by a factor of well over 100 per cent. In very short periods of time an area can go from a state of undercapacity, with unused facilities, to one of overcrowding and congestion (or vice versa). Moreover, occasional visitors are unlikely to have much knowledge of local hazards and to know how to respond to them. Hazard and risk scenarios vary with place and season. In the summer, valleys with campsites may be flashfloodable. In the winter, Alpine ski slopes may suffer avalanches and rockfalls. In the spring, snowmelt may cause debris flows and slush avalanches to run down onto roads and houses. Tsunamis and storm surges may threaten tourist coasts, and earthquakes may put hotels at risk. Few tourists wish to know about such problems and hence the burden of protection falls on local authorities. This can be a very vexing problem. For example, it is not uncommon in Alpine areas for the maximum concentration of visitors and winter-sports enthusiasts to coincide with the peak avalanche risk.

Once again, the first step is to create scenarios for hazards, vulnerabilities and risks. As far as possible, this means assessing the location and timing (or at least seasonality, where this is a factor) of possible hazard impacts. Where on a coast will tsunamis or storm surges produce the deepest flooding or most destructive waves? When is the hurricane season? Where on a mountain are the avalanche tracks or debris-flow paths, bearing in mind that such phenomena tend to be repeated or remobilized? What does the pattern of isoseismals in past earthquakes suggest about future seismic hazard? Are there route corridors in the area along which hazardous materials are transported, or nearby locations where they are stored or processed?

The situation is likely to be more fluid for vulnerability and risk. A compilation is needed of the locations and capacities of tourist accommodation: hotels, pensions, bed and breakfast establishments, inns, motels, residences, campsites, chalets, and so on. Where tourist presences are registered in such establishments, one can gain an idea of the seasonal shift in numbers, bearing in mind that the tourist presence is subject to fluctuation with economic conditions, weather patterns, scares and whims. If more detailed information is needed, it will be necessary to convene a group of helpers and spend a tourist season conducting censuses of the number of people who visit specific places and attractions, and the number of vehicles on roads at specific times. The overall objective is to build up a profile of aggregate preferences and habits among tourists and to relate these to hazards, so as to understand the shifting pattern of risks.

In addition to general tourism – mere presence in the area – one needs to bear in mind that there will be more specialized kinds of both visitor and visit. Day trippers may arrive by coach, car or boat. They may visit the sites, have lunch and depart, such that they are not noted in hotel records, however numerous they may be. Sanatoria may be frequented by convalescent people, shrines and sanctuaries by pilgrims, convention centres by conference goers, and even laboratories by visiting scientists. All need to be factored into the totals of transient visitors. Some may have special needs or be handicapped in some way; all can be regarded as a category distinct from permanent residents of the area.

The last step in constructing the risk profile will be to consider the hazard to facilities. Are ski-runs and hotels located on potential avalanche tracks? Are well frequented beaches at risk of tsunamis? Are hotels and lodgings (seismically very vulnerable structures in some places) likely to collapse in earthquakes? Every effort should be made to ensure that structural mitigation is carried out, although it should, of course, be compatible with restrictions on environmental impact. Where structural measures are not likely to resolve the problem, or where they are not likely to be applied, the emergency planner should design non-structural procedures. Surveillance of the hazard, alarm and alert systems, and evacuation processes, need to be put in place. Scientific forecasting organizations, such as seismological laboratories and meteorological services, need to be made aware that there is a specific problem of safeguarding transient visitors. Hazard forecasts must reflect this need, for example, by intensifying surveillance and shortening the time between information bulletins at times of maximum tourist presence.

To ensure that tourists respond adequately to emergencies, some attempt will have to be made to explain to them the risks and the procedures for hazard avoidance. Publicity must be simple and effective. In order to make the message acceptable, it is wise to couch it in terms of having a safe vacation rather than as a direct appeal to tourists to interest themselves in hazards. Material can be distributed in hotel rooms and through tourist information offices. Signs can be erected where they are most likely to be read. For example, the mouth of a valley frequented by hikers and campers may bear signs that read "In case of flashflood, climb valley side slopes". It is opportune to observe tourist behaviour and design measures to correct that which is inappropriate. This may involve active policing, for example to ensure that curious people do not return to the shore when the beach has been evacuated after a tsunami warning. In this context, alarms such as sirens must have a clearly understood function that engenders the right response from transient visitors. Where specific areas are subject to particular dangers, restrictions on access may be necessary. These will require extra enforcement: visitors who have little or no experience of particular hazards sometimes tend to lack the motivation to believe warnings and interdictions. In this respect, special difficulties will be experienced with the threat of low-probability events of high consequence, such as volcanic eruptions that occur after long periods of quiescence. The visible evidence of past impacts is not clear enough to remind people that the risk is genuine.

Leisure time has increased enormously in recent decades in the Western world. So have opportunities for travel. This has fuelled the rapid, sometimes explosive, economic growth of tourist areas. Hazards may have remained the same in such areas, but vulnerability and risk have tended to increase greatly. Roads and buildings may have been constructed with little regard to hazards. The transient population may have increased enormously; even the resident population, which may have grown substantially, may have come to include many people who have not been there long enough to have a clear idea of the risks, especially if hazards occur infrequently. This means that emergency plans for such areas must be adapted continually to changing circumstances. The process of adaptation refers to all stages of the emergency-planning procedure, from the construction of hazard scenarios to the arrangements for post-disaster relief. There is a significant risk that resources for emergency planning and management will, as a result of the continuous growth of vulnerability, constantly lag behind the needs associated with risk.

Plans for emergency management in tourist areas obviously require the participation of representatives of the industry. In seminars, advisory sessions and training courses, such people need to be encouraged to participate in the planning process. Consultation and feedback to the emergency planners are essential, not merely because the experience of tourist operators can reveal unexpected risks and solutions to them, but also because the industry will, to some extent, determine the limits of what can be achieved by emergency planners. Measures that are likely to reduce profits or drive tourists away are sure to be unpopular and will be resisted by the operators. On the other hand, a tragedy with many tourists as casualties can seriously damage the image of an area. This is a good argument to use when one has to convince sceptical representatives of the industry to participate in the emergency-planning process and the eventual plan.

6.7 Planning for libraries and archives

In terms of emergency planning, the vulnerability of libraries and archives is in a special category. They represent accumulations of information, the fruit of years of work and, often, generations of concentrated expertise; and they may be both precious and difficult or impossible to replace. The greatest risk is posed by large natural hazards, such as floods, earthquakes and hurricanes, although it is not unknown for libraries to be subjected to terrorist attacks, usually as an attempt to harm a community's cultural identity. Structural collapse will obviously damage the contents of such institutions, although not necessarily irreparably. Flooding may do more harm, specifically to basements and lower floors. In violent storms, the loss of a roof, or even a few windows, can allow wind and rain to enter the building and thus do incalculable harm to its contents. Earthquakes can shake the contents of shelves onto floors and damage them. Merely recomposing sequences of books, let alone repairing them, can be an arduous task in such circumstances.

Besides the building itself (which in some cases is an historical monument in its own right), the items at risk fall into four main categories:

- materials such as books, periodicals, documents, maps, photographs, plans, designs, prints, microforms and specimens
- rare, precious and fragile materials, such as antique books and documents
- equipment such as computers and storage devices
- records, such as catalogues.

All such materials should be arranged in such a way as to minimize the risks. For example, if 20 per cent of the building is floodable, this area should be devoted to low-vulnerability uses (e.g. a reading room or reception point) and the highest-value or least-replaceable materials should be kept in the safest place. This obviously requires a detailed assessment of hazards to the building. Moreover, where possible, there should be back-up copies of records and documents. Computer disks and tapes, photographs, files and microform records should be kept as duplicates at another site and in a safe place.

In an emergency, there needs to be a plan to put immediate safeguards into place. If flooding is a problem, there should be stocks of sandbags and squads of workers to deploy them around the exit points. If there is enough lead time,

251

and if planning is adequate, it may be possible physically to remove library materials. This requires both stockpiling of containers and meticulous arrangements to fill, remove, transport and temporarily house them. It needs to be done in such a way as to minimize damage to the materials and, where possible, to conserve the structuring given by classification systems. When time is limited, materials should be evacuated in order of importance, value or irreplaceability.

The library or archive director who assumes the worst may wish to consider planning for the restoration and conservation of damaged materials. After floods or storms, books may need to be dried out, which is a difficult process that requires drying each page if the book is to be saved from excessive warping and from the growth of mould. It requires huge stocks of rice paper and climatically controlled indoor conditions.

Special attention needs to be given to fire control. Sprinkler systems and other indiscriminate use of water could easily do more harm than good. However, reading materials tend to be highly flammable and easily damaged by flames, heat or smoke. The answer is to be able to contain fires and stop flames and smoke from spreading. This is a technical problem for architects and fire-service advisers. In this context, smoke detectors are important early-warning devices. Fire response needs to be planned carefully and tested by exercises. Fire-damaged books and documents may be rendered brittle by heat and charring. Very little can be done to repair such damage, but important material can sometimes be microfilmed or photographed before it disintegrates (digital cameras may help expedite this process). Smoke and fire are, of course, inimical to computerized archive systems; their damage can sometimes be reduced by maintaining positive air pressure in computer rooms. This is often done anyway to reduce the presence of dust in the atmosphere.

Recovery from disaster in a library or archive requires a team leader who will assign tasks to a series of work crews and coordinate the work and flow of materials. The first operation is to salvage goods from the library or archive and remove them to a safe place – dry, fireproof and capacious. Next, an inventory of damaged materials should be compiled. Where possible, simple means should be used to limit further damage, for example by hanging wet books or sheets of paper on lines of string (assuming that this would not damage them further). Equipment and supplies for salvage and repair work must be assembled and crews trained to use them. As work progresses it is important to keep a photographic and perhaps written record of salvage operations. This may have a bearing on subsequent conservation work with respect to specific items.[*]

---

[*] For further details the reader is referred to the excellent manuals on disaster planning and recovery for libraries and archives by Buchanan (1988), Fortson (1992) and George (1994).

6.8 Protecting fine art and architecture

Cultural heritage – consisting of works of art, architecture, landscapes, archae-ological sites, and so on – forms a category that may require special protection against disasters, because the works or sites are fragile, complex and unusually vulnerable to unexpected impacts. Emergency protection may be expensive and require considerable resources, but it is almost invariably cheaper than the complex work of restoration of artefacts and sites damaged by disaster. More-over, unique heritage may be irretrievably lost if it is not protected properly.

6.8.1 Works of art

Many of the planning provisions outlined in the previous section also apply to the repositories of works of art, such as galleries, museums and storage ware-houses. Even more than books, works of art may be highly sensitive to extremes of heat or humidity and thus require controlled environments. Structural col-lapse can damage them seriously or irreparably; heat and flames may ruin or consume them, and even stone can be damaged by fire. Earthquakes are par-ticularly threatening, as they can overturn furniture, shake paintings from walls, or throw artefacts off shelves so that they smash on the floor beneath. Hence, in seismic zones, works of art need to be secured, whether they are on show or not. In all cases, measures also need to be taken to safeguard and duplicate catalogues and inventories of works of art.

Although social hazards, such as acts of vandalism or terrorism, may be quite unpredictable, natural hazards may suggest fairly clear scenarios, through which the disaster planner can learn from the sad history of damage to works of art in the past. It is often necessary to exhibit art treasures in par-ticular settings, and these too may be vulnerable, for example, to flooding. As usual, the situation needs to be surveyed and a census made of works at risk. Where opportune, the general municipal or regional disaster plan should include a section on the procedures for saving particular works of art at risk. Squads of accredited and clearly identified rescuers should be able to pass quickly through security systems in order to gain access to works at risk. This may mean making special arrangements to open galleries and museums when disaster is about to strike, which means involving curators, superintendents, security guards and other museum workers in the counter-disaster effort. Procedures for entering secure environments and removing vulnerable works should be carefully worked out in advance. This may involve means of unlock-ing doors and cabinets, disabling alarms, detaching works of art from walls or other settings, crating them and moving them – all without compromising security or creating a risk that "rescuers" are thieves in disguise. The problem may be acute and require many helpers and considerable ingenuity when a work of art is very large and heavy. In all such cases, the evacuation procedure cannot be improvised but must be planned meticulously in advance. With fragile works it probably cannot be tested until disaster is imminent.

In summary, the problem of protecting works of art can be divided into two parts: strategies that can be followed before disaster strikes, when there is enough lead time to be able to protect the works, and post-disaster recovery work. Regarding the former, efforts need to prioritized on the basis of the importance of works of art and the vulnerability of buildings that contain them (which partially governs the magnitude of the threat to the works themselves). Emergency work should be carried out on the basis of established priorities and a register of art treasures at risk. Work should not be undertaken if warning time is inadequate. A basic choice must be made *a priori* between protecting works *in situ* and removing them to safe places.

In situ protection methods, such as sandbagging and the use of protective coverings, require stockpiling of materials, study and rehearsal of the means of deployment and assembly, and organization of manpower. The translocation of works requires attention to questions of access to cabinets, the means of dismantling works or removing them from walls, the use of lifting gear and carts where necessary, stockpiling and use of boxes or other protective containers, routes to be taken with each work and means of transporting it. In both cases, the question of access and how to get past security systems needs to be tackled, as does the location of places of temporary safe storage, with associated questions of custody, security, climatic control and other protection requirements. It is helpful to draw up a computer record of each work of art, or group of works, giving details of the protection strategy that has been designed for it, and how this is to be put into effect.

The disaster planner also needs to evaluate the probable extent of post-disaster work on the recovery and restoration of damaged art treasures. Granted that most aspects of this will be imponderable before the worst has happened, nevertheless, restorers, curators, site superintendents, custodians and other workers need to be alerted to what they may have to accomplish in a future emergency. Open discussion of the scenarios for damage to works of art may lead to better efforts to conserve them and mitigate the risks.

6.8.2 Heritage sites

The protection of works of art is a subset of the wider question of how to protect heritage sites in general from the effects of disaster. This very important topic is beyond the scope of the present work, but it is worth devoting some space to the protection of architectural works and archaeological sites. The first task is to list them and subdivide them by category, age or function. Works of ancient, historic or modern architecture will be vulnerable to some of the following: structural damage or collapse; water scour, wind damage, fire damage, rising damp, accelerated weathering, cracking or superficial damage; and loss of details, ornaments, mouldings or statuary. Buildings, sites and monuments should be subjected to a pre-disaster vulnerability survey. This should identify not only particular susceptibilities to loss but also signs of particular weakness and decay. On the basis of the survey, mitigation measures can be suggested, such as fire-suppression systems, physical barriers and structural bracing.

Whether or not pre-disaster mitigation is carried out, damage in disasters is probably inevitable, at least in the more severe impacts. However, no historic building need be demolished for this reason alone, although it may be expensive to reconstruct. The disaster planner should either arrange to stockpile or to locate sources of wooden baulks and steel scaffolding to buttress damaged buildings. One or more architect, surveyor or structural engineer should be assigned the task of supervising the work, and a construction company or municipal workers should be ready to carry it out. In an ideal case, reconstruction plans would be at least partially drawn up for each listed heritage building before it is damaged by disaster. This is not inconceivable, in that many forms of damage are predictable. Although there is likely to be little support for efforts to devote time and money to such an exercise, it may be worth trying to estimate the probable future need for reconstruction work at each site, as this will help to determine the magnitude of vulnerability.

Lastly, where the area of the disaster plan contains major works of art or architecture, or major archaeological or historical sites, attempts can be made to involve national and international bodies in disaster mitigation. For example, the International Council on Monuments and Sites (ICOMOS) is actively engaged in promoting the protection of cultural heritage against disasters of all kinds. It has both national and international committees. The United Nations Educational, Scientific and Cultural Organisation (UNESCO), based in Paris, is also heavily involved in measures to protect heritage sites.

6.9 A plan for the mass media

Information is one of the most vital commodities in disaster and it tends to be in short supply precisely when demand is greatest. The mass media – radio, television, newspapers, magazines and Internet-based newsfeeds – are one of

255

the principal links between the disaster–response community and the general public. It is therefore imperative that the emergency plan make full provision for involving the media, so that the public are properly informed.

For survivors, the public and even, to a certain extent, the suppliers of national and international aid, television, radio and newspapers will usually be important sources of information about disasters. Public opinion both influences journalism and is heavily influenced by it. The news media's interest, or lack of interest, in a disaster can have the effect of turning relief on or off like a tap, according to the public's motivation to contribute to appeals for donations. In this respect, academic researchers who have studied the role of the media in disasters are divided: the pessimists regard journalists as foes, always ready to distort information for the sake of a good story; the optimists argue that the media are willing to collaborate responsibly with the disaster-relief community to improve the quality of broadcast or published information.

To know how to integrate the mass media into a disaster plan, it is first necessary to understand their role in disasters. Although the present work is not intended to be a treatise on such topics, a short description is warranted. News about disaster comes from four sources:

- reporters sent into the field
- journalists based at the nearest centre of government
- press-agency syndicated reports
- official communiqués.

The news itself is a perishable commodity: values change daily, or even by the hour. It constitutes a "frame of reference" that transforms events into publicly discussable phenomena, simultaneously a recorder and a product of social realities. In disasters, news tends to mirror the event far more than it elucidates the underlying causes. Questions of newsworthiness determine the importance to a journalist of a story of or particular elements of it, and these tend to be based on criteria of impact, social context, human, political or social interest, novelty and familiarity.

Some very serious problems have been experienced with news-media coverage of disasters. Political or ethnocentric bias may be apparent in news stories. Myths and rumours may be propagated, especially where journalists have little familiarity with the subject of disasters or the places where they occur. Media coverage of catastrophes tends to be unsystematic and unsustained, with scarce attention to the accuracy of details, and poor interpretation of the drift of events. Despite this, massive publicity given by the media to disasters can increase the "convergence reaction" and bring huge numbers of people unnecessarily into the affected area: sightseers, hordes of reporters, dozens of television crews, well intentioned but unorganized volunteers, and so on. Transportation, accommodation, communication and information systems can be overloaded by heavy media usage at a time when they are needed for search and rescue. Moreover, the demands of the media for information may induce authorities to make unwise statements, especially shortly after the impact, when the situation is far from clear. Finally, negative reports in the media can affect the credibility and authority of disaster managers. The solution is to design and implement a specific plan to collaborate with the news media and involve them in mitigation and relief efforts. This may even lead to courses on civil protection for journalists, or at least setting up a pool of accredited representatives of the news media.

Generally, the media want to convey accurate information to the public. Journalism is a competitive business in which rival channels, stations and newspapers must be monitored constantly and reputations must be protected. Reporters will collaborate willingly with disaster planners and emergency managers who adopt positive, honest and welcoming attitudes to them. The reporters can be induced to visualize their roles in mitigation and relief work.

Collaboration between the media and the authorities can help stop rumours, dispel myths, avoid confusion, inform and educate the public, and convey official information efficiently to general recipients. Honourably treated, the news media can be induced to convey information on the seriousness, duration, scope and effects of disaster, and on the progress of rescue, relief and recovery efforts. They can be persuaded to promote mitigation and public education efforts. However, needs differ between media: radio needs information quickly; television requires a series of eye-catching immediate visual images; and newspapers need more detailed information and graphics. All news-media organizations work to strict deadlines and depend on obtaining regular or constant supplies of information. They work through both formal and informal networks, which must be properly understood if the media are to be managed successfully. One also has to recognize the degree of stress and pressure under which reporters habitually work. There is often little time to verify the accuracy of stories as they break.

The disaster plan should designate an emergency **public-information officer** (PIO) to liaise with the media. This person must have skills and experience in writing, public speaking, the use of computers and audiovisual equipment, public relations and how the news media work. The PIO must develop strong links with news-media representatives, especially editors and reporters, and must show no favouritism. The PIO's role is not limited to the post-disaster emergency phase, but also involves promoting mitigation and planning initiatives, and publicizing the work that civil-protection authorities do in times of quiescence. The overall aim is to protect people from disasters by raising their general level of consciousness of the problem and supplying them with the information that they will need to order to reduce their own risks.

Besides informal contacts, the principal ways of conveying official information about disasters to the news media are as follows:

- news releases and situation reports
- fact sheets and summaries of background information
- biographical sketches and photographs of the main people involved
- face-to-face and telephone interviews
- audio or video clips
- press conferences
- briefing sessions
- training courses
- feature stories
- brochures
- public service announcements.

It is essential that the PIO has a good understanding of the media's needs and, vice versa, that the media comprehend the PIO's needs. During an emergency the PIO must use all possible means of conveying information to alert and warn the public, supply accurate data, and monitor news-media reporting. Accurate and generally intelligible information should be supplied in order to quell rumours and misassumptions, and to stimulate appropriate public responses. Monitoring of reports in the media (perhaps by compiling a daily round-up of press cuttings and video clips) will enable PIOs to check the impact of news that they have given out the day before and adjust future news releases in order to optimize public response to them. To accomplish all of this, the PIO must know who the media are, where they are located, what audiences they reach, how frequently they are published or broadcast, who is in charge of each news organization, and how to reach editors and reporters at any time. He or she must also have an idea of what news outlets the majority of the public habitually monitor, and when they are most likely to do so. A list of appropriate types of information to supply to media representatives during emergencies is given in Table 6.13. Table 6.14 is a list of misassumptions about disaster that are commonly perpetuated by news reports.

The public-information component of the emergency plan should include some or all of the following:

- an explanation of who is responsible for generating and who for distributing official information about disasters in the local area
- how the release of public information and access to it will be organized
- guidelines and checklists for dealing with the media
- forms and logs for record keeping
- pre-scripted pro forma for media releases and emergency public-information messages.

Appendices should give lists of local, regional and national media, editors

Table 6.13   Information typically required by the news media during emergencies.

Initial requirements
- a general explanation of what has happened
- a status report on the latest developments
- access to the main people in charge of emergency operations and centrally involved in the story
- the chance to take still pictures or video sequences of the scene of the disaster (e.g. of the main places where buildings have collapsed or people have been killed)

Continuing requirements

Casualty figures
- the number of dead and where bodies are being taken
- whether anyone of prominence is among the dead and injured
- the number of injured people, the nature of injuries and where they are being treated
- the stories of people who escaped serious injury or death

Property damage
- what buildings were lost, and at what cost
- how important the damage is to the community
- whether such damage ever occurred in the past, and if so, when

The event
- what it was that caused the disaster
- interviews with eye witnesses

Emergency response to the disaster
- who gave the alarm
- information on predictions, warnings, premonitory signs, and so on
- who was first to respond, and what they saw when they arrived
- how many persons, pieces of equipment and departments responded to the disaster
- how emergency managers are handling the situation and how well they are coping
- whether any acts of heroism have occurred
- what is being done to protect the community against further hazards and risks

---

Source: Partly derived from "Effective media relations", R. E. Churchill, in *The public-health consequences of disasters*, E. K. Noji (ed.), 122–32 (Oxford: Oxford University Press, 1997).

and journalists, whose names, addresses and contact numbers should be kept constantly up to date. The PIO should of course have both a good working knowledge of the disaster plan and a clear idea of how to explain its provisions to journalists and the public. He or she should also have the contacts and skills needed to connect media representatives with civil-protection personnel and scientific experts when more informed comment is needed.

The relationship between scientists and news reporters deserves special attention, as they can easily misunderstand each other. Scientists are trained to provide hard accurate information that can be very useful to journalists and ultimately to the public. However, scientific information is not always easily understood by non-specialists, few reporters have scientific training, and it may be difficult to distil science into easily understandable certainties. Controversy among scientists may be very healthy in academic and research circles, but it can be ruinous to the lay person who wants some incontrovertible truth on which to found beliefs and actions. Moreover, the media do not always distinguish adequately between bona fide scientists and charlatans, especially

Table 6.14 Typical myths and misassumptions about disaster, which are often perpetuated by the news media.

Myth	Disasters are truly exceptional events.
Reality	They are a normal part of daily life and in very many cases are repetitive events.
Myth	Disasters kill people without respect for social class or economic status.
Reality	The poor and marginalized are much more at risk of death than are rich people or the middle classes.
Myth	Earthquakes are responsible for very high death tolls.
Reality	Collapsing buildings are responsible for the majority of deaths in seismic disasters. It is not possible to stop earthquakes, but it *is* possible to construct anti-seismic buildings and to organize human activities so as to minimize the risk of death.
Myth	People can survive for many days when trapped under the rubble of a collapsed building.
Reality	The vast majority of people brought out alive from the rubble are saved within 24, or perhaps even 12, hours of the impact.
Myth	When disaster strikes, panic is a common reaction.
Reality	Most people behave rationally in disaster. Although panic is not to be ruled out entirely, it is of such limited importance that some leading disaster sociologists regard it as insignificant or unlikely.
Myth	People will flee in large numbers from a disaster area.
Reality	Usually there is a "convergence reaction" and the area fills up with people. Few of the survivors will leave and even obligatory evacuations will be short-lived.
Myth	After disaster has struck, survivors tend to be dazed and apathetic.
Reality	Survivors rapidly get to work on the clear-up. Activism is much more common than fatalism (this is the so-called "therapeutic community"). In the worst possible cases only 15–30% of victims show passive and dazed reactions.
Myth	Conditions in the disaster aftermath are really much worse than they look because the authorities are concealing the worst of the damage and casualties from the public.
Reality	This is almost never the case.
Myth	Looting is a common and serious problem after disasters.
Reality	The phenomenon of looting is rare and limited in scope. It mainly occurs when there are strong preconditions (i.e. a disaster is hardly necessary to start it off), as when a community is already deeply divided.
Myth	Disease epidemics are an almost inevitable result of the disruption and poor health caused by major disasters.
Reality	Generally, the level of epidemiological surveillance and healthcare in the disaster area is sufficient to stop any possible disease epidemic from occurring. However, the rate of diagnosis of diseases may increase as a result of improved healthcare.
Myth	Disasters cause a great deal of chaos and cannot possibly be managed systematically.
Reality	There are excellent theoretical models of how disasters function and how to manage them. After more than 75 years of research in the field, the general elements of disaster are extremely well known, and they tend to repeat themselves from one disaster to the next.
Myth	There is usually a shortage of resources when disaster occurs and this prevents their effective management.
Reality	The shortage, if it occurs, is almost always very temporary. There is more of a problem in deploying resources well and using them efficiently than in acquiring them. Often there is also a problem of coping with a super-abundance of certain types of resource.

Table 6.14 (continued)   Myths and misassumptions about disaster.

Myth   Any kind of aid and relief is useful after disaster, provided that it is supplied quickly enough.

Reality   Hasty and ill considered relief initiatives tend to create chaos. Only certain types of technical assistance, goods and services will be required. Not all useful resources that existed in the area before the disaster will be destroyed. Donation of unusable materials or manpower consumes resources of organization and accommodation that could more profitably be used to reduce the toll of the disaster.

Myth   Medical personnel with any kind of background are needed in disaster situations.

Reality   Routine and simple medical needs will probably be covered by local personnel and only medical workers who are trained in disaster work will be needed from outside the area.

Myth   In order to manage a disaster well it is necessary to accept all forms of aid that are offered.

Reality   It is much better to limit acceptance of donations to goods and services that are actually needed in the disaster area.

Myth   Unburied dead bodies constitute a health hazard.

Reality   Not even advanced decomposition causes a significant health hazard. Hasty burial demoralizes survivors and upsets arrangements for death certification, funeral rites, and, where needed, autopsy.

Myth   Disasters usually give rise to spontaneous manifestations of antisocial behaviour.

Reality   Generally, they are characterized by great social solidarity, generosity and self-sacrifice, perhaps even heroism.

Myth   One should donate used clothes to the victims of disasters.

Reality   This often leads to accumulations of huge quantities of useless garments that victims cannot or will not wear.

Myth   Great quantities and assortments of medicines should be sent to disaster areas.

Reality   The only medicines that are needed are those used to treat specific pathologies, have not reached their sell-by date, can be properly conserved in the disaster area, and can be properly identified in terms of their pharmacological constituents. Any other medicines are not only useless but potentially dangerous.

Myth   Companies, corporations, associations and governments are always very generous when invited to send aid and relief to disaster areas.

Reality   They may be, but, in the past, disaster areas have been used as dumping grounds for outdated medicines, obsolete equipment and unsaleable goods, all under the cloak of apparent generosity.

Myth   Technology will save the world from disaster.

Reality   The problem of disasters is largely a social one. We already have considerable technological resources, but they are poorly distributed and often ineffectively used. In addition, technology is a potential *source* of vulnerability as well as a means of reducing it.

Myth   Famine victims usually die of starvation.

Reality   They mainly die as a result of the nutrition-infection complex, i.e. malnutrition and famine related diseases such as typhus and cholera.

Myth   The situation will return to normal in a few weeks' time.

Reality   The social, economic and psychological effects of disaster, and often the damage caused, can persist for years, or even decades.

Source: Partly based on "The nature of disaster", E. K. Noji, in *The public-health consequences of disasters.* E. K. Noji (ed.), 17–18 (Oxford: Oxford University Press, 1997).

Box 6.3 The Eschede train crash

The incident

The Eschede train accident took place at Eschede, Germany, on a morning in June 1998. It involved the northbound Munich–Hamburg intercity express train 884, "Wilhelm Conrad Roentgen", of the Deutsche Bahn railway company. The train is an intercity express of 12 passenger coaches, weighing about 50 tonnes each, with electric locomotive units at each end. It is 410 m long, has a capacity of 759 passengers, and is capable of travelling at 280 km per hour. As many passengers alighted at Hannover, the train is slightly less than half full when it reaches Eschede, a town about 100 km south of Hamburg. It nevertheless contains a party of 80 schoolchildren.

At just before 11.00 h local time the train, travelling at about 200 km per hour, hit one of the concrete support pillars of a 40 m-long overpass bridge and demolished it. The front locomotive detached from the carriages and continued for a further 2 km before being brought to a stop by the driver, who was not injured. All 12 coaches derailed: the first three came to a halt 350 m after the bridge and the fourth skidded into a nearby wood. The front part of the fifth carriage tore away from its rear and came to rest 100 m in front of the fallen bridge, which crushed the rest of this coach. The next carriage ended up sideways in front of the bridge and the other six were compressed and telescoped against the bridge or propelled over it by the momentum of the rear locomotive, which fetches up against the wreckage. A car fell off the bridge.

The accident was probably caused when a steel tyre fractured on a wheel of the fourth coach, which caused it to derail as it went over points (i.e. a switch) and to jack-knife into the bridge support 300 m later. In the first 14 hours of the rescue operation, 76 bodies are recovered. After 48 hours the final death toll is 102, including 5 people whose injuries prove fatal after they have reached hospital; 69 people are seriously wounded and 18 moderately injured; about 40 are lightly injured.

The conditions are ideal for the rescue operation. The weather is warm and hazy, with negligible winds. No southbound train encountered the wreckage before movement was halted on the line. Access was possible from either side of the crash site, and there was plenty of hard level open space in the immediate vicinity for helicopter landings and vehicle staging. Eschede town is near enough for the first emergency responders to arrive on site within 10 minutes of the crash. A warehouse next to the overpass bridge on the western side was an ideal site for an advance medical post. A second advance medical post was set up on the eastern side in tents. An auxiliary medical post for people who are not seriously injured was set up in the gymnasium of a school about 500 m from the site.

Current was switched off in the overhead electrical lines of the railway about 10 minutes after the crash. The first responders begin work at the site shortly before this happens.

Within 15 minutes of the crash the Eschede emergency dispatcher alerted all relevant local responders, including the general hospital of Celle, nearby, whose medical personnel arrived on site within 20 minutes.

Air space was closed over the area and access restricted to emergency aircraft. Helicopter landing space was designated in fields on either side of the railway. Ground space was divided into a western sector, including the carriages that came to rest forwards of the collapsed bridge, and an eastern zone, which included the bridge and the remaining carriages. At the height of operations about 1200 rescuers were at work on the perimeterized site, including substantial contingents of military personnel from the Bundeswehr (army) and German border patrols.

Haematomas, cerebral traumas, crush injuries and internal haemorrhagic lesions are the principal causes of death, and most of the people who die did so at the time of the crash. About 50 doctors reached the site. Of the 87 significantly injured patients, 27 were transported by helicopter and 60 by road ambulance. Some 23 medical centres received patients. General evacuation of injured patients was achieved from 65–100 minutes after the crash.

Box 6.3 (continued) The Eschede train crash

Problems encountered during the emergency operation

In retrospect, the emergency operation went very well in several vital respects. There was good coop-eration among police, firemen, military personnel and others. However, there were problems.

The first of the problems involved communications. Landline and cellular telephones were ren-dered inoperable, as Germany does not have a system that guarantees access to preferential and emergency users during periods of overload. Moreover, the computers were not powerful enough to cope with the level of telephone traffic. Two-way radios (tuned to the 2 m frequency) and the one gen-eral emergency radio frequency also became overloaded. Some commanders had to resort to couriers and runners in order to maintain communication. Others had to relay messages through military radio channels rather than civilian ones. Communications collapsed between air units and civil forces on the ground. At Eschede the emergency telephone operator was overwhelmed with work and lacked ade-quate support. Finally, the Casualty Bureau had only ten telephone lines, one for each agency and a single line for relatives, which proved totally inadequate. All of this underlines the need to have powerful, multiple, dedicated communications channels and to be able to switch between them at will.

Some 32 helicopters reached the site, but there was a lack of coordination of landing facilities. The problem was eventually solved by setting up a temporary "control tower" using an army radio set. On the other hand, pilots were fully aware of the location of the 54 hospitals with helicopter landing pads located within a 90 km radius of the site.

Full-scale triage did not prove necessary, but it should have been practised anyway and in any case was inadequate. The injured and dead were taken to the same building. Bodies were packed into freight containers. No records were made of who was dead or who had been injured, nor of where the injured were taken. Eventually, a plea had to be broadcast on television asking hospitals to say whether they had any Eschede victims, and police had to make systematic inquiries of medical centres throughout the region. One patient was eventually located in a hospital 300 km away from the crash. Meanwhile, the mass migration of doctors from the Celle hospital, the nearest to the site of the crash, left it seriously deficient in medical personnel. Had patients not been transported farther afield by heli-copter, there would have been problems at Celle.

Difficulties were experienced with managing personnel and vehicles. Too many rescue personnel reached the site and, as a result, the situation in the vicinity of the two field-command posts (the dis-aster area was divided into operational sectors) was excessively chaotic. Personnel were not used adequately in relays, which might have kept their energy up and limited cases of critical-incident stress (CIS), which were observed among emergency workers. Finally, emergency personnel and vehicles needed to be better distinguished by labels or colours that indicated what their roles were. For exam-ple, emergency medical doctors, of which 23 were present on site, were not immediately identifiable as such. A national or international system of identification was needed.

In synthesis, many of these problems could be solved by planning, exercising and preparing for future disasters. It is obviously important to learn the lessons of what did not go well in the past and to act upon them in the interests of improved response capability. Eschede underlined the importance of restricting access to the sites, managing movements and personnel well, and guaranteeing the flow – and the recording – of information.

Sources of information: Esther Bärtschi (2000), "Sciagura ferroviaria di Eschede (Germania) del 3 giugno 1998: analisi critica dell=intervento", *N&A mensile italiano del soccorso*, Year 9, **100**, 2–15; syndicated press reports of UPI and AFP.

in matters as uncertain as earthquake prediction. Lastly, scientists may be antagonistic to the media if they perceive their work to be misinterpreted or interrupted by journalists. A skilled PIO can reduce or eliminate many such problems by interpreting between the worlds of science and journalism.

In summary, public information should be a vital part of the emergency plan, as it determines a significant amount of public reaction to the plan and its operation. In the end, emergency planning is about protecting the public, who have a right to be properly informed about what is being done on their behalf.

There are other problems concerning the mass media: their role in the convergence reaction that follows disaster and their role in over-stimulating public responses. Regarding the former issue, after a train crash near Selby in the UK, in early 2001, the narrow rural roads leading to the site filled up with news-media vehicles, which obstructed the movement of emergency services. The December 1988 air crash at Lockerbie (Scotland) led to an invasion of 2000 media personnel in a severely traumatized community of 6500 inhabitants. In other events, news-media helicopters have obstructed emergency-service flights over disaster areas. These are problems that resolved in collaboration with the news media, both at the disaster-planning stage and at the height of the emergency. However, if they can be sorted out with respect to local journalists, they may be more difficult to resolve for national and international news teams.

The second issue is that news reporting of disasters can motivate the public to overwhelm emergency agencies with enquiries. The solution is to get them to broadcast or publish appeals for specific forms of collaboration (for example, *not* to telephone the local police station).

6.10 Psychiatric help

Recently much psychological research has been conducted in the wake of disasters. However, the field is still young and relatively undeveloped. Moreover, the presence of psychiatric counselling teams among post-disaster relief efforts is also relatively new. It is, however, very worthwhile, as both victims and relief workers may need help in maintaining their mental health.

In places where no psychiatric support services exist for disaster work, the emergency planner may wish to ascertain whether an appropriate structure can be created. Can local mental-health professionals be persuaded to volunteer? Are they willing to organize themselves into a crisis-intervention group and act as unpaid volunteer workers when disaster strikes? Do they have, or can they be given, appropriate training in disaster psychiatry and counselling? Can appropriate facilities be found for them, such as consulting rooms?

If such a group exists, or can be created, it should be fully integrated into the emergency-planning structures. The psychiatrists should be treated as normal fully fledged disaster-relief workers. They should be encouraged to work in collaboration with other relief units and especially with medical doctors, who are often the first professionals to detect the need for psychiatric help, either because patients confide in them or because mental wounds are as evident as physical injuries. Doctors should be encouraged to refer patients to the psychiatrists, where appropriate; and, likewise, the psychiatrists may need to refer patients to medical doctors if physical ailments are being concealed. Full integration of the psychiatric unit into the emerging response structure is essential for these reasons and because, in order to do their work effectively, the psychiatrists will need to gain the confidence of other emergency workers and of victims and survivors. Although their official status may make them seem forbidding to some people who need their help, to others it will act as a guarantee of bona fide intentions.

Separate approaches may be needed in the two branches of post-disaster psychiatric work: the care of victims and of relief workers. Both groups may need help. Victims may be a prey to acute depression (however, it is not certain that disasters increase suicide rates), panic attacks, regressive behaviour, withdrawal or irritability. They may suffer psychosomatic effects such as skin rashes or loss of bodily control. Acute stress may occur and can increase the incidence of physical ailments, such as heart attack, stroke and high blood pressure. In extreme cases, patients may suffer from Wallace's **disaster syndrome**, a spontaneous withdrawal from external stimuli and a gradual process of re-establishing contact (although even in the worst case no more than 15–30 per cent of survivors are likely to manifest any trace of the symptoms).* The care of victims requires methods of ascertaining who is in need of psychiatric help, given that the problem may be partially hidden by the diffidence of sufferers and their unwillingness to seek aid. It will also require a means of gaining their confidence and establishing the psychiatrist as a point of reference and source of support. Psychiatric assistance to victims may be prolonged – in severe cases for more than a year – and hence the unit should plan for a sustained presence.

Relief workers can manifest various kinds of post-traumatic stress disorder. These include critical-incident stress, which is common in inexperienced people who are suddenly confronted with horrifying situations of death and destruction. Tiredness mixed with stress can lead disaster managers into the state in which they overrate their ability to solve problems (the **Jehovah complex**) or try to solve too many problems at once (the **magna mater complex**). In all cases, stress, depression or psychiatric imbalance resulting directly from the

---

* See pp. 565–6 in *Natural disasters*, D. E. Alexander (London: UCL Press, 1993).

disaster will tend to reduce the efficiency of relief workers. On occasion it can mean that serious mistakes are made.

Psychiatric monitoring of the relief effort is thus warranted. Where possible and appropriate, team leaders should be in the front line of this. When workers start to show signs of yielding to stress or begin to behave irrationally or anti-socially, they should be withdrawn from frontline service, encouraged to rest, and offered counselling if necessary. However, given the demands of the victims for psychiatric help, and the probable small size of the psychiatric unit, there may be little opportunity to give individual assistance. One possible solution is to plan to end the emergency phase with a group seminar or debriefing session at which the psychiatrists are present. In this, participants in the relief effort can be encouraged to share their experiences and seek mutual support as they externalize some of their grief, shock or anxiety, and start to release accumulated tension. By interpreting reactions to the disaster in a timely way, the psychiatrist can facilitate this as a controlled and constructive process.

Among both victims and relief workers, special efforts need to be made to locate and diagnose subjects at risk: old people who have suffered bereavement or other catastrophic losses, young and inexperienced volunteer workers who are having their first taste of mass casualty situations, and so on. All participants in the emergency need to be involved in this process of seeking out those who need help.

Drugs to control psychiatric conditions should be prescribed with extreme care after disaster, in order to avoid dependencies, side effects and interaction effects between medicines. It is particularly important to be cautious if medical records are not available for the patient in question. A related problem is that of substance abuse (prescription drugs, narcotics, solvents, alcohol), which may increase in the stressful conditions that follow a disaster. This requires even closer cooperation between medical doctors and psychiatrists, especially as it is difficult to detect the symptoms until they are chronic (and even then the causes may be misinterpreted in the absence of experienced help). Substance abuse can also lead to violence, especially domestic strife, and increases in this may be noted by police. Hence, psychiatrists may consider cooperating with police forces in order to search for the underlying causes of post-disaster increases in wife- or child-battering incidents. If there is a serious endemic problem of drugs or alcohol, then it may be useful to involve in the disaster plan organizations such as Alcoholics Anonymous, detoxification centres and counter-drug associations.

Nowadays, many jurisdictions impose a legal requirement on employers to provide their employees with counselling when they are subject to serious stress in the course of their duties. Failure to offer psychiatric help when and where it is needed could be regarded as contributing to psychological injury and render the employer liable for damages. "Employer" in this case may refer to the heads of many different kinds of emergency service.

6.11 A note on the integration of plans

This section will add a few specific comments to the points made in §4.4 on the integration of disaster plans at various levels of government and among various organizations.

A basic choice must be made regarding whether to make specialized disaster plans autonomous or part of the general municipal or regional plan (Fig. 6.14). The advantage of autonomous plans is that they do not depend on the provisions or resources of the parent plan, but there is a risk of duplicating functions and activities. Integration between plans enables more resource sharing, but can over-complicate liaison between different working groups and possibly result in some equivocation or duplication in the command structure.

Whatever strategy is adopted, it is essential that participants in all plans have a clear idea of where their responsibilities lie and where they end. Besides careful specification of tasks in each plan, and the institution of training courses, it is often helpful if personnel who fulfil specific roles have a clear idea

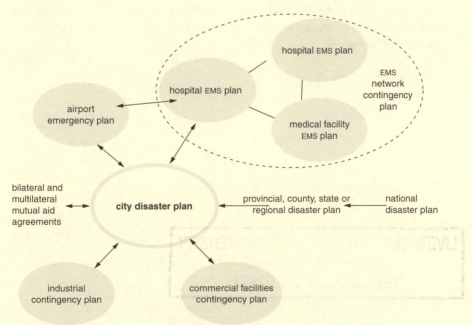

Figure 6.14   Integration of local, regional and national emergency plans.

of what other participants' roles involve. This can increase the level of respect for the roles and functions of one's co-workers and help ensure that people are not expected to carry out tasks that they could not possibly accomplish.

When thinking about these problems, it is essential to distinguish between the internal and external workings of an organization (such as a hospital or industrial plant) in a disaster. Usually, internal workings can be subject to self-contained planning, but external workings demand some degree of integration with the wider reality of emergency management. Integration between plans can therefore be partial, and selected participants in the specialized plan can be empowered to work with outside forces.

Cooperation between the personnel of different plans requires adequate flows of information on what is happening and what is being accomplished. This cannot be left to chance: there is too great a temptation to ignore the work of organizations other than one's own. Hence, specialized plans need a liaison officer who will provide updates on the functioning of the plan, either bilaterally to opposite numbers in other organizations or via a centralized agency for information collection, probably the main local emergency-operations centre.

Lastly, municipal emergency planners are often sources of expertise and may be able to act as consultants to organizations that need specialized emergency plans. This would tend to favour integration of plans by homogenizing methodologies. Regional or other government authorities at the intermediate level also have a potential or actual role in fostering standardized emergency planning methods and offering resources in common to both municipal and specialized planners.

CHAPTER 7

Reconstruction planning

The inevitability of disaster makes it imperative to consider the need for recovery and reconstruction once damage has been done. Although it may be impossible to predict the need for reconstruction that will be generated by a future disaster, the procedures should be studied and some preparations made before the event. This will improve the efficiency of the recovery process and reduce suffering in the aftermath of the next disaster, when conditions will be chaotic, time will be at a premium, and there will be relentless demand for solutions to pressing problems, such as the need for replacement housing and the reconstruction of damaged buildings.

The following account refers to situations in which the next disaster is expected to do considerable damage to the built environment in all its forms (the archetypal example would be a major earthquake in a heavily populated area). Emergency planners will have to design temporary measures for the housing of displaced survivors, strategies for the repair of basic services, and long-term programs for reconstruction. It is a vital principle that the last of these should improve current disaster-prevention and mitigation measures, so as to reduce vulnerability to disasters.

7.1 Temporary measures

According to some widely used temporal models of disaster, the emergency phase is followed by a period of recovery of basic facilities, which is followed in turn by reconstruction (Fig. 7.1). In practice, the phases overlap and hence recovery must begin before the emergency is over, especially in communities where survivors have a strong determination to overcome the difficulties caused by disaster. The period of restoration, or initial recovery, is characterized by a series of needs that must be fulfilled. Immediate problems must be solved, such as the provision of food, shelter and other basic necessities to those

269

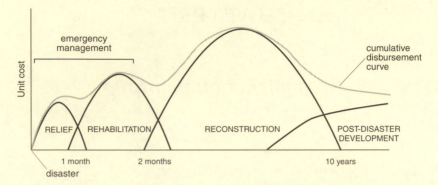

Figure 7.1 Phases of recovery in the aftermath of disaster.

who have lost them in the catastrophe. Public safety must be ensured, if necessary by cordoning off areas where damaged buildings may collapse and by maintaining evacuation orders, however much opposition there is to them among the evacuees. The safety of rescue squads and other groups of workers must also be ensured. Basic services, such as electricity and water supply, food delivery and public transportation, must be restored (see §7.2), Productive activities must be restarted, if necessary in temporary premises. Lastly, measures must be taken that provide a firm basis for full recovery, reconstruction and economic relaunching of the disaster area. Accordingly, it is worth dedicating some planning activities to the reconstruction phase. Although much of what will need to be done cannot be foreseen in advance of the next disaster, some of it can, and advance planning and decision making will, wherever it is possible, shorten the time required to restore services and facilities.

As noted in §5.8, the disaster plan should designate safe areas for the erection of temporary camps for homeless survivors. To summarize the requirements for such a site: it should be firm, compact, level, ample in size, well drained and not likely to become waterlogged or excessively muddy during wet periods. It should lie within reasonable distance of a safe and adequate supply of water for drinking and washing, not be located close to mosquito-breeding grounds (e.g. ponds and wetlands), have adequate all-weather access to metalled roads, and where possible should not be close to commercial or industrial areas likely to produce noise, odours, air pollution or traffic jams. In addition, it is helpful if the site has some protection against the elements, such as a screen of trees that would act as a windbreak. Obviously, the site should not be subject to landslides, debris flows (alluvial fans and piedmont areas are to be avoided for this reason), avalanches, floods or earthquake shaking.

As an absolute minimum, there should be a space ratio of at least $40\,m^2$ per person (all inclusive), which translates to at least 4 ha per 1000 people. However, the minimum value will vary (generally towards higher values) with the type of accommodation to be installed at the site. Single units (tents, trailers, container homes or prefabs) should have at least 2 m of space between them. Some planners recommend 8 m between tents, which allows people to pass without tripping over pegs and ropes. Walkways and access roads should be treated against dust in dry weather and drained well during wet periods (i.e. drainage ditches should be dug where necessary). Elevated duckboards may be required if the ground becomes very muddy. Generally, camps should have a wide general-access road and be planned to facilitate expansion, if space is available, without loss of form or function. Cleaning, garbage collection and disinfestation should follow a regular schedule and, like sanitary facilities, refuse-disposal areas should be segregated from living quarters. Sanitation and fire-prevention rules should be made known to people who live in the camps and should be rigidly enforced.

Generally, temporary camps should not contain more than a thousand people each. This is necessary in order to facilitate day-to-day management and disease-prevention activities. Service buildings, such as clinics, churches and community centres, should be grouped together and away from living areas. Wherever possible, housing units should be assigned so as to preserve a sense of community. This can be accomplished by assigning the dwellings to families on the basis of the original street addresses of homeless survivors. In sum, disaster planners and managers should be prepared to deal with the mass of engineering, logistical and social problems that setting up temporary camps for survivors involves, and the disaster plan should designate appropriate areas on the basis of needs predicted by hazard, vulnerability and risk scenarios.

Two interlocking needs arise in areas damaged by disaster. One is to ensure public safety by restricting access to areas that are dangerous and the other is to ensure circulation of traffic in order to facilitate the relief effort and to permit reconstruction to take place. Rapid structural survey and field survey of continuing hazards must be planned for, so as to indicate which areas need to be cordoned and placed off-limits to general access. This involves devoting some thought to formal means of communication between the surveyors and disaster managers, how areas are to be closed to access, who will enforce orders and carry out any surveillance that may be necessary, and who will be allowed into the restricted areas and under what circumstances. Here, the safety of emergency workers becomes paramount. Although most of these considerations will have to wait until after the disaster, some preparations can be made beforehand by running through the processes involved and training key workers.

It is a good idea to run courses on how to conduct disaster work for utility personnel, construction companies and the police, as these groups will have to cope with some of the most precarious and dangerous situations caused by hazards and damage. The emphasis should be on promoting procedures that ensure the safety of such workers (indeed, funds may be available to train personnel in safety enhancement). It is very easy for large disasters (and even some small ones) to result in fatalities among those who come to help: rescue workers or construction labourers crushed by falling masonry, electricians electrocuted, gas-company workers asphyxiated or killed by explosions, and so on. Such people may have to work in unfamiliar conditions and with unfamiliar hazards, which account for the increase in casualty risks.

Damage to the built environment will make it necessary to shore up structures that have not collapsed (or not entirely) but are in a precarious state. Braces, collars, rods, baulks, counterpoises, scaffolding and other forms of buttressing will be needed, and so a source of these should be located in advance of the disaster. It is useful to identify bridges that may collapse in floods, earthquakes or debris flows, and to ascertain whether they might then be replaced by Bailey bridges. Military engineering corps, or other such units, should then be sounded out regarding their ability to construct Bailey bridges or fords, and how rapidly they can do so in an emergency.

After disaster has struck, wreckage should be removed where it poses a hindrance to relief and recovery efforts. This will require cranes, bulldozers, dumptrucks, diesel fuel stocks, drivers and operators. However, prior estimates of need should not be based on scenarios that involve the indiscriminate bulldozing of rubble. In the recovery phase, clearance activities should be kept to a minimum and the rest left until the reconstruction period, when it should be properly planned.

As noted elsewhere, temporary premises may be needed after disaster for some of society's basic functions. Industries, artisans and shopkeepers may need them in order to restart production or sales. Schools, colleges, clinics and municipal offices may need to be transferred. If the spatial extent of the prevailing hazard scenario is restricted, as would be the case, for example, in many flooding situations, a survey of property outside but near to the hazardous area may provide some prior answers about where to transfer operations, even though the eventual pattern of needs may be impossible to predict. Where the hazard is geographically more diffuse, part of the demand for prefabricated buildings may come from the industrial, commercial, agricultural, scholastic and municipal sectors. It is helpful to involve urban planners in the recovery process. In this context, it represents a sort of parallel planning process, similar to that pursued by the planning of permanent facilities. Like other professionals who become involved in disaster work, urban planners may need training (see Ch. 8) so that they appreciate the differences between needs related to disasters and those that pertain to normal urban growth.

Following disaster, the disaster planner must be prepared to address the needs of the productive sector. Whereas the emergency phase usually involves social welfare, rather than market forces, commerce will soon return, and indeed the sooner this happens the better it will be for the area's economic recovery process. Shops, factories and workshops may be severely affected by disaster. Hence, the commercial, small-business and industrial sectors will need to be surveyed in terms of losses of premises, equipment, inventory, workers and jobs. Some of the potential solutions can be worked out in advance, especially if factories or other commercial organizations have created their own disaster-recovery plans.

In many cases, the key to recovery from disaster is funding, which endows people and organizations that have sustained losses with the flexibility to choose solutions to their problems. The financial side of recovery is one that is very often left until the post-disaster period and then tackled using *ad hoc* methods. A more provident and efficient approach would be to ask some "what if . . . ?" questions before disaster strikes, however uncomfortable they may make administrators feel, and design some countermeasures. Can conventions be negotiated with the insurance industry to provide affordable policies against disasters, to compensate not merely for damage but also for economic effects such as lost production? Fiscal stringency may militate against using grants to stimulate recovery from disaster. Low- or zero-interest loans are often more effective, especially if repayments can be fed directly back into the reserve of funds from which the loans came, in order to maintain it. The advantage of designing such systems before disaster rather than after is reaped in savings of time, as eligibility criteria and the necessary bureaucracy will already have been worked out.

7.2 Restoration of services

As noted in the §7.1, restoring basic services or assuring their continuity provides the basic framework within which reconstruction work can be planned and carried out. The basic services include electricity, gas, water mains, sewerage, water treatment, telephone lines, health services, postal deliveries, school classes, food and goods deliveries, and municipal functions such as planning and cadastral offices.

When preparing for disaster, the emergency planner should work closely with utility companies. These should be expected to have, or to create, disaster-recovery plans. The criteria for the success of such plans are the speed and efficiency with which a basic service can be restored after a given level of damage and disruption has occurred. In the case of mains water supply, the process may be prolonged if damage is extensive, as it may be in a large earthquake.

Functioning supplies of water can be hyperchlorinated as an interim measure to give them some protection against faecal contamination. Water quality needs to be tested frequently and at various points in the distribution network in order to identify sources of contamination. In the meantime, water-quality advisories and injunctions on water use can be issued. But the search for leaks may take months. Moreover, massive leaks can cause floods or landslides that aggravate existing damage and require more reconstruction, unless the leakage can be stopped quickly. A complete survey of water supplies will include inspection of the drainage basin and its land uses, geology and vegetation, and potential sources of water pollution. At water-treatment plants, the intake, pumping rates, treatment processes, storage facilities, water pressures and head losses will need to be surveyed, and the operational records of the plants must be scrutinized. Interconnections between different water supplies should be examined, together with actual and potential cross connections that could lead contaminants to be back-syphoned into the system. Lastly, the integrity of laboratory facilities for water-quality monitoring should be checked.

One common cause of contamination is the intermixing caused when parallel water mains and sewers break at the same point. Leakage from sewerage systems and from damaged septic tanks can also contaminate shallow aquifers, and the failure of wastewater treatment plants can lead to the pollution of surface waters. Close inspection and prompt repair are needed.

Buried pipelines are mostly vulnerable to earthquakes, landslides and subsidence. Above-ground utilities may succumb to hurricanes, windstorms, icestorms, snowstorms or floods, or in a limited way to explosions, fires or transportation crashes. For example, hurricanes can take the roofs off telephone exchanges and drench the switching equipment with water, unless the buildings are constructed to withstand force 12 winds. In each case, main trunk lines will obviously need to be restored before consumer connections.

When restoring electricity lines, care needs to be taken to ensure that live cables neither come into contact with water nor arc as a result of being close to one another. Lines that have been severed by storms but remain live can send current into pools of water on the ground and make them potentially lethal. The explosion and fire risks associated with the repair of gas mains may require evacuation before work can start. It should in any case be carried out in liaison with fire brigades. Telephone services will require restoration of landlines, exchanges, switching gear, cellular repeater towers and satellite uplinks. Temporary facilities for public access can be provided while this is going on, especially for cellular telephones.

In synthesis, for utilities the process of recovery should begin with a census of damage to key facilities: exchanges, antennae, repeaters, generating stations, transformer substations, pumping stations, gas holders and trunk lines. It should then look for breaks in networks, proceeding from the main lines hierarchically down to the subsidiary feeders and distributaries. Geographic information systems (see §2.3.1) can be used as mapping and analysis tools to aid response; the rest should be prescribed by the utility recovery plan.

As noted in §7.1, schools, clinics and municipal offices may need temporary relocation to safer premises; thus, their shelter requirements are similar to those of homeless survivors (§5.8 and §7.1).

7.3 Reconstruction of damaged structures

After a major disaster, affecting a wide area and a large population, reconstruction may take 10–25 years to complete. It is a complex process, often dogged by unforeseen setbacks and sometimes characterized by a "boom and bust" economy in which the stimulus provided by the reconstruction effort gradually runs out and gives way to economic stagnation. In an ideal world, reconstruction plans would be formulated, at least in outline, on the basis of damage scenarios before disaster strikes and these become reality. In a few cases this has been done, but it is likely to remain rare, for seldom do the resources and political willpower exist to support pre-disaster reconstruction planning.

Time is necessary to the social component of the planning and execution of post-disaster reconstruction. Not only must plans be drawn up, but there must also be adequate consultation and ratification. Hence, one should beware of "instant" reconstruction, which suggests that democratic processes in planning may have been ignored. Nevertheless, careful attention to the precepts of emergency planning may result in quicker, more efficient and equitable plans than would otherwise be the case.

In many ways, reconstruction planning is akin to any other form of urban and regional planning. However, the usual emphasis of these is on expansion of facilities, coupled with some renovation; post-disaster planning instead is concerned to renovate some facilities on a very large scale and create substitutes for others where reconstruction to original standards is not possible or desirable. In this process some very hard choices have to be made. Unless damage is slight and very self-contained, it is seldom possible to restore facilities to exactly how they were before disaster struck. For example, modern demands for space are such that most urban landscapes are reconstructed to lower densities than those that had been lost. Hence, reconstruction almost always involves some mutation of the original form, function or appearance (one might also say mutation of the *genius loci*) of places; anything different would represent the deliberate creation of an anachronism. It may nevertheless require that certain features of symbolic value be reconstructed as they were or be left in ruins to remind the population of the disaster. But the mutation can be either left to chance or deliberately programmed in such a way as to accelerate the processes of socio-economic change. Decentralization of cramped inner-city sites is a case in point.

Post-disaster reconstruction planning has several objectives. First, it seeks to identify, in a coordinated way, buildings and facilities that are to be restored, modified, modernized or demolished. In this it must strike a balance between the "start again", or *tabula rasa*, approach of complete demolition and site clearance, and the preservationist ethic of *in situ* restoration–reconstruction. At the same time, planning must facilitate coordination among property owners, especially where parts of a site, city block or building have different owners, and reconstruction is impossible without adequate agreement between them, especially on cost-sharing matters.

Where possible, reconstruction plans should seek to ensure continuity and availability of funds, together with equity of access to them. Reconstruction funding is a complex and sometimes rather haphazard process. It often involves protracted negotiation between local, regional and national governments, with the lower levels arguing for more funds and the upper ones trying to restrict disbursements. On occasion it may necessitate special laws or decrees. Logically, it should be easier to obtain or justify funding if there is a clear idea of what the funds are to be used for, and reconstruction plans can furnish this, together with detailed estimates of costs. Adequate planning is necessary in order to keep the reconstruction process going and stop it from flagging. In the end it will be necessary to make the difficult transition from replacement reconstruction to post-disaster development and economic relaunching. Good planning of the former can provide a firm base for the latter. However, to avoid a boom and bust cycle, planners may need to take action against profiteers. This may mean regulating the supply of building materials, and perhaps their market prices, employing a system of competitive tendering for contracts, and ensuring a steady but not reckless availability of credit and mortgages. Some degree of intervention in market processes will probably be inevitable, whatever the prevailing political ideology.

The process of reconstruction planning and management involves several stages and is as follows. The first step is to attract funds for planning and redevelopment. Besides efforts to extract money from government coffers, it may be possible to encourage the participation of the private sector in reconstruction planning. The next stage is to acquire the input data on which to base plans. Three types of site survey are required. The first involves mapping and compiling lists of the damage caused by the disaster in terms of cost, type, location and seriousness. The second involves mapping and investigating the geology, soils and geotechnical features of the area, if necessary with the aid of borehole data on underlying sediments. This will give a picture of site stability and the foundation-bearing capacity of the land. Thirdly, natural and technological hazards need to be mapped so as to create a microzonation map for use in mitigation planning (see §7.4).

In the next stage, the disaster zone is divided up, at various scales, into areas for restoration reconstruction, redevelopment, urban expansion and new sites for replacement urbanization (e.g. for permanent rehousing of homeless survivors). This will require a mixture of greenfield and brownfield sites and some adjustment of urban density. Key buildings and facilities need to be identified and detailed reconstruction plans must be prepared for them. Once the mix of different sites has been determined, the general reconstruction plan should lay out the pattern of urban services, such as access roads, street lighting, parking spaces and utility corridors. It should add any structural hazard-mitigation measures that are needed. Detailed planning should proceed from this point with the integration of the general urban plan (perhaps drawn at a scale of 1:10 000) with plans for particular sites (probably drawn at geographical scales larger than 1:1000). During this process, the cost of reconstruction should be estimated and summed. Alterations to the property cadaster should be noted.

Once the draft plans are complete, the safety of sites should be verified by investigating hazard, vulnerability and risk as they will manifest themselves in the reconstructed environment. It is obvious that a reconstruction plan should seek to reduce vulnerability, risk and the impact of future disasters, not perpetuate or increase them. Each aspect of the plan needs to be systematically subjected to expert evaluation in the light of these factors. Once this is done, the reconstruction plan should be presented to the general public at a public hearing. Interested parties should be given at least a month to lodge objections. These should be evaluated by a planning commission and the plan modified as necessary to accommodate both safety considerations and valid objections. It should then be formally approved and put into effect.

The time required to carry out all of these steps varies considerably from one situation to another. If major geotechnical and geological evaluations are required, damage to be repaired is complex and extensive, and funds are short, the planning process may take several years, during which the disaster area will appear to stagnate. Even highly efficient and streamlined planning is likely to take many months. It is very important that the process be democratic, that there be fair access to funding and credit, fair apportionment of sites and materials, and plenty of opportunity to give a thorough evaluation to the reconstruction program as set forth in the plan. This requires adequate publicity and special efforts to involve the public and property owners. Absentee owners represent a special problem because of the difficulty of keeping in contact with them to resolve questions related to the reconstruction process. Finally, the success of reconstruction planning often depends on the efficiency of political processes and the lack of the sort of political polarization or opportunism that can stultify initiatives. Generally, local administrations with good political connections to central and regional government (the main sources of funds) tend to reconstruct more quickly than those with poor connections.

7.4 Development and mitigation

Economic development can be a sign that reconstruction after disaster is complete and many of the problems created by the impact have been overcome. On the other hand, economic stagnation may signify inability to relaunch the productive capabilities of the area after the setback caused by the disaster. As noted in the previous section, providing a firm base for the post-reconstruction period should be a prime objective of reconstruction planning. Moreover, both reconstruction and post-disaster development should be considered as opportunities to institute a wide variety of measures designed to reduce hazards, vulnerability and risks, and to mitigate disasters. Indeed, in areas subject to recurrent natural or technological hazards, all new development and redevelopment should be evaluated in terms of whether it contributes to vulnerability or mitigates it. This should be done *before* it is undertaken and the results of the evaluation should lead to improved safety.

In terms of risk mitigation, planners should evaluate development or redevelopment proposals according to the following criteria:
- Would the proposed project exacerbate existing hazards? For example, lining a stream channel with concrete in order to accelerate flow through it might reduce flood hazards in the immediate vicinity but could increase them farther down stream.
- Would the project create new risks or cause secondary hazards? For instance, an earthquake could lead to a toxic spill at an industrial plant, or a flood might cause a dam to be overtopped by excess water, or even to burst.
- Would the development project put people at risk through the possibility that structures might collapse?
- Alternatively, to reduce the risks to manageable levels, would mitigation be excessively expensive?
- Or if it were not carried out, would the potential damage in a disaster be excessive or very costly?

Clearly, some analysis will be required of probabilities of occurrence and the costs and benefits of mitigation. Mechanisms used in order to inhibit unwise development and encourage mitigation practice will vary with local, regional and national planning laws. One very clear requirement is that hazard mitigation should be well integrated with urban and regional planning practice.

The wide variety of mitigation measures can be classified broadly into structural, semi-structural and non-structural approaches (some examples relating to flood hazards are given in Table 7.1). A river-bank levee is a form of structural control of flooding, whereas the preservation of natural flood-retention areas is a semi-structural measure, and the use of insurance against flood damage is a non-structural approach (see §7.4.1 and §7.4.2). Historically, many countries have started off by relying exclusively on structural protection, often a single measure at the outset and a series of prescriptions later on. The countries and regions with more advanced policies have then progressed to a mixture of structural and non-structural measures (semi-structural approaches tend to be rather limited and are included with structural ones hereafter in this chapter). Hence, increased reliance on non-structural approaches has become a hallmark of evolving hazards-control policy. Recently, however, there have been signs that some countries with poorly designed mitigation practices may be going straight to a non-structural approach without first engaging in the structural-engineering rite of passage. The historical lesson of structural mitigation practices is that more and more of them tend to be required, and over time costs rise steeply but returns (**benefit–cost** factors) diminish.

Table 7.1 Flood hazards and their abatement (after Cooke & Doornkamp (1974: 117) and other sources).

Critical flood characteristics	Adjustments to floods	Flood hazard abatement	
		Structural	Non-structural
Depth	Accept the loss	Levées, dykes,	State laws
Duration	Public relief	embankments, flood	Zoning ordinances
Area inundated	Abatement and	walls	Subdivision
Flow velocity	control	Channel capacity	regulations
Frequency and	Land evaluation and	alteration (width,	Building codes
recurrence	other structural	depth, slope and	Urban renewal
Lag time (time to peak	change	roughness)	Permanent
flow)	Emergency action	Removing channel	evacuation
Seasonality	and rescheduling	obstructions	Government
Peak flow	Land-use regulation	Small headwater	acquisition of land
Shape of rising and	Flood insurance	dams	and property
recession limbs		Large mainstream	(creation of floodway
Sediment load		dam	and public open
		Gully control, bank	space)
		stabilization and	Fiscal methods:
		terracing	building financing
			and tax assessment
			Warning signs and
			notices
			Flood insurance

The following sections will not offer an evaluation of mitigation measures, which can be obtained from other texts and will depend in most cases on local circumstances. The aim here is to explore some of the planning implications of designing a set of mitigation techniques for any particular hazard, vulnerability or risk situation.

7.4.1 Structural measures

The range of structural measures used to control hazards is extremely wide. Seismic engineers and architects have evolved complex codes of practice to make buildings resistant to earthquake shaking. Hydraulic engineers have designed many measures to reduce or contain flooding. Avalanches are commonly controlled by batteries of structures in wood, steel, concrete, earth or rocks. Industrial and mining processes are protected by containment structures, and transportation systems by alarms, automatic brakes and fire-control devices. In most cases, improvements in safety have followed the terrible lessons of particular disasters, in which death, injury and destruction have shown the need for greater safety and suggested ways in which it might be attained.

Structural measures can be evaluated in terms of costs and benefits, usually expressed over the finite design life of the structure in question. Costs should be calculated as the sum of the following:
- construction and continued maintenance, repair and administration of the structure
- environmental impact, assessed in many different ways with respect to flora, fauna, landscape aesthetics, and so on
- any possible restrictions on activities or amenities that result from the presence of the structure (e.g. restrictions on gravity-fed irrigation of agricultural fields caused by a levee).

Benefit is assessed in terms of both damage and casualties avoided by adopting the mitigation measure and can be given a monetary value in order to compare it with costs, so as to ascertain whether net benefits are positive and mitigation is therefore economically justified. Benefit–cost analyses require careful consideration of which factors to include and which to leave out of the equation. In this respect, different choices can radically alter the balance.

Although structural protection measures differ substantially from one hazard to another, there are various standard precautions that can be taken when developing a new urban area that is judged to be susceptible to some form of natural or technological hazard. These are especially important where floods, earthquakes, landslides or fires are a problem. The density of development should be relatively low, with plenty of space between buildings, or at least between groups of buildings. Streets should be wide and unobstructed, and cul-de-sacs should be short and not crowded with buildings. Residential areas should be well separated from industrial zones and main transportation corridors, and houses should be set back from the immediate source of any hazards that are on or adjacent to the development. Multiple access points are required in case the main routeway becomes blocked, and refuge or evacuation areas (perhaps with structural protection) may be needed. Water supplies, storage facilities and hydrants should be constructed in numbers sufficient for fighting fires. Urban landscapes should be designed to ensure access by emergency vehicles, even when roads may be cluttered with evacuees.

In areas of flood or earthquake hazard, all new or reconstructed urban areas should contain zones designated for the evacuation of the population and the regrouping of emergency vehicles. These should be well signposted from the urban centres and main routeways. They should be safe, level and well drained (sports pitches are often the most appropriate local areas for these uses, especially if there is floodlighting). Emergency-vehicle parks should have adequate room for entering, exiting and manoeuvring with trailers or large vehicles, even when the park is fairly full. Supplies of sand and gravel can be stockpiled to spread over muddy areas in wet weather. Helipads should be firm, level areas with floodlights at four corners and a white H painted within a circle of about 2 m diameter. Trees or other obstacles such as power lines (which are usually invisible from approaching helicopters) should not obstruct the site in any way.

Hazards can seldom be controlled adequately without employing at least some structural protection. However, the structural approach has drawbacks. To begin with, it may adversely modify the hazard. Thus, artificial levees may contribute to river-bed aggradation to the extent that flood hazard increases rather than diminishes, unless the levees are raised again, a solution that cannot be repeated indefinitely. Structural protection thus often interferes with the equilibrium of the process that it seeks to contain. Hence, attempts to control beach erosion using groynes tend to destabilize processes of longshore sediment transportation, which creates a need for second- or even third-generation structures in pursuit of an equilibrium that cannot be regained.

Structural protection can easily lead to an over-reliance on technology that may fail, perhaps because of a unique combination of circumstances. No structural measure taken against hazards is ever totally infallible (Fig. 7.2). Moreover, some measures that offer only partial protection may impart a false sense of security. Thus, although the probability of disaster may decrease because of the application of protection measures, risk may increase paradoxically, as, for instance, the protected area is newly urbanized and more items are

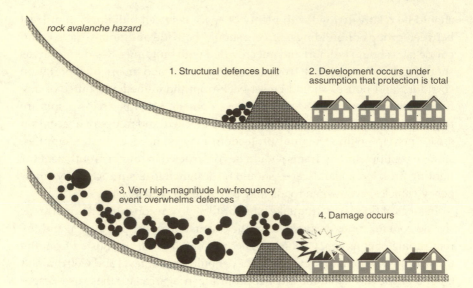

Figure 7.2 Structural measures do not offer 100 per cent protection against major hazards.

placed in the path of the hazard. Moreover, understanding of both the hazards and the engineering methods employed to contain them may progress such that the measures adopted in earlier periods are superseded. This is one reason why structural protection often seems to involve spiralling costs.

The outcome of this is that some structural protection is usually necessary but, per se, it is seldom an adequate solution. The disaster planner should evaluate proposals for structural mitigation in the light of several questions:

- What are the long-term costs likely to be, especially in terms of maintenance, as these can, on occasion, exceed initial construction costs?
- Are there hidden environmental or economic costs? For example, do avalanche barriers or landslide-prevention structures degrade the scenic value of a mountain valley?
- Does the measure constitute a serious threat to safety if it should fail through neglect, misuse, modification or an unexpectedly powerful hazard impact? For example, rockfall barriers constructed of masonry or stone can contribute debris to landslides if they collapse.
- Will development or other forms of risk taking increase in response to the mitigation measure? For example, will the presence of a rockfall barrier encourage development behind it on the probably mistaken assumption that it is proof against the highest-magnitude event that could occur at the site?
- If this is a potential problem, what percentage of protection can the structural measure be expected to provide, and can development be restricted on that basis?

These are questions that may need social and economic enquiries in addition to structural, mechanical or civil engineering feasibility studies and design exercises. The disaster planner is thus urged not to take too narrow a view of structural mitigation and not to try to justify it in terms of only one set of criteria, such as engineering feasibility.

It is generally recognized that structural protection is best combined with selected non-structural measures in a comprehensive program of hazard reduction. This involves prevention or attenuation of hazards (usually through structural measures), and the avoidance or attenuation of impacts (usually, although by no means exclusively, by means of non-structural protection). We will now examine some of the options that do not rely on engineering.

7.4.2 Non-structural measures

Again, the intention in this section is not to offer a guide to non-structural protection but to consider some of the planning implications of using it.

Non-structural mitigation measures can be subdivided into four types:
- land-use control
- norms and regulations
- financial measures
- emergency civil protection.

The last of these has been dealt with extensively in Chapter 4, so this section will concentrate on the other types.

Land-use control mainly involves placing restrictions on land use or designating areas for particular uses. In the North American system, zoning and subdivision regulation can be used on the basis of accurate knowledge of hazards to indicate which areas to designate in which way (assuming that the public and local administrations support such measures). European urban and regional planning tends to be more prohibitionist than its North American counterpart, which makes it well suited to ensuring that land uses are compatible with hazards. In contrast, while both continents regulate new development significantly, European practice probably offers less opportunity than that of North America to regulate existing land uses. In addition to specifying the uses of areas of land, these can be acquired by compulsory purchase in order to ensure freedom from vulnerability and risk, assuming that land-acquisition programs have enough political and public support to succeed.

Norms and regulations can be used both passively and actively, either to prohibit actions or to mandate others. Buildings codes, sanitary regulations and similar sets of norms exist to ensure public safety by prohibiting dangerous or injurious practices. With careful enforcement they can also be used to encourage good practice: development that fails to meet minimum standards should be demolished or upgraded compulsorily.

Financial measures fall into several groups. Grants and credit can be used to stimulate good mitigation practice, either by making them available in ways that encourage protection or by restricting access to them when it is feared that their unwise use would increase vulnerability and risk taking. Taxation can be used selectively to deter unsound development, discourage the settlement of unsafe areas, or, through tax relief, to provide incentives for mitigation. In many ways, taxation is the most widely used disaster-relief and mitigation measure, as all citizens are subject to it; hence, we all pay for safety and damage repair. Lastly, insurance is becoming increasingly popular as a means of putting income aside to pay for future losses. However, it involves problems of establishing equitable and affordable premiums. If policy holders were to pay in strict actuarial proportion to the risks of natural hazard losses that they bear, premiums would be excessively costly, for example, in the lower floodplains of rivers that frequently overflow their banks, or on perennially unstable slopes. Many natural and technological hazards are not evenly or randomly distributed and particular locations tend to be affected repeatedly. Thus, the US National Flood Insurance Program, which was created by the federal government because private insurers withdrew from the flood-insurance market, faces the problem of reimbursing some claimants over and over again. In fact, there is some element of subsidy in all natural-hazard insurance schemes, in which policy holders with small risks subsidize those whose exposure is great.

One consequence of this situation, and of the rising value of natural hazard losses (Fig. 1.3), is that both private and publicly sponsored insurance against disasters is in crisis. For the insured it is becoming hard to obtain and for the insurers hard to underwrite, as it is difficult to find sufficient capital. In many cases, natural hazards and some technological risks are subject to provisions separate from comprehensive property insurance, which often specifically excludes the risks in question. There may instead be riders, additions to policies or separate policies. Despite their cost to the insurees, these are becoming increasingly popular, which has led the insurance industry (more properly, the re-insurance industry, which pools the risks of insurance companies) to scramble for means of financing schemes. At the world scale (see Fig. 7.3) it is currently unclear whether major-hazard insurance will become more common or will begin to limit itself, as companies, and perhaps even governments, withdraw from insuring certain risks of disaster.

To make hazard insurance schemes more equitable, affordable and viable, they need to be imposed in such a way as to discourage voluntary risk taking and to reward mitigation efforts. It is therefore sensible to combine them as much as possible with hazard-mitigation plans. In countries that have begun to

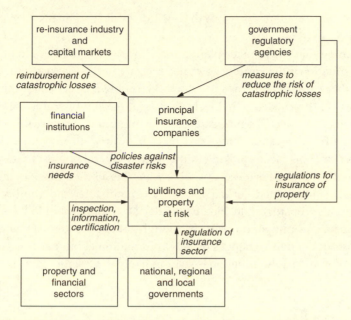

Figure 7.3 The international system for insurance against catastrophic loss.

organize themselves from this point of view, national provisions may predominate. However, there may be scope for regional or local arrangements in which, either individually or as consortia, insurance companies are induced to offer policies at controlled premiums in return for regulation of the risks. Thus, the city of Florence (Italy) has an agreement with a company that offers flood insurance on the proviso that its payouts are limited by depth of flooding. At depths of over 80 cm, when damage becomes much more costly, government money replaces private-sector funds.

Several questions hang over non-structural measures. One is the degree to which control and prohibition are possible in any given society. Although the aim of ensuring public safety by mitigating hazards is thoroughly laudable, the measures taken to achieve it can easily come into collision with questions of rights and fundamental liberties, to the extent that there may be a lack of enforcement. Hence one needs to be careful to compare the situation on the ground with the more theoretical picture offered by the regulations.

Another question is the degree to which mitigation measures are compatible with one another. This is particularly important for non-structural mitigation techniques, as these almost always need to be used in groups rather than singly. One may ask, for example, whether regular insurance includes coverage against hazards, and whether such insurance can have a percentage levied on it to contribute to emergency medical services or finance civil-protection activities. It is sometimes possible to conclude local agreements with insurers to help achieve these aims.

Assuming that the disaster planner has some choice in these matters and that the situation is not entirely governed by stark necessities or political expediency, there are several ways in which the problem of choosing mitigation measures can be tackled. Obviously, technical feasibility studies are a vital prelude to any choice. The next stage is to examine and compare several combinations of structural and non-structural measures in order to determine which is best. This means looking at them from several points of view and in the light of various users' needs. In some ways it is also a process of searching for possible unexpected consequences that might alter the validity of the strategy under consideration. Finally, the disaster planner should be prepared to change hazard-mitigation policy (although not to the extent of being inconsistent) as vulnerability changes or new information about hazards comes to light. In this context the situation is seldom likely to be static.

CHAPTER 8

Emergency-management training

Training is an important adjunct to planning and management and is relevant to all phases of the disaster cycle. With regard to *mitigation*, students need to know how to choose, combine and implement disaster-reduction measures. For adequate *preparedness* it is a question of how to write, implement and test emergency plans and warning systems. In *response*, one must learn how to manage the early phases of emergencies, and *recovery* demands knowledge of how to manage programs of repair and reconstruction .

Procedures can be designed to ensure that emergency-training courses meet these objectives. First, one needs to ask some fundamental questions. Is a formal course the best method of ensuring adequate training in emergency management? Assuming that such a course *is* appropriate, what knowledge and skills do the trainees need to acquire? Secondly, a detailed assessment of needs should be made and the objectives of the course defined. Teaching methods are then developed and materials amassed in such a way as to meet each objective of the course. Once the course has been taught, its efficacy can be evaluated by observing the results and eliciting feedback from participants. Then the course can be revised or modified so that it more closely fits the needs of participants.

Although the main emphasis is on training selected individuals (usually local or regional government employees) to manage the emergencies generated by sudden-impact disasters, in fact the constituency is rather broader. Emergency managers are supplemented by emergency and mitigation planners (if these are not the same people as the former group). An analogous figure is the director of emergency medical services, who may require slightly – but only slightly – more specialized training. Police and military officers with specific roles in the civil-emergency sector may need instruction, as may public-information officers. It is also clear that a wide variety of professional workers can benefit from emergency-management training: these include medical personnel (doctors, nurses, technicians), veterinarians, firemen, engineers, architects, scientists (such as geologists and hydrologists), lawyers, policemen, judges, pilots, ships' commanders and mass-media representatives. Moreover,

287

there are the various categories of volunteer and auxiliary who participate in emergencies: in addition to generic disaster-response volunteers, there are paramedics, ambulance drivers, mariners and divers.

Each of these groups has a different role to play during a disaster and each will require a different form of training. However, all those who must function in a leadership role will have a common set of educational needs. Thus, whatever the speciality in question, good disaster managers have the following characteristics. They will have obtained a solid grounding in hazards and risks, forms of vulnerability to disaster, and mitigation options. They will be thoroughly familiar with their legal obligations and be able to make good use of laws that facilitate relief and aid. They will have received a full training in techniques of disaster planning and mitigation; hence, they will know how to write a comprehensible and functional disaster plan. They will be well versed in procedures of emergency management and rapid communication, and will have been trained in information management, in the sense of both using advanced information technology to manage disasters and being knowledgeable about public and media relations. In sum, the well trained disaster manager will be able calmly but rapidly to analyze complex situations, make decisions firmly, and manage people and resources under pressure. These characteristics therefore need to be translated into specific training goals. Moreover, where possible it is wise to select applicants for places on training courses in relation to their promise as managers, degree of motivation, mental flexibility, resistance to stress in the workplace, and aptitude for quick creative thinking.

Training methods fall into three main categories: academic (i.e. largely theoretical), professional and practical. The most traditional form is simple classroom instruction, but as the emphasis of training is practical, it needs to be supplemented by participatory activities and exercises. These can include presentations by trainees, group discussions and interactive lessons on computers. More effective results can be achieved through personal and group research exercises that lead to the preparation of theses or research reports. The practical aspect can be emphasized by hands-on training in the use of equipment and software. Site visits can be made to emergency-operations centres, scenes of past disasters, or places of exceptional risk, and field exercises can be used to develop logistical skills.

A basic dilemma in teaching emergency management concerns the distribution of parts of the course among different teachers. It is usually unlikely that a single teacher will have the knowledge and experience to teach all aspects of a comprehensive course. Hence, there is a tendency to use a wide variety of instructors, whose involvement is usually limited to only a few hours. Although this has the merit of exposing the students to a variety of viewpoints and competencies, it can easily lead to fragmentation and lack of coherence.

Continuity must be maintained and the best ways to achieve this are either to minimize the number of teachers or to appoint a tutor who will constantly moderate the students' experiences and integrate the various teachings. Alternatively, if resources allow, the training experience can be subdivided into a series of thematic short courses.

Disaster exercises can be divided into the tabletop variety, which are conducted in seminar rooms or virtually in front of the computer screen, and field simulations, using full-scale equipment and real people. They can both be excellent ways of testing policies, plans and procedures, and can bring to light weaknesses such as shortages of personnel or equipment. Timing, coordination, communication, roles and responsibilities can all be taught. Moreover, poor performance on the part of trainees can be evaluated and measures taken to rectify it during the exercise. In addition, field exercises are an invaluable source of general publicity for mitigation efforts.

Although emergency-training needs, and the means of satisfying them, are not especially difficult to identify, there is no firm consensus on what needs to be done. In part this is because counter-disaster training is in its infancy in many parts of the world. In part it is because there is no centralized authority to oversee training standards. Furthermore, disasters are increasingly international events: many transcend international boundaries (e.g. Alpine floods or snow-avalanche episodes, and Caribbean hurricane disasters). Others may be purely domestic in scope of impact, but the relief effort will elicit participation from as many as 70 different countries. Moreover, it is increasingly common for disaster managers to be sent on international assignment. Hence, a set of standardized international protocols is needed for emergency-management training. These should identify the qualifications necessary and should specify minimum standards for the form of training, the subject matter of lessons, and the minimum number of hours of instruction to be devoted to each topic.

Despite the need for standardization, it is clear that training can be implemented in several different ways. For the general all-embracing emergency-management training course, three basic curriculum models have emerged: the traditional (cause-and-effect) model which relies upon a linear progression from causes to impacts to responses, a more modern and adventurous concept-based approach (or key-themes model), and scenario-based methods.

8.1 The cause-and-effect model

The first and most traditional method involves presenting subject material in a linear progression from causes to impacts and then to responses. Typically, the course begins with a review of basic concepts, including magnitude, frequency and intensity of hazards and their impacts. Hazard analysis is subdivided into categories based on the natural, technological and social agents of disaster. Next, principles of emergency planning and management are taught. Special topics can include how to direct emergency teams, how to manage an emergency-operations centre, and how to cope with stress during a disaster. The last of these topics may lead to a more general survey of sociological and psychological aspects of disaster. This in turn provides an introduction to the important question of how to manage public information and liaise with the mass media. Next, disaster mitigation methods may be taught in the context of planning and managing recovery and reconstruction. Lastly, skills are developed in computer use, cartography, image interpretation, fieldwork and, where possible, field exercises. The traditional linear approach has the advantages of being straightforward and graced with an easily comprehensible structure. However, of the three models, it is the one least conducive to innovative teaching methods.

The skills that a disaster-management training course should aim to develop are as follows. First, trainees will require a basic knowledge of disasters. This includes a systematic treatment of the "how", "why", "where" and "when" of physical hazards, a penetrating analysis of human vulnerability, and an exposition of **risk assessment** (risk = hazard × vulnerability). Secondly, a basic knowledge of mitigation techniques is required. These can be divided into methods of prediction and warning (including the social corollaries of technological systems), short-term preparedness methods and long-term mitigation options. These aspects can be taught in a traditional classroom setting, although self-study, small-group projects, workshops and discussion groups can also be used.

Thirdly, technical skills to be acquired include the ability to use computers effectively (with particular emphasis on emergency-management software, Internet and e-mail capability), use of advanced telecommunications, map reading and cartographic skills, and the acquisition of at least some familiarity with geographic information systems and how remotely sensed images are interpreted. Fourthly, managerial and communication skills need to be developed. These include the ability to work effectively with the mass media, to construct plans, and to manage relief workers, victims and the public. In the interests of encouraging foresight and vision, it is especially important that trainees learn to build and explicate scenarios of hazard, risk and disaster impact, for these are essential bases for planning. Legal skills include the knowledge of obligations and rights under the law, while analytical skills can be developed by using problem-solving methods and algorithms and by teaching research methods that will be used to gather data and information.

8.2 The concept-based approach

The second method is structured around the key concepts of emergency management. This involves linking terms with the skills they relate to. Thus, analysis of the term should lead to acquisition of the skill. Characteristic examples of the terms, and of the skills linked to them, include the following:

- hazard – how to analyze events
- vulnerability – how to analyze situations
- risk and exposure – how to analyze and manage probability
- prediction and forecasting – how to understand the relationship between technology and its users
- warning – how to understand the relationship between technology and the socio-economic system
- planning – how to manage resources
- evacuation – managerial capacity
- management – the ability to direct operations
- logistics – a sense of timing and organization
- scenario construction – the ability to predict events and outcomes.

This model is relatively good at encouraging trainees to regard disasters as complex problems that require broadly interdisciplinary solutions based on lateral thinking and coordinated strategies. It can be combined with the third model, to broaden the themes and deepen the analysis by use of scenarios that pose specific problems for students to resolve.

8.3 Scenario-based methods

Scenario methodology is a versatile means of training emergency personnel, as it bridges the gap between abstract classroom instruction and practical training during real disasters. This section will describe how the method can be used in the classroom and will assess its potential as a means of preparing trainee emergency managers for the difficult, chaotic and stressful situations that they will eventually have to face on the job.

It is highly appropriate to teach emergency managers how to use scenarios as a basis for formulating disaster plans, as the scenario indicates the amount and nature of resources needed to combat the hazard and how they must be deployed. Scenarios are also useful as a means of testing students' abilities to respond effectively to practical problems and hence can be used both to screen applicants for courses and to examine them at the end of their training. In addition, the methodology is particularly useful for illustrating the limitations of particular situations. Thus, for example, students who would send in helicopters to rescue stranded people may need to be reminded of the constraints imposed by low cloud cover, strong crosswinds, congestion of flight paths, or simply the high ratio of maintenance time to flying time for such machines.

It could be argued that the most realistic scenarios in emergency training are the disasters of the past. One can develop these by documentary research and get students to work systematically through the stages of their management. Moreover, disaster managers gain credibility as well as expertise from their experience of real events. However, despite the obvious value of learning the lessons of past events, and especially those of past errors, there is much value in purely hypothetical scenarios, not so much in substitution for the analysis of past disasters but as a parallel means of training. Hindsight tends to condition attitudes to emergency management in subtle ways. It can restrict the solutions to problems by imposing a *de facto* set of outcomes on the situations under analysis. It may therefore be more efficacious to make scenarios hypothetical, although thoroughly realistic, and thereby free them from the constraints of preconditioned outcomes. A good way to do this is to base the scenarios loosely on actual emergencies, so that participants will find them realistic.

A basic method of using scenarios to teach emergency management involves providing participants with a minimum of necessary information about the starting situation (or "boundary conditions") and assigning them one or more roles and one or several tasks to perform or objectives to reach. The scenario proceeds by discussion as a strategy is worked out to solve the problems and reach the objectives.

In the simplest approach, the scenario refers to a relatively restricted geographical area, perhaps a small or medium-size town, a short space of time (perhaps the first few hours after disaster has struck), a relatively clear-cut role such as that of emergency coordinator or disaster manager, and an uncomplicated set of objectives, for example, to organize the rescue of people trapped beneath the rubble of collapsed buildings. More sophisticated approaches may involve increasing any of the following:

- the size and complexity of the geographical area involved
- the number of roles, and hence the degree of interaction and role playing among the participants
- the number, range or specificity of objectives
- the timescale or duration of the scenario
- the nature of responses required of participants (from passive to active involvement).

Scenarios induce participants to think through the consequences of decisions and actions. However, they must be devised and utilized in such a way that the students are forced to face up to such consequences. Despite the risk that the more assertive members of a class will dominate the process, it is advisable to conduct a scenario by group discussion, which facilitates collaboration and the exchange of ideas. In emergency management, people with different personalities and backgrounds will contribute different perspectives, and the sum total of these may be required if difficult problems are to be solved effectively. However, it can often be productive to divide a class up into small discussion groups, to pose specific questions and to take notes on the gist and direction of the ensuing discussion. The scenario can be stopped at intervals in order to examine how it has progressed, and at the conclusion a debriefing session may help participants understand what has happened during it.

The basic building blocks of a scenario include the nature of the disaster impact (what, where, when and who?), ground rules and basic logistical factors, the roles of participants, the objectives they should try to reach, and complicating factors or setbacks. The entire starting framework can usually be written out in somewhere between three paragraphs and three pages (500–1500 words of text: see the example on pp. 294–296). Good results can often be achieved by giving students a one-page starting scenario, allowing them five minutes to read and digest it, and allotting 20–25 minutes to the ensuing discussion. As emergency management is at least partly about anticipating the unforeseen, this method has the advantage of facing students with the sudden impact of a novel situation. In a real emergency they would have to think on their feet and hence the scenario demands a rapid but reasoned reaction.

Such is the flexibility of the scenario method of training that it can be used creatively in a variety of different ways. To begin with, updates can be given to discussants, which either introduce chance factors or provide basic information withheld at the outset. In fact, in real emergencies the ability to make decisions is limited by the piecemeal arrival of essential information, and this important factor needs to be built into scenarios that aim to simulate such conditions. The course instructor can hold back critical information and release it at strategic moments in order to boost the discussion or alter participants' views of the developing situation. Information on the speed or intensity of impacts can be altered in order to fit, or indeed to stimulate, participants' capabilities as managers. Skilful use of the ability to supply information selectively at different points in the discussion can help accustom trainees to the need to make decisions in the absence of critical data, as they will have to do in real emergencies.

Depending on how a scenario is structured, it can be used for a variety of specific teaching objectives. One can create a situation of conflict or of difference of opinion in order to teach participants how to mediate. Alternatively, one can describe a situation of more or less chronic uncertainty in order to explore the processes of making decisions in the absence of necessary information, as is common in the early stages of emergencies. Another form of scenario may describe a situation in which resources (manpower, equipment and vehicles) are scarce, which is again common in the initial phases of response to disaster. One can create a less-than-optimal situation in which things do not go according to plan, which is valuable as a means of teaching students how to cope with the chaotic aftermaths of disaster. A further approach is to create an "overload" situation that can be tackled effectively only by a sort of triage process, in which managers must rank tasks by priority and learn to do first what has the most beneficial effect, as there is no time to accomplish all the necessary tasks. In fact, as shortage of time for decision making is a common aspect of sudden emergencies, it is useful to face students with scenarios containing problems that must be resolved within a specified period, for example, nine or ten minutes for a one-page scenario.

Scenarios can also be used for teaching a variety of skills in addition to decision making. They can be employed to train participants to keep systematic records of decisions made, actions taken, equipment utilized, manpower deployed and tasks authorized. As in many cases emergency managers are rotated on the job, records are vital to the continuity of management and should be kept meticulously, despite the stresses and distractions of the work. Short scenarios can be used, with appropriate forms, to teach participants to write emergency communications messages, perhaps as responses to requests formulated in the scenario. Indeed, while teaching communications processes, it is helpful to develop a longer scenario based entirely on emergency messages, as a disaster manager in an emergency-operations centre would receive most information in this form. Hence, the flexibility of the scenario approach comes in handy again: time is a purely notional commodity that can be compressed and drawn out as needed. Lastly, scenarios can be extraordinarily useful for getting students to draw mental maps of disaster areas and the dynamics of situations that develop in them. Field research has shown that the ability of disaster-relief coordinators to visualize emergencies spatially (through response maps) is critical to their success as managers. This skill can be taught by turning the verbal descriptions in scenarios into sketch maps. The maps drawn by different students can be compared and analyzed in class time.

Space does not permit a range of examples to be presented, but one simple case is offered as an illustration of how the scenario method of teaching emergency managers works. This brief narrative will then be analyzed in terms of its teaching potential. It runs as follows.

[Fire in a road tunnel]   National Highway 997 passes in a tunnel 12.1 km long under Mount Indomitable (Fig. 8.1). The mountain marks the boundary between two administrative provinces; hence the northern entrance to the tunnel is in one jurisdiction and the southern entrance is in another. The highway consists of four carriageways in the open air of the valleys but is reduced to two lanes in the tunnel, with a total width of 7.5 m within the tunnel. Between the two lanes and the walls of the tunnel are walkways that have a maximum width of 70 cm. The tunnel, which is lit by sodium lamps, is fitted with air-quality sensors that activate large fans that extract polluted air and draw in fresh air along small, oblique ventilation shafts. There are no pressurized cabins, but every 1500 m along the tunnel there are hydrants (the water pipes run under the road surface), extinguishers and emergency telephones, which are all located in well lit alcoves. There are traffic lights at both entrances to the tunnel, but there is no direct means of monitoring conditions inside it. Moreover, there is no service tunnel and no manhole exists that would provide vertical access to the inner sections.

The only alternative route across Mount Indomitable is by a narrow road, with many hairpin bends, which climbs steeply and winds over a high pass.

The incident occurs at 18.05 h, local time, on a Wednesday in July, during a period of heavy tourist traffic. No restrictions are in force regarding the circulation of heavy goods vehicles, of which many are also on the road. The heavy traffic carried by the tunnel at this time has given rise to considerable pollution and the ventilator fans are fully operational. At km-6 (kilometre posts are measured from the north side) a large truck laden with cans of paint crashes into a vehicle carrying rolls of newsprint. The tunnel is full of cars and trucks with various kinds of loads.

At 18.19 h the first group of people from the margins of the area where

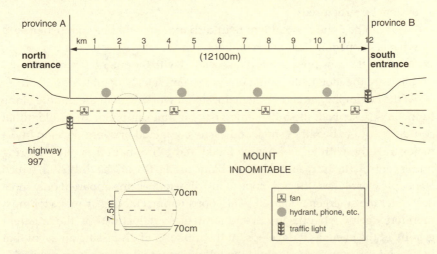

Figure 8.1 Sketch map illustrating hypothetical scenario of road-tunnel fire.

the crash has occurred reach the north entrance to the tunnel. Some of them are in a state of shock and others are extremely upset. They say they heard a loud bang and saw flames and smoke in the distance towards the middle of the tunnel.

This basic description of the beginning of what could turn out to be a major tragedy is based on information from several tunnel fires that occurred in central and southern Europe in 1998 and 1999, which called into question this form of road transportation and the design of security measures associated with it. Tunnel fires pose severe problems for emergency services. To begin with, access to the burning area is restricted. Smoke and flames are drawn along the tunnel, which offers very little opportunity for the intense heat of a fire to disperse. Most road tunnels do not have elaborate security systems and all too commonly no specialized plans exist for dealing with emergencies that occur.

As shown in the above description, the scenario offers both more information than an emergency manager who is first on the scene would probably obtain right away (e.g. the exact time of the incident) and less information than he nor she needs to organize an effective response. As no details of the available emergency forces are offered, the strength, nature and location of these relative to that of the tunnel must be assumed, although these are details that could instead be supplied so as to expedite the development of the scenario.

Pertinent questions for discussion include the following:

- Exactly what sort of emergency operation needs to be mounted at this stage?
- On the basis of the information furnished, what special measures and precautions are likely to be needed?
- How should field operations be organized and how should their command system be structured?
- What can be done to obtain more information on what has happened, and what is happening, inside the tunnel?
- What are the coordination needs associated with the emergency operations?
- What are the inevitable limitations of the emergency operations?

At the beginning, unknown elements are at work deep in the middle of the tunnel. As described, the initial emergency is an example of an incident that could turn out to be rather small and easily contained or could develop into a major tragedy, with tens of people killed and a fire that takes days to bring under control. (In the case of the Mont Blanc tunnel fire of March 1999, in which 50 people were killed, the full scope of the disaster became apparent only three days after the emergency started.) The coordinating emergency manager must therefore decide what resources and strategies are justified by the apparent gravity of the event as it develops. In the interests of discussing an event that in the end turns out to be serious, the following updates have been devised.

[Update – 18.29 h]   The temperature in the tunnel has risen to the point at which, from km-4 (north end) and km-10 (2 km from south entrance), it is impossible to move farther in without wearing heat-resistant protective clothing. It seems likely that people have been trapped in the central part of the tunnel. A truck driver who reached the north entrance on foot says that he has abandoned a tanker full of formaldehyde solution in the vicinity of km-3. He also notes the presence at km-2.5 of a tanker full of petroleum spirit with no driver in the cab. Because of the problems of turning or backing up large trucks, relatively few drivers have been able to remove their vehicles from the inner part of the tunnel. No further traffic movements appear to be taking place in this area and no driver able to escape has remained at the wheel. The tunnel is full of dense black smoke. Electricity supplies have been interrupted.

[Update – 18.45 h]   A huge explosion occurs in the interior of the tunnel.

The question then becomes one of adjusting the response to the newly gathered information on the seriousness of the incident and, of course, seeking to protect fire and rescue crews while they are in the tunnel.

A scenario such as this requires students to develop a response strategy that is rapid and effective. It must be flexible enough to cope with sudden unexpected changes in circumstances and it must be prudent, as the lives of rescue workers are at stake. Above all, the disaster manager must not seek to solve all problems by applying a set of ideal measures and there must be no self-recrimination if the response is less than perfect, as conditions are unlikely to allow neat, well rounded solutions. During the progress of the disaster, the emphasis is strongly on rapid answers to immediate problems: blocking traffic, ensuring the early arrival of emergency vehicles, rescuing people from inside the tunnel, monitoring conditions that cannot be predicted or even observed adequately, and gathering basic information from people who have fled the inner part of the tunnel. The order in which priorities are established could be critical to the success of the operation. It is also necessary to establish command posts at either end of the tunnel and to ensure good communication between them. Finally, one may have to cope with the possibility that the explosion that happens 40 minutes after the crash has killed or injured rescue workers. The scenario has deliberately been conceptualized as a tough situation to resolve.

Scenarios are a form of communications model and as such they are part of a continuum defined by the number of roles and the degree of complexity of simulation processes (Fig. 8.2). At one end of the continuum is the Delphic questionnaire method, which is a means of eliciting information from a panel of experts. It assumes that only one role or approach is defined, at the end of the

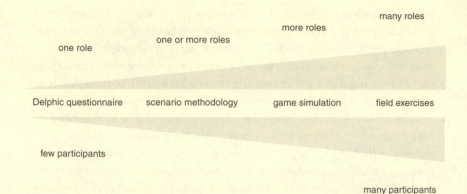

Figure 8.2 The spectrum of methods for emergency management training.

Delphic process, by collective strategies or policies. Because it involves the controlled release of information during the exercise, the method is an efficient means of eliciting opinions from the experts, but it probably has only limited value in teaching. At the other end are field exercises, which are by nature very heterogeneous in composition and aims. In the middle are scenario methods, neither heterogeneous nor unduly restrictive.

As noted above, instructional methods based on scenarios usually involve only one or a few roles to be assigned to participants. If the scope of the exercise is broadened and the number of roles is increased, one can create an emergency-simulation game. Designed to facilitate negotiation between participants and to develop managerial skills, this will begin with a scenario that both sets the scene for role playing and establishes the parameters and limitations of the game. The scenario develops by the processes of decision making, action and negotiation as the participants play their roles and thus carry the game forward. The potential for teaching emergency management by this method is considerable, but few examples of disaster games have been described in the general scientific literature. Strong potential also exists for game simulation using virtual reality and computerized scenario building. Again, few examples appear in the literature.

More attention has been given to the most sophisticated form of emergency simulation, field exercises. Where civil-protection structures are well developed, such activities are usually conducted regularly. They involve multiple role playing on a large scale. Once again, a scenario is used as the starting point of the simulation, often based on the worst future impact to be expected in the area. The scenario teaching methods that form the subject of this paper can help trainee emergency managers to prepare for both field exercises, which need to be conducted efficiently in order to shorten emergency-response times, and real emergencies.

In essence, the scenario method has considerable potential for further development as a means of teaching the principles of emergency management, especially as it is easily integrated with other forms of model and simulation used in this field. Moreover, scenarios can help bridge the gap between the theoretical studies associated with a degree in, for example, Earth sciences or sociology, and the need to solve problems in extreme situations created by sudden emergencies. Scenarios are thus a valuable tool in creating the necessary mental reorientation. Typically, recent graduates who are starting careers in emergency management tend to think about problems in the slow and well modulated way that colleges and universities have taught them. They tend to examine a problem from all sides and search for the best possible solution regardless of constraints. But emergencies rarely allow such luxuries. Instead, they require quick and dirty solutions in which experience and training combine to furnish the inspiration needed to arrive at the best-case answer by quick judgement and without going through long intermediate stages of hypothesis testing. Scenarios can help accustom students to the process of applying theoretical knowledge to practical problems under emergency conditions. Thus, they merit further study and development as a training method.

A corollary of the scenario method is the use of case studies, which can help trainees to learn to analyze events or situations in depth. An event can be transformed into a case study only if it can be turned into a generalization – i.e. if it has something to contribute to general knowledge beyond the unique account of what happened. If an appropriate event can be found, there are several ways to treat it. The simplest one is to construct a narrative account of what happened. This can be organized around a particular question or activity, and must have a strong conceptual basis if it is to bring forth a usable generalization. An alternative is to use the factual material to help answer a series of open questions representing what needs to be known or deduced from the event. On the other hand, the explanatory case study combines an accurate rendering of the facts of the case with alternative explanations of the event, which are then evaluated so that the most plausible or reliable explanation is selected. Finally, a case survey is a synthesis of various case studies. It is used to search for similarities and differences between these events in the light of lessons to be learned. Whichever type of case study is utilized, it should end with conclusions, and perhaps recommendations, that build upon the analysis of the material used.

8.4 Trends in disaster education and training

One can conclude from this brief survey of training methodology that, in the absence of generally agreed guidelines, individual decisions must be made about the choice of methods, strategies, workloads and course content. Most teaching methods, and even possibly the three curriculum models, may be complementary rather than mutually exclusive. But whatever the eventual strategy, it is axiomatic that emergency-management training must be severely practical, although it should always include an adequate basis in theory, for in a crisis this may provide an essential conceptual framework for the harassed manager. In each case, procedures are required to evaluate the effectiveness and efficiency of training methods. The ultimate goal here is to increase the degree of professionalism of disaster managers.

Several present trends indicate the probable future development of training for emergency management. Disasters will continue to cross international borders and so, therefore, must training. This will involve sharing expertise and experiences until the day when widely accepted standards are created that make qualifications derived from training courses internationally valid. During this process it is clear that information technology will play an ever greater role; therefore it should be utilized to the full in training courses. Although information offered over the Internet is often ephemeral and unreliable, a vast stock of resources for learning about disasters and their management is being assembled on the World Wide Web and it deserves to contribute to the learning exercise. Distance learning is becoming increasingly popular and is now often based on the Internet. Courses may be freely accessed or they may require prior payment in order to obtain a password. Other courses are accessible to all comers, but examination and certification are available, at cost, only to people who fulfil a set of study requirements and take an on-line examination.

One of the positive byproducts of training is that the role of emergency manager is gradually becoming accepted by society. This means that the greatest future challenge for instructors and trainees alike is to create a body of professional disaster managers who are capable of involving the public and the private sector in disaster mitigation and preparedness. This is how the public's acceptance can be confirmed and built upon to make communities safer.

CHAPTER 9

Concluding thoughts

In countries that are at peace and have stable systems of government, the average citizen does not devote much thought to disasters or to how to prepare for them. Fair enough: one does not want to encourage a morbid interest in death and destruction. However, every effort should be made to create a social environment in which protecting oneself and one's family, friends and colleagues against disaster is considered normal, not unusual. But until there is a fundamental change in the popular sense of what civil protection is about, we will have to accept that mitigation is in the hands of experts and professionals, not ordinary people. In some future world it may be shared more evenly between disaster planners and lay people, such that citizens take more responsibility for their own safety, but in most countries there is still a very long way to go before that is the case. Indeed, even official disaster planning is still somewhat of a novelty. Training courses are in short supply: a comprehensive list published in 2000 suggested that even in the USA, which has taken such matters to heart, only 1 per cent of universities and colleges offered diplomas, certificates, degrees or postgraduate courses in disaster management, and less than 4 per cent offered training on disasters as part of other qualifications. Moreover, there are as yet few guarantees that education and training courses meet adequate standards of rigour, comprehensiveness and appropriateness of content. Indeed, in this field there are no recognized universal standards or protocols: there is no ISO specification for emergency-management training.

On the other hand, both public and official interest in disaster prevention are growing fast. These are breathless times, characterized by an accelerating pace of social, economic and technological change. They pose a challenge to the aspiring emergency planner: how to make life safe in an increasingly complex world, how to provide some very fundamental security in times of rising uncertainty, and how to adapt the planning process to both present-day changes and future developments that can hardly be foreseen right now.

Given these considerations, what makes a good disaster planner? Apart from knowledge of the theory and reality of catastrophes, and experience in managing them, there are some other basic ingredients. One needs a talent to translate a complex reality successfully into a simple strategy. This requires considerable selectivity in the choice of details and a knack for divining not so much what to include in the plan as what to leave out of it. One also needs the ability never to become bogged down in the details, never to lose sight of the grand plan. Were it not for the fact that military operations tend to be too authoritarian for civilian uses, one might also say that the disaster planner should have the talents of a good general. Levity apart, one cannot over-emphasize the importance of plain common sense in both disaster planning and emergency management. No amount of theory or training can dispel the high probability that what does not feel right probably is not right. Hence, training, experience and common sense are complementary ingredients in the formation of a disaster planner or manager. We can add to these aspects the need for good managerial and interpersonal skills, given that a plan will not succeed unless it can be presented in a way acceptable to users.

Formidable though these requirements may seem, the aspiring disaster planner should not be daunted by them. Although one cannot afford to make serious mistakes during a fully fledged disaster, before that moment there will be ample opportunity to revise and adjust each element of the plan.

Disasters can be both unique and repetitive. Either irregularly or on a pre-dictable cycle they can recur at intervals, and despite the singular character of each event there is much common ground between them. As a result, planners can build on invaluable experience gained in previous events. But these also indicate the need for constant vigilance and continuous improvements in pre-paredness. In no sense can a disaster plan be written and put aside until it is needed: disaster planning is a continuously evolving process of adaptation to changing needs and circumstances. This may mean that the political and finan-cial backers of the plan need repeatedly to be convinced of its importance. When disaster has not struck for a long time, there is tendency to forget how serious and immediate a problem it can become. It can prove fatal in more ways than one to drop one's guard against natural and technological hazards.

Although the prediction of hazard impacts is becoming an increasingly refined and successful science (and art, for that matter), prophecy needs to be practised with extreme caution in this field. However, present trends suggest the following. The amount of fixed capital in the world is increasing steadily, especially in hazardous areas, and hence so are vulnerability and the need for protection against hazards. Moreover, an ever-wider variety of economic activ-ities are being pursued, which complicates the matter; and both personal mobility and leisure activities are increasing, which puts more people more at risk. Although there is an increasing emphasis on safety in transportation, greater volumes of traffic and the spread of mass travel by aircraft and car to new places are causing the rate of accidents and fatalities to rise again through-out the world. Furthermore, the number of dangerous chemicals being pro-duced continues to increase, and their manufacture, storage and transportation pose ever more challenges to safety. Yet, on the other hand, there is little evidence that the frequency and magnitude of natural hazards is increasing, although there are fears that storms and floods may become more intense as a result of the greenhouse effect. The current increase in the size of natural dis-asters is therefore an artefact of human vulnerability, not changing natural extremes. All this adds up to greater, more complex and insistent challenges to emergency planners and managers.

It is debatable as to whether the world is truly turning into the riskier place that it seems to be becoming. Much depends on the scale of the analysis and the factors taken into consideration. Some risks are diminishing and some are increasing. There are rises in vulnerability but there are also gains in mitigation. However, it is clear that risks are becoming more complex and sophisticated. They will pose unprecedented social and technical challenges. Fiscal strin-gency and the application of rigid efficiency criteria to government present the emergency planner with the need to accomplish more with fewer resources. Yet it can be shown that a well run, efficient disaster-management agency offers great value because it saves taxpayers money by reducing losses (the benefit–cost ratios of good hazard mitigation are almost invariably positive, and sometimes spectacularly so). This is an incentive to any rational government, which is not overwhelmed by fiscal ideology, not to strip the resources from such an agency. It is also a strong incentive to avoid that other great pitfall, the bureaucratic management of disaster. To repeat: complex situations need to be attacked with simple strategies, and that goes for the paperwork as well. It is especially true, as countless examples show, that more paperwork does not mean less fraud or less waste of resources.

Even though the overall strategy for planning against and managing disas-ters may be simple, the technology behind it is becoming steadily more com-plex. Information management and communications are the key sectors. For example, extraordinary degrees of personal mobility in communications are being imparted by the use of cellular and satellite telephones. Even further advances arise when personal computers can be connected by mobile tele-phone from the scene of a disaster to vast numbers of users via the Internet. The Internet's ability (or that of parallel intranets) to survive disaster impacts and transmit data, text, audio messages and visual images of all kinds in real time, or nearly so, has opened up extraordinary new prospects for disaster manage-ment. For example, one can plan to have an **anchor desk** at a site more or less remote from that of the next disaster, which broadens the geographical scope of management. During the emergency, information is transmitted to the anchor desk from the emergency-operations centre, or directly from the incident-command post (see §4.1.2). It is sorted at the anchor desk and sent to other locations to be analyzed by appropriate experts, who return their judgements and advice for prompt transmission to the EOC or ICP. This opens up the interesting prospect that, in future, disasters might be managed in part by panels of emergency coordinators and other experts who are not physically present in the disaster area (or even in close physical proximity to one another) but who receive a constant stream of information to work on. Truly, the concepts of space and place are changing radically in this new information age.

However, it is unwise to place too much reliance on technology. Electronic gadgetry is labour saving and powerful, but it cannot work well if the social environment in which it is used does not facilitate good results. Disasters are, in the end, about people – people as victims, survivors, rescuers, carers, managers and planners. Let that not be forgotten in the race to equip ourselves with the latest hardware. Machines do not solve problems (although they can also certainly create them), people do. Hence, in the final analysis, the emergency planner should first and foremost strive to create an environment in which people can work constructively together in a state of mutual understanding and collaboration. This is the key to solving the disasters problem.

Glossary

Many terms commonly used in emergency preparedness and civil protection are explained in the text of this book. With the reader's wider interests in mind, the glossary is intended to be a list of supplementary definitions, including those of terms routinely encountered in the field but not specifically mentioned in this work.

acute toxicity A toxic effect that occurs within 14 days (usually within 24 hours) of exposure to a single dose, or rapidly repeated multiple doses, of a hazardous substance.

advance medical post See first-aid station.

advisory A message to say that a hazard is in the early stages of approaching, and warnings may follow.

aerial photography Systematic acquisition of standardized visual images of terrain and human settlement from a chosen altitude in the lower atmosphere. Most scientific and logistical aerial photography is vertical, panchromatic and stereographic. This means that images (usually greyscale) are similar to visual maps of the land surface and they overlap between successive prints, which have a slightly different incidence of light. Viewed as pairs, they can give a stereographic effect that displays topography and features in three dimensions.

aftershock Part of a sequence of small earthquakes that follow a large main shock. Aftershocks usually diminish rapidly over the first few days after the main shock, although some anomalously large ones can occur.

agency A division of government or a non-governmental organization with a particular function in post-disaster work. Government agencies may have statutory responsibilities for particular areas or tasks.

agent-generated demands Emergency needs arising directly as a result of the phenomenon that caused the disaster (e.g. the need for sandbags to prevent flooding). See also *response-generated demands*.

air attack The use of aircraft directly to suppress wildfire, usually by dropping water or spraying fire-retarding chemicals onto the flames.

alarm A visible or audible signal that warns of danger.

alert A pre-disaster call to place emergency forces on stand-by, pending mobilization.

all-hazards planning Most hazardous places are threatened by more than one set of risks. Besides the presence of several classes of extreme phenomenon, there may be secondary hazards (e.g. earthquake-induced landslides). It is often more efficient to plan for all the major hazards to be expected in a region rather than separately for single ones. This allows economies of scale to be achieved and risks to be tackled comprehensively.

Sources: Emergency Management Australia; California Specialized Trailing Institute; FEMA Emergency Management Institute; Pacific Emergency Management Center; Simeon Institute.

anchor desk   A workstation in contact by Internet with the emergency-operations centre. It receives images, data and situation reports in real time, sends them by Internet for analysis to experts at remote sites, collects the results and forwards them to the EOCs, all very rapidly.

anthropogenic hazard, anthropogenic disaster   See *social disaster* and *technological disaster*.

aseismic   Not seismic; either not subject to seismic risk or not built to resist earthquakes.

assembly area   A location designated for the assembly of people affected by an emergency. They may be victims awaiting evacuation or emergency personnel about to start work at the site of a nearby incident.

background levels   In risk analysis, inherent natural or normal levels of risk in the absence of any specific factor (e.g. an approaching hurricane or a poorly constructed building) that might raise them.

base camp   The location designated in the disaster plan to provide sleeping facilities, food, water and sanitary services to emergency-response personnel.

basic life support   Medical term for basic measures to protect the airway, assist breathing and maintain circulation of a seriously injured person without the use of drugs, defibrillation or advanced techniques. This is usually achieved by mouth-to-mouth resuscitation and heart massage.

benefit–cost analysis   See *cost–benefit analysis*.

black-box model   A model with internal workings that are not specified. The modelling process therefore relates exclusively to the connection between inputs and outputs.

boiling liquid expanding vapour explosion (BLEVE)   The result of the failure of a closed tank that contains a liquid at a temperature above its boiling point. The explosion will come with a pressure wave and possibly shrapnel as the vessel is torn apart.

boundary conditions   Starting values and limiting conditions in simulation and scenario modelling.

briefing   Advice given to emergency personnel about the incident or disaster whose effects they are about to tackle and regarding the tasks that they will perform.

building failure   The partial or total collapse of an architectural or engineering structure (building, bridge, etc.) or its sudden, unexpected and total inability to fulfil the function for which it was designed. Failure occurs when the load (usually the dynamic, or transient, load) on a building exceeds its strength, as expressed in terms of rigidity (stiffness) and flexibility (ductility, or ability to absorb the forces that cause deformation). Large sudden-impact disasters such as earthquakes or hurricanes can cause widespread building failures, and this is one of the principal causes of human casualties in disasters.

cache   A set of tools, equipment or supplies stored at a designated location with the intention that it be used in specific types of incident.

call-up   The resource-deployment command.

cardiopulmonary resuscitation (CPR)   Basic life support for injured people by means of expired air resuscitation and external cardiac compression.

casualty   Death or injury (mortality or morbidity) in disaster. Injury can be divided into physical trauma (e.g. fractured bones) and psychological trauma (e.g. post-traumatic stress disorder). Casualty incidence rates are monitored by epidemiologists.

catastrophe   In this book the terms *disaster* and *catastrophe* are used synonymously, because definitions of the two terms are not sufficiently well developed, precise or subject to consensus that the terms can rigorously be distinguished from one another. However, some authors regard a *catastrophe* to be more cataclysmic than a disaster and to affect a larger area. Jurisdictions affected by catastrophe, or so it is argued, are more thoroughly overwhelmed by it that they would be in the case of a mere disaster. As there are no quantitative measures of the distinction, or even adequate functional ones, I regard it as unsafe.

choropleth map   A map that represents different ordinal classes of phenomenon (e.g. 0–9, 10–19, 20–29, or slight, moderate, severe) as shaded or coloured areas.

civil defence   The progenitor of *civil protection*. It began with early twentieth-century arrangements to protect civilian populations against the effects of warfare, and particularly with 1940s air raid precautions (ARP), a quasi-military means of organizing the public against aerial bombardment and radiation emissions caused by nuclear attack. Civil defence reached its apogee in the 1950s in response to the threat of a nuclear exchange. Given the changing nature of the strategic threat, the need to tackle other hazards, and the impossibility of protecting people against general nuclear war, it has largely been superseded by civil protection.

civil protection   The process of protecting the general public, organizations, institutions, commerce and industry against disaster, by creating an operational structure for mitigation, preparedness, response and recovery. Military forces do not play a central role in civil protection, which is in the hands of administrative authorities, such as municipal, provincial, state or national governments.

command   Authoritative direction of personnel and resources in the performance of tasks associated with emergency management. Laws, bylaws, constitutions, or other legal instruments usually govern who commands whom.

communicable disease   A disease caused by bacteria or viruses that can be spread from one person to another, by human contact, promiscuity or sharing facilities. It is both infectious and contagious.

complex emergency   Although all emergencies are, in a sense, complex, since the early 1990s a specific type has been given this name. It occurs in poor developing countries, although not exclusively. Warfare, civil strife and guerrilla insurgencies combine with natural hazards, such as drought or floods. Civil government is in a state of collapse and the only rule is the force of arms. The result is a long drawn-out emergency that cannot be ended merely by providing relief but requires the restoration of socio-economic stability and peace to break out. In many complex emergencies, relief efforts become caught up in local conflicts and the neutrality of emergency-relief units is thus hard to maintain.

computer-aided dispatch   The use of computer technology to help dispatch personnel, vehicles and materials to emergencies and to monitor the results. Recorded messages, automatic telephoning, vehicle-tracking systems, and computerized inventories are all possible components of computer-aided dispatch systems.

container home   A development of the building-works site office and the goods container. Designed to be transported on a flatbed articulated truck or railway wagon and offloaded by crane onto a level site, it has two small bedrooms, an integrated kitchen, a living area, WC and shower unit, and flexible connections to utilities. A standard design is mass produced to provide medium-term shelter to homeless survivors of disaster. Versions can be adapted for use as clinics, municipal offices, and so on. Modular in form, it is suitable for use as several connected units. The design life is about ten years. Performance is poor in hot weather and high winds, but good in cold conditions. Floor area is about 25–30 m^2.

control   The direction of management and rescue activities in an emergency. Authority for control is specified in legislation and emergency plans. It is allied with the processes of directing and assigning tasks to emergency workers, and assuming responsibility for failures. Many control functions affect multiple organizations.

control area   A site with a perimeter around it to prevent the entrance of unauthorized personnel or equipment.

convergence reaction   The frequently observed tendency of many people to converge upon a disaster area in the immediate aftermath of a catastrophic impact. Rather than being deserted by survivors, residents and others, disaster areas tend to fill up with victims, survivors, relatives, relief workers, scientists, politicians, journalists, VIPs, sightseers, utility workers, and so on.

cost–benefit analysis   In terms of disasters, the benefits of inhabiting areas at risk, of carrying out various activities in them and of putting oneself at risk, set off against the costs of damage and losses in disaster (including estimated future costs of the event) and of mitigation works.

cost effectiveness   A measure of efficacy obtained by quantifying the costs and benefits of

an activity and comparing them in a cost–benefit ratio. Cost effectiveness does not necessarily require a predominance of benefits over costs, nor can all benefits always be quantified
as readily as costs can.

creeping disaster   A disaster with an impact that develops progressively or cumulatively
over a long period of time, usually years or decades, although possibly months. There is no
obvious identifiable point at which the impact suddenly becomes chronic or reaches crisis
point. Slow subsidence or extremely slow landsliding beneath an urban area can be an
example of a creeping disaster.

crisis   In disasters, a point at which normal mechanisms for coping (personal, organizational
or institutional) suddenly cease to function as a result of the seriousness of the impact. New,
more appropriate (and often more Draconian) procedures must be instituted with alacrity.

critical-incident stress (CIS)   A form of *post-traumatic stress disorder* that occurs in frontline
emergency-relief workers who experience frightening, horrifying or otherwise dramatic
conditions and which lead to a strong psychological reaction, often of depression and inability to cope. CIS can usually be abated by rest and counselling. Debriefing also helps,
although it may instead exacerbate the condition.

culture   The accumulation of beliefs, traditions, customs and historical memories that collectively define how a given group or population lives. It determines attitudes and actions
and hence disaster planners should view it as both a constraint upon the range of possible
emergency actions and a positive factor, in that cultural identification can be reinforced in
order to promote social solidarity and resistance to disaster.

damage assessment   Investigation of damaged property and quantification of the value of
losses. Monetary estimates usually depend on the cost of repairing the damage, which in turn
depends on the adoption of particular techniques and components whose values are known.

database   A collection of information (in numerical or textual form or as images) that has been
classified and stored in such a way that it can be accessed selectively by potential users. Databases can be stored on cards or as other hard-copy archives, but increasingly they are kept
as digital records accessible via computer networks or the Internet.

debriefing   The process of collectively analyzing, discussing and digesting the lessons of an
incident or disaster at the close of the emergency phase. It takes the form of an informational
meeting and discussion among participants in the relief effort.

decision-support system   A computer model that uses pre-determined logic and strategies
to help the decision-making process to become explicit and more rational. Information can
be supplied from a databank in order to inform the decision, and rules can be set to guide
it. See also *expert system*.

Delphi technique   A means of acquiring expert opinion on a problem (the name derives from
the Oracle at Delphi in ancient Greece). A panel of experts is convened and supplied with
a questionnaire. The answers that they give are analyzed and used to compile a new questionnaire, which is submitted to the panel. This method, which involves no dialogue between
participants, avoids domination of the discussion by more assertive members and leads to
a written record of opinions.

desktop exercise   An indoor simulation of an emergency-response situation. Usually a
scenario is explicated by group discussion, and strategies are worked out collectively. Also
known as a *tabletop exercise*.

digital elevation matrix (DEM) or **digital terrain matrix (DTM)**   A three-dimensional block
diagram based on spot heights, which are usually regularly spaced and are stored as {x, y,
z} values in a matrix. It can be shaded, rotated or viewed from different perspectives and with
different light sources.

digital terrain matrix (DTM)   See *digital elevation matrix (DEM)*.

disaster epidemiology   A branch of medicine and public-health studies primarily directed
towards saving lives in disaster impacts, preventing post-disaster disease epidemics, and
maintaining the general health of the affected population. It is practised by medical doctors,

public-health experts and statisticians, all acting as epidemiologists. See *epidemiology*.

disaster medicine An umbrella term for the work of medical, surgical, nursing, paramedical and public-health experts in saving lives, treating the injured and caring for the survivors of disaster.

disaster syndrome A psychological defensive reaction pattern first observed systematically by the psychologist Anthony F. C. Wallace in the aftermath of the tornado that struck Worcester (Massachusetts) on June 9, 1953, and killed a total of 94 people. The syndrome consists of one stage of spontaneously withdrawing contact from the world around the subject person and three stages of progressively re-establishing it. To victims of the syndrome the disaster appears as a sort of symbolic "end of the world".

dispatch The process of sending emergency responders to the scene of an incident or disaster. It is usually carried out from a control room or emergency-operations centre and may be conducted manually by operators with telephones or radios, or using computer technology and recorded messages.

dose rate In risk analysis, impacts per person per unit time. The concept stems from the analysis of radiation hazards.

duty log A record of events and actions taken over a disaster manager's or emergency-service dispatcher's period of duty during the emergency phase of a disaster.

earthquake intensity A measure of the perceived shaking effects of earthquakes and of the damage they cause to the built and natural environments. Intensity scales are not perfectly correlated with the magnitude of seismic waves and energy expenditure of earthquakes. Although it partly depends on these factors, seismic intensity is also a function of local and site-specific geological conditions, and the resistance of buildings to dynamic displacement. In the Americas the modified Mercalli scale is used; in Europe, the Mercalli–Cancani–Sieberg (MCS) and Medvedev–Sponheuer–Kárník (MSK) scales are used; in Russia the scale is GOST, and in Japan the Japan Meteorological Agency (JMA) scale is current. Apart from the seven-point JMA scale, the others have 12 divisions, which vary incrementally from imperceptible shaking to cataclysmic damage. Different scales are used, and scales are periodically revised, because building typologies differ from place to place and over time. As the most important feature of the scales is damage to buildings, they must reflect the propensity to damage of each construction type.

earthquake magnitude A standardized measure of the size of the largest seismic waves created by the movement of a geological fault in Earth's brittle crust. Charles Richter's 1935 magnitude scale is now known as local magnitude (M_L), and is used in conjunction with other scales, such as those for body waves (M_B) and surface waves (M_S). As the scales are logarithmic, earthquakes with high magnitude are very much larger than those with moderate or small magnitude. Seismic magnitude is an indirect measure of the amount of physical energy released by an earthquake.

electronic distress beacon (EDB) A battery-powered beacon used to alert authorities to a distress situation such as a sinking ship or crashed aircraft.

elements at risk The people, buildings and structures, infrastructure, economic activities, and services exposed to hazards.

emergency An imminent or actual event that threatens people, property or the environment and which requires a coordinated and rapid response. Emergencies are usually unanticipated, even though they can, and should, be planned for. It is implicit that the consequences of ignoring an emergency or not dealing with it properly are avoidable casualties or damage.

emergency management Short-term measures taken to respond to particular hazards, risks, incidents or disasters. Resources and manpower pertaining to government, voluntary and private agencies are organized and directed on the basis of a plan that anticipates needs and coordinates efforts by assigning tasks to particular responders, organizations or field units.

emergency mapping The cartographic depiction of selected aspects of disaster impact or the subsequent emergency-relief effort. As the situation can change rapidly because of increases in the knowledge of damage and casualties, or to developments in the relief effort, there is a certain imperative to emergency mapping. It is therefore best carried out using computers with programs, such as geographic information systems, that produce refresher graphics.

emergency medical services (EMS)   The organization of facilities, personnel and equipment for the benefit of people who have been injured in disasters and for longer-term care when regular medical systems have been damaged or are insufficient. Doctors, nurses, paramedics, hospitals, clinics, ambulances, air ambulances, communications systems, pharmaceutical supply systems and epidemiologists are involved in EMS.

emergency medical technician (EMT)   A medical specialist with particular skills in pre-hospital emergency medicine. Ambulance crews may contain one or more EMTs.

emergency-operations centre (EOC)   The "nerve centre" of disaster-relief housing in a room or rooms, an apartment or a building located outside the main disaster area, but usually not at great distance from it. Here, emergency operations are directed. The EOC must have good communications channels to hospitals, centres of government, and field-operations posts or directly to operations in the field. The EOC can consist of a single chamber or it can have two or more rooms. These may include a conference room for determining strategy, a situation room for accumulating and analyzing information, a communications room for transmitting and receiving information, a media-briefing room and a room for resting and recuperating in. Generally, small jurisdictions have only one- or two-room EOCs, perhaps consisting of a conference room and a general operations room.

emergency-position-indicating radio beacon (EPIRB)   A compact, buoyant, self-contained radio transmitter that operates at frequencies of 121.5 megahertz or 406 megahertz. It may use satellite-based communication.

emergency-support function (ESF)   A functional area of emergency response, such as transportation, logistics or communication, designated by an emergency-coordination agency. In an emergency-operations centre one or more staff members will have responsibility for coordinating each ESF. They will liaise with personnel in the field and with ESF coordinators at other locations.

endemic disease   A disease that is usually present in a given geographical area under normal conditions. Hyperendemic diseases are particularly persistent and resistant to eradication by programmes of control.

epicentre   The point on Earth's surface directly above the focus (hypocentre) of an earthquake.

epidemiology   The systematic study of particular diseases, medical and sanitary conditions and classes of injury in human populations. See *disaster epidemiology*.

ETA   Estimated time of arrival.

ETD   Estimated time of departure.

evacuation   Precautionary, temporary, planned removal of people or movable items that, if left in place, would result in avoidable casualties or damage.

event tree   A flowchart that expresses a sequence of events, usually connected with industrial processes and their possible breakdown or failure. Event trees are used in risk analysis to identify modes and pathways of failure.

exercise   A field or desktop simulation of an emergency and its management. Exercises are used to train personnel, test plans, work out logistics, identify needs and weaknesses in emergency-management provisions and demonstrate capabilities.

expert system   A computer-based logic system in which formal logic and a database are provide answers to particular sequences of questions. It can be used to help make decisions in complex situations such as emergencies.

exposure   The length or proportion of time that a person, building or other entity runs a risk. Fixed capital (e.g. houses, bridges, factories) is permanently at risk of non-seasonal hazards. Seasonal hazards, such as floods and hurricanes, vary the proportion and intensity of exposure. Routine behaviour problems also cause exposure to vary. For example, if the principal risk arises from the collapse of bridges during earthquakes, people will be most exposed to it during daily commuting to or from work or on other forms of regular journey.

failsafe design   In effect, designing buildings and structures to fall down safely, or machinery to break safely. Failsafe design tries to ensure that, when things go wrong (e.g. when an earthquake causes part of a building to collapse), the worst consequences do not occur (e.g. people survive the collapse because they are able to take shelter). Much of failsafe design procedure involves preventing one failure from leading progressively to another (e.g. when one bolt shears through, all the others will follow suit).

failure mode   The manner in which a component, piece of equipment or physical system fails, as expressed by the consequences when this happens. Failure mode and effects analysis (FMEA) is used to analyze patterns of potential failure modes in a technical system. Failure modes, effects and criticality analysis (FMECA) ranks the patterns of failure in terms of their potential seriousness and likelihood.

fan-out contact system   A progressive means of alerting or calling-up emergency personnel. A contacts B and C, B contacts D and E, C contacts F and G, and so on. When carried out by telephone, this is sometimes known as a *telephone tree*. Cross checking and alternative arrangements are needed to avoid breaks in the sequence as a result of the absence of any of the participants.

fault tree   A flowchart used in engineering to define possible failure patterns in equipment. The fault tree represents combinations of system states and causes of failure that can contribute to a specified major failure (called the *top event*).

Federal Emergency Management Agency (FEMA)   An agency of the US Federal government charged with reducing losses caused by disaster and protecting critical infrastructure. The head of FEMA is a member of the Cabinet of the US President. Many of FEMA's emergency-management structures and programmes of emergency management and hazard mitigation are regarded by other countries as models to emulate.

field command post (FCP)   See *incident-command post (ICP)*.

field exercise   An emergency-management simulation exercise conducted out of doors and designed to test plans, acquaint personnel with tasks and work out logistics. It is based on a reference scenario that details what will happen during the simulation.

field hospital   A mobile medical unit in which wards, operating theatres, etc., are usually housed in tents. It mainly supplies back-up care in disasters.

fire appliance   A generic term used to describe any fire-service vehicle used to fight fires or tackle other emergency problems.

fire containment   The construction during a fire emergency of a barrier to stop the flames from spreading.

first-aid station   A field medical post, usually either in the open air or in a tent, where triage is performed on patients and their condition is stabilized. Significantly injured patients are taken from the advance medical post to hospital.

first-wave protocol (FWP)   The procedure for assigning particular proportions of patients after primary triage to different medical centres.

first-wave ratio (FWR)   The relative capacities of medical centres with respect to their initial ability to receive and care for the victims of a mass-casualty incident.

flashflood   A flood that rises rapidly (i.e. in a few hours or less) with little or no advance warning. Most flashfloods result from intense rainfall over a small area; some are caused by dam failure or by the sudden release of water from behind an ice barrier or landslide that blocks a stream.

flood stage   The height of water in a river channel at which water starts to spill over the banks and inundate the floodplain. It correlates with a particular flow rate, or discharge in metres per second.

floodplain   Level land that may be inundated by floodwaters when a river overflows its banks. In some countries, notably the USA, floodplains are defined as the area with a 1 per cent chance of being flooded in any year. In reality, floodable land may extend beyond such zones. Equivalent areas exist on low-lying coasts, where storm surges may result in flooding by sea water.

311

floodproofing   Design or alteration of a particular building or structure to resist flooding or reduce its impact.

focus (of earthquakes)   The point on a fault at which fault movement begins and from which earthquake shaking spreads dynamically outwards. It is also known as the *hypocentre*. Shallow-focus earthquakes do most damage and have foci located less than 50 km beneath Earth's surface. The point on the surface directly above the focus is the *epicentre*.

forcing function   The "motor" of a simulation model: the subsystem component that replicates the fundamental process that causes things to happen.

forecast   See *prediction*.

frequency   With respect to natural hazards, the number of events of a given or minimum size per unit time (e.g. the number of earthquakes of magnitude greater than 5.9 per 100 years). The reciprocal of frequency is *period*, the average interval between events of a given size. For example, the *return period* (or *recurrence interval*) of an earthquake of magnitude Mw = 6 may be 75 years (i.e. the frequency is 1.33 per 100 years). As average values are used, the measure takes no account of irregularity of occurrence. Highly irregular events do not easily yield an accurate measure of frequency and period.

games simulation   An interactive group simulation with a set of rules based upon simplifications of real situations. The game proceeds by a combination of set rules and constraints, negotiation and interaction among the participants, and chance (e.g. the throw of a die). It enables participants to sharpen their interpersonal and managerial skills and their abilities to deal with unexpected developments.

general systems theory (GST)   A general theory of physical, environmental and social systems, which defines the concepts of boundaries, energy flows, subsystems, thresholds, energy-flux regulators, and so on. When it was first formulated in 1950, GST was intended to be a unifying concept in the sciences because of its wide applicability. However, this has made it a somewhat blunt tool of inquiry, although it is still used, for example in hydrology and electrical engineering.

geographic information system (GIS)   A computer graphics and analysis software package that allows spatial (i.e. geographical, territorial) information to be digitally encoded (digitized) and displayed with respect to {x,y} coordinates. Many different variables can be encoded and displayed in relation to each other as selected *overlays*. Spatial analysis can be carried out on the variables and their mutual relationships. GIS is particularly useful in the mapping and study of risk, vulnerability, hazard, impact, response and mitigation. However, it is only a tool and not a substitute for analysis and policy.

global positioning system (GPS)   A navigational system based on 24 satellites in orbit 20 000 km above Earth's surface that transmit a set of signals to ground-based receivers that enable locations to be determined precisely.

grey-box model   A model whose internal workings are specified by certain simplified processes or procedures that enable the modeller to gain an understanding of how real processes work and thereby to produce a satisfactory output response to input data.

ground truth   Field confirmation that what is apparent in aerial photographs or satellite images has been interpreted correctly. For example, if dark or mottled tones on the images appear to represent areas burned by wildfire, ground truth is sought by going into the field and looking at the areas to ensure that this is the case.

hazard   A natural, technological or social phenomenon that threatens human lives, livelihoods, land use, property or activities. Some hazards may result in a single disaster impact, others are recurrent on a regular (e.g. seasonal) or irregular (i.e. random) cycle. The majority are recurrent rather than unrepeatable events. Many types of hazard impact can be characterized by a magnitude-frequency relationship in which the larger the impact the lower its frequency of occurrence.

hazard watch   The period in which signs of impending disaster impact intensify without reaching the point at which the timing of the impact can be definitely forecast. The watch may give rise to a generalized alert, or even a mobilization, although more commonly it puts emergency forces on stand-by. If the hazard abates rather than intensifies (i.e. the danger passes), the watch may terminate in a *stand-down* or *all-clear*. Otherwise, it will be upgraded to a full-scale warning and alarm, with complete mobilization of the emergency forces.

HAZMAT team A hazardous-materials response team of professional workers who have specialized equipment and expertise for clearing up toxic spills, fighting catastrophic pollution, neutralizing uncontrolled chemical or physical reactions, and preventing fires and explosions that can spread dangerous substances.

helibase The main location for parking, refuelling, maintaining and loading helicopters involved in emergency work.

helispot A location designated as a safe landing or take-off area for helicopters.

hypocentre See *focus*.

IDNDR See *International Decade for Natural Disaster Reduction (IDNDR, 1990–2000)*.

incidence The number of new cases of a disease or injury recorded in a given population in a specified period of time, usually 24 hours.

incident A sudden event, usually resulting in an *emergency*, that requires a response from one or more agencies. Incidents are more restricted in scope and consequences than are disasters.

incident-command post (ICP) or field-command post (FCP) Mobile communications and command centre stationed on the periphery of the main disaster area and in close communication with both the scene of field operations and the emergency-operations centre.

incident-command system (ICS) A means of organizing disaster-response forces in a non-hierarchical manner, so that units work alongside one another under the aegis of a coordinator, but are not subject to a vertical chain of command. The ICS method works best where units can easily be assigned different tasks, with minimal overlap, and where communication between units is excellent, so that information on the disaster and relief efforts is widely shared among emergency workers.

incident commander The person who is responsible for directing emergency operations at the site of an incident or disaster.

intensity of impact In broad terms, number of deaths, physical and psychological injuries, and people rendered homeless by the disaster; scope of destruction and damage; effect on industrial and commercial productivity, and on employment; effect on human activities, and scale of donations, extra taxation and other financial reparations.

International Civil Defence Organization (ICDO) An intergovernmental organization based in Geneva that works to develop and coordinate civil protection. It collaborates with governments to mitigate and prepare for disaster.

International Decade for Natural Disaster Reduction (IDNDR, 1990–2000) Proposed in 1984 by Dr Frank Press of the US National Academy of Sciences, the IDNDR was instigated under United Nations' sponsorship in order to mitigate natural disasters and improve emergency preparedness throughout the world. Various international conferences and projects were organized and more than 100 countries set up national IDNDR committees and programs of research, training and education in the field of natural disasters. International programs included Project Radius, for the reduction of seismic vulnerability in urban areas. The IDNDR has been superseded by a more permanent arrangement, the *International Strategy for Disaster Reduction (ISDR)*, again under UN auspices, with headquarters in Geneva.

International Federation of Red Cross and Red Crescent Societies (IFRC) A worldwide grouping of national humanitarian organizations that work to prevent disaster, reduce suffering and provide emergency aid.

International Strategy for Disaster Reduction (ISDR) See *International Decade for Natural Disaster Reduction (IDNDR)*.

intranet A computer network that connects the local area networks (LANs) of a single organization. If there is added outreach, it can be termed an *extranet*.

313

involuntary risk   A risk that is borne because there is no reasonable alternative, and because the bearer is unable or reluctant to forgo the benefits associated with taking the risk.

isoseismal   A contour line that divides areas affected by an earthquake according to seismic intensities. This, the MM=IV/V isoseismal, divides areas affected to modified Mercalli intensity IV from those affected to intensity V.

Jehovah complex   Among emergency managers and disaster-relief workers, the illusion that they can solve any problem with which they are faced. It results from loss of judgement attributable to stress and exhaustion. See also *magna mater complex*.

lahar   A mudflow composed of water and volcanic deposits. Primary lahars occur during, or as a direct result of, eruptions; secondary lahars have other causes (e.g. the collapse of a crater wall and sudden drainage of a crater lake).

land-use control   The processes of urban and regional planning, and of regulating public health, that impose restrictions and interdictions on the uses to which particular plots of land are put. It is one of the principal instruments of non-structural hazard mitigation, because it can be used to prevent development, or require structural mitigation, at sites where there is a demonstrably strong risk of damage and casualties. However, it is often less effective where development has already taken place.

landing zone   See *helispot*.

LC$_{50}$ The concentration of a toxic substance that is estimated to kill 50 per cent of a population of laboratory animals in a specified short period of time.

LD$_{50}$ The dose of a toxic substance, in mg per kg of body weight, that is estimated to kill 50 per cent of a population of laboratory animals in a specified short period of time.

levee   A raised linear earthen defence against flooding, usually along the banks of a river channel.

lifelines   Vital communications and essential services that are liable to be compromised in disaster. They include the transportation networks along which emergency vehicles and evacuees will travel, main utility corridors for the distribution of electricity and water, and the medical-assistance infrastructure. It is one of the prime objectives of disaster planning to preserve lifelines during impact or restore them quickly afterwards.

liquefaction [potential]   Liquefaction is the spontaneous transformation of a solid material to one that behaves as a liquid. It is common in geological formations that contain lenses or layers of saturated sand and in "sensitive" clays (those whose microscopic structure is highly sensitive to sudden rearrangement during disturbance). It is usually caused when the material cannot disperse the oscillatory stresses caused by earthquake shaking as quickly as they are generated. As the formations subject to liquefaction can be identified by geological fieldwork, susceptibility (*liquefaction potential*) can be mapped.

logistics   The planning and execution of field operations, such as transportation, warehouse management, and deployment of rescue squads.

long term   In the aftermath of disaster, the period of reconstruction and post-disaster development, or of persistent effects. A major disaster may create effects that last for 10, 25 or, in a very few cases, 50 years.

macroseismic field   The area that experiences the visible effects of earthquakes in terms of casualties, damage and disruption. It is usually defined as the area enclosed by the III/IV isoseismal.

macrozonation   The division of a region or large area into zones that express future hazards, vulnerabilities or risks on the basis of accumulated information about where disasters have struck in the past and what distributed spatial effects they have had. Appropriate scales are 1:50 000 to 1:250 000. See also *microzonation*.

magna mater complex   Among emergency managers and disaster relief workers, the desire to take all problems onto their own shoulders without delegating responsibility. Like the

314

Jehovah complex, it results from loss of judgement attributable to stress and exhaustion.

magnitude   The physical strength of the hazard impact. It is often expressed in terms of kinetic-energy liberation or some surrogate measure (e.g. height of waves). It correlates with damage and effects of impact, but only loosely, as the physical force meets with a resistance inversely proportional to the vulnerability of the items it impacts.

main shock   The largest earthquake in a particular sequence.

mass-casualty incident   A disaster in which many people are killed or seriously injured. There is no threshold number of casualties, but the total will probably exceed the capacity of ambulance services and medical centres to deal with injured people using normal procedures. Emergency plans will be put into action, including triage to determine who is treated first.

medical transportation area   The section of the triage-post area where casualties are loaded into ambulances for transportation to hospital. Also known as a *staging area*.

medical triage officer   A doctor, nurse or other qualified medical worker who performs triage on patients at the advance medical post or triage station.

medium term   In the aftermath of disaster, the period that extends from the end of the initial emergency to the point, usually years later, when reconstruction is firmly established. It is centred on the period of initial recovery and repair of damaged services.

mental map (perceptual map)   A memorized or sketched map of the physical or human landscape. People tend to distort spatial relationships in the way they remember them. For example, they recall more detail about nearby or familiar places than about distant or unfamiliar ones. This has an important bearing on emergency management, as decisions are often taken in stressful situations on the basis of information that is far from complete.

microzonation   The division of a small area into zones that express future hazards, vulnerability or risks on the basis of field survey of past impacts and present dangers. Appropriate scales are 1:2500 to 1:25 000.

mitigation   Medium- to long-term activities designed to reduce the impact of future disasters. Methods are divided into structural mitigation (e.g. building levees along a river to reduce flooding), semi-structural (e.g. allowing floodable areas to exist along a river floodplain in order to contain floodwaves) and non-structural (e.g. flood-damage insurance). Most modern mitigation strategies involve combinations of methods.

morbidity   In disaster, physical or psychological injury or illness. See *casualty*.

mutual aid   Voluntary provision of services and assistance by one jurisdiction to another. Mutual aid is usually governed by a compact or agreement that specifies its nature, limitations and the circumstances under which it will be requested or offered.

natural disaster   The impact of an extreme natural phenomenon on the human system (lives, livelihoods and activities). For disaster to occur, the impact must exceed the ability of the human system to absorb or reflect it without suffering considerable, albeit temporary, disruption and losses.

natural hazard   An extreme natural phenomenon that threatens human lives, livelihoods or use of land. Many geophysical phenomena are not hazards but resources in their less extreme forms. For example, rainfall is essential to water supply, but when it is excessive it can cause floods, and when it is lacking, drought may occur. Thus, extremes can be viewed in terms of both excess and deficit with respect to long-term average values.

non-structural mitigation   Measures used to reduce the impact of future disasters without the use of engineering or architectural techniques. Non-structural mitigation includes insurance coverage, land-use and planning measures, and emergency management.

organizations (in disaster)   Discrete social systems characterized by high degrees of internal interaction. Sociologists divide them into *adapting* (changing function to meet the needs of the disaster), *emerging* (newly created to meet the needs), *expanding* (absorbing volunteers or conscripted or convoked members), *extending* (enlarging their brief to meet needs created by the disaster), and *redundant* (not useful to the relief effort). Sudden, unexpected and intense impacts can put organizations into crisis, although usually temporarily.

peak discharge The maximum flow rate, or discharge, in a stream channel during a flood. If it exceeds channel capacity (the maximum discharge that the channel can hold), overbank spillage flooding will occur.

period See *frequency*.

planning map A map on which prescribed or suggested land uses or planned processes are shown. It serves a prescriptive, rather than a representative, function.

post-traumatic stress disorder (PTSD) A psychological condition induced by close or prolonged contact with profoundly disturbing conditions. PTSD tends to be greatest in people who have little psychological preparedness (through lack of either training or experience) for disaster and where the worst consequences of the impact are sudden, unexpected and very graphic. PTSD does not usually develop into a permanent psychological illness, although it can lead to long-term states of depression.

prediction A statement to the effect that a particular impact will occur in a particular area, with a particular magnitude and set of effects, during a particular time interval. Predictions are the responsibility of bona fide scientists and are usually given in probabilistic terms. The word is virtually synonymous with *forecast*.

prefabricated dwelling ("prefab") A small building that is manufactured in sections, transported to a site and erected there on a prepared foundation, usually a concrete pad. Light-walled prefabs are usually built with wooden frames that have glass-fibre infill, heavy-walled prefabs are made in sections of concrete. A light-walled prefab may last 10–15 years or more if it is properly looked after. Such buildings are widely used as medium-term shelter for homeless survivors of disaster. Their sites require considerable urbanization work in order to provide stability, access and utility services.

preparedness Short-term actions taken to reduce the impact of impending disaster. The actions include warning, evacuation and stockpiling of emergency supplies.

probable maximum precipitation (PMP) The greatest quantity of precipitation per unit time that is expected to occur in a given area with a given recurrence interval (e.g. 500 years). It forms the basis of civil-engineering calculations concerning runoff and water yield.

probable maximum flood (PMF) The flood resulting from the probable maximum precipitation and the worst flood-producing catchment conditions that can realistically be expected.

protocol A set of standard procedures for carrying out a task or achieving a well defined goal. The term is widely used in emergency medicine and in computing.

public information officer (PIO) The disaster manager's spokesperson. The PIO's task is to liaise with the news media and representatives of the general public and private organizations. He or she must ensure that news and information passed on to the public via these sources is not only timely but also accurate, and that unfounded rumours are supplanted with correct information. The PIO provides a vital link between the public and the disaster-relief community. This is especially important, as public sensitivity to the problem of catastrophe, and therefore mitigation and preparedness activities, depends on the availability of appropriate information, as does the warning process on many occasions.

pyroclastic flow A ground-hugging flow of solid particles suspended in gas caused by an explosive volcanic eruption. Pyroclastic flows can be directional (*nuées ardentes*) or non-directional (some *ashflows*) and may have temperatures as high as 600°C and speeds of more than 100 km per hour.

radionuclide An atomic nucleus undergoing radioactive decay.

reaction time The time that elapses between genesis an incident and the departure of the first emergency crew. It encompasses notification of the emergency-operations centre and dispatch of the crew.

recovery The phase of restoration of basic services after disaster (e.g. electricity, mains water supply, transportation and healthcare). Reconstruction is planned during this phase.

recurrence interval   See *frequency*.

release   In risk analysis, the rate at which a hazard strikes, which is usually expressed in terms of frequency and return period, or with respect to the trend of cumulative impacts.

rendezvous point   A pre-arranged location for mustering personnel and assembling vehicles and equipment when an emergency strikes.

response   The immediate and short-term reactions of the disaster-relief community to an emergency situation. It includes search-and-rescue operations medical assistance, and the early provision of food and shelter for survivors.

response-generated demands   Emergency needs arising from the damaged social fabric, rather than directly from the agent that caused the disaster (e.g. the need for shelter for homeless survivors). See also *agent-generated demands*.

return period   See *frequency*.

risk   The potential interaction of hazard and vulnerability for a given exposure of the items at risk. It is often expressed as an index (a decimal proportion of 1, certain loss) that can easily be converted to a percentage when required. Risk is taken on voluntarily (as with dangerous sports) or involuntarily (as with the journey to work), although the distinction is somewhat arbitrary. Risk is communicated, perceived, analyzed and managed. All human activities involve some degree of risk, but people's attitudes to the risks they run are almost never completely logical and rational. It is extremely difficult to compare different risks and reduce them in relation to their objective seriousness. Hence, risk-reduction strategies tend to embody a degree of arbitrariness which is a function of public preferences. Risk is also an entirely hypothetical concept, in that, when damage or losses occur, the risk that they would happen ceases to exist.

risk analysis   The systematic study of risk conditions and the probable impacts of future events, incidents and disasters.

risk assessment   The determination of priorities for risk reduction by evaluating and comparing risks against each other, against targets for reduction, or against particular standards or criteria.

risk communication   The communication of information about particular risks to the public (individuals, groups, organizations, institutions) and monitoring of the public's response.

risk estimation   The process used to produce a measure of the level of risks being analyzed. Risk estimation consists of the following steps: frequency analysis, consequence analysis and their integration.

risk evaluation   The process of judging the acceptability of risks or ranking them in terms of priority for reduction.

risk management   The process of applying measures that reduce risks, having identified, characterized and analyzed them.

scenario   A hypothetical sequence of events, based closely on observation of real circumstances and on logical projection of their consequences. It is used to develop a detailed picture of how hazards, impacts, relief efforts and so on will probably develop over time. It is thus an important basis for planning.

search and rescue (SAR)   The process of searching for trapped or lost people, locating them, rescuing them and, if they are alive and injured, giving them initial medical treatment prior to transportation to hospital. In urban settings it is known as *urban search and rescue (USAR)* or *urban heavy rescue (UHR)*. USAR usually requires work to be concentrated at particular sites (e.g. collapsed buildings or transportation crashes) where fires have to be extinguished and rubble or wreckage moved. Rural or regional SAR often involves wide searches for missing people, covering much ground or sea area. The main aim of SAR and USAR is to rescue people alive and save them from death. They therefore need to be carried out with great rapidity and efficiency: few people can survive under rubble or in hostile open-air conditions for more than a day.

secondary hazards, secondary disasters Threats, risks and impacts that result from the impact of the primary hazard (e.g. an earthquake) on a particular source of risk (e.g. a dam that might collapse).

sensitivity analysis The process of systematically varying the numerical values input into a digital model in order to gauge the scope and nature of output responses.

short term In disasters, a period of a few days to a few weeks that stretches from the impact to the point at which repair and recovery are firmly established processes. It is centred on the initial post-disaster emergency.

simulation exercise A disaster drill based on a reference scenario. The simulation can be conducted indoors as a theoretical *desktop exercise* or in the open air as a logistical *field exercise*.

social disaster The impact on the human system (lives, livelihoods, activities and property) of riots, crowd crushes, terrorist outrages, etc. Together with technological disasters, this category falls under the umbrella heading of *anthropogenic disaster*.

span of control The ratio of supervisors to supervised people or units within an emergency-management organization. The optimum ratio is considered to be 1 to 5.

structural mitigation Engineering and architectural measures to reduce the impact of disasters. Examples include river levees, anti-seismic buildings and avalanche barriers on slopes.

tabletop exercise See *desktop exercise*.

technological disaster Severe impact upon human lives, livelihoods or activities caused by the malfunction, breakage or misuse of a piece of modern technology. Transportation crashes, radiation emissions, toxic spills, and chemical or gas explosions are included in this category. Together with social disasters, the category comes under the umbrella heading of anthropogenic disaster.

technological hazard A threat posed by the use of advanced modern technology, for example, related to the emission of toxic substances (radionuclides, toxic chemicals, etc.) into the atmosphere or water supply or onto the land.

tephra The solid, fragmentary products of a volcanic eruption: ash, lapilli, blocks and bombs.

thematic map A map of the distribution of a particular phenomenon or characteristic (e.g. solid geology, landslides, coasts undergoing erosion).

triage The process of classifying injuries on the basis of immediate need for medical attention coupled with an assessment of the benefit that the patient is likely to derive from a given type of treatment. The highest priority is given to seriously injured patients who stand a good chance of recovering if treated properly. The lowest priority goes to dead, moribund and slightly injured victims. Triage is practised to determine who receives priority in first aid, transportation to medical centres and treatment on arrival. Tags have been designed so as to designate patients by priority level and accumulate basic data on their condition.

tolerable risk A risk that people who are exposed to risk are expected to bear it without excessive concern. Tolerable risks may be accepted reluctantly, but accepted all the same.

top event See *fault tree*.

trauma Physical or psychological injury.

unit A small group or element within an emergency-management organization, which is dedicated to a particular function or set of activities. In the field a *unit* may be synonymous with a *squad*.

unity of command The idea that each person within an organization reports to only one designated leader, who has sole authority over that person's activities.

unreinforced masonry building (URMB) A structure built of brick, cement block or stone, with or without lime or cement mortar. URMBs tend to perform poorly in earthquakes because they are weak, inflexible and heavy enough to generate relatively large inertial displacements. Brick, weak stone, cement or mortar can be pulverized by the hammer effect of vertical earthquake motions. However, URMBs can be strengthened by adding ring beams to the tops of walls and cementing soldered steel grids to walls. These are measures that increase the unitary strength and ductility of masonry (i.e. its ability to absorb the temporary displacements caused by dynamic loading).

urban search and rescue (USA)   Sometimes known as *urban heavy search and rescue*. See *search and rescue*.

verisimilitude   Correspondence between model (or scenario) and reality. A good model is a functional reduction of complex conditions of reality, which by selectively simplifying them preserves the essence of the reality and thus makes it easier to understand.

vernacular buildings   The workaday tradition in architecture. Ordinary, often mass-produced, buildings constructed according to the usual local practices.

volunteer emergency worker   A person who voluntarily engages in emergency-relief work, usually as part of a group or organization. Unorganized volunteers can be a drain on emergency workers. It is best that volunteers be trained members of accredited organizations that have agreed roles during emergency situations.

vulnerability   Susceptibility to loss, damage, destruction or casualty in future disasters. It is strongly related to poverty, lack of mitigation, lack of political power, and marginalization, but it cannot be predicted completely by any of these factors. Hence, its origins tend to be complex.

warning   An alarm signal or message coupled with a recommendation or order to take action (such as mobilize or evacuate). The warning is a more advanced stage of mobilization than the hazard watch and it pertains to conditions in which there is a high probability, or virtual certainty, that disaster will occur relatively soon (within minutes, hours, or at the most a few days). Warnings can be subdivided into the technological processes of conveying the message (i.e. communications systems) and the social process of informing the people and ensuring that they understand and act upon the warning message. Warnings are separate from predictions. Whereas the latter are the responsibility of scientists, warnings are the preserve of civil administrators.

wildfire   Environmental fire, forest fire, bush or brush fire. Unplanned, uncontrolled outbreaks of fire caused by natural or anthropogenic causes. Lightning is an example of the former, arson of the latter.

window of opportunity   During the aftermath of a disaster there is often a stronger-than-usual demand for safety and mitigation measures. In this period, the political and economic conditions will favour civil-protection initiatives in particular ways related to the (negative) lessons of the recent disaster.

zonation   The division of a geographical area of land (e.g. a valley, town or region) into homogeneous sectors with respect to particular criteria (e.g. intensity of hazard, degree of vulnerability or risk, dependence on a particular operations centre).

Bibliography

Disasters: general texts

Alexander, D. E. 1993. *Natural disasters*. London: UCL Press.

Burton, I., R. W. Kates, G. F. White 1993. *The environment as hazard* (2nd edn). New York: Guilford Press.

Cooke, R. U. & J. C. Doornkamp 1990. *Geomorphology in environmental management* (2nd edn). Oxford: Oxford University Press.

Hewitt, K. 1997. *Regions of risk: a geographical introduction to disasters*. Reading, Massachusetts: Addison Wesley Longman.

IFRCRCS 2000 [annual]. *World disasters report 2000*. Geneva: International Federation of Red Cross and Red Crescent Societies.

Mileti, D. S. (ed.) 1999. *Disasters by design: a reassessment of natural hazards in the United States*. Washington DC: National Academy of Sciences (Joseph Henry Press).

Smith, K. S. 2001. *Environmental hazards: assessing risk and reducing disaster* (3rd edn). London: Routledge.

Disasters: communications

Burton, A. 1990. *A basic guide to disaster communications*. Olympia, Washington: Emergency Response Institute.

Cate, F. H. 1995. *International disaster communications: harnessing the power of communications to avert disasters and save lives*. Washington DC: Annenberg Washington Program.

Coile, R. C. 1997. The role of amateur radio in providing emergency electronic communication for disaster management. *Disaster Prevention and Management* **6**(3), 176–85.

Creasy, C. 1994. Emergency communications. *Fire Engineering* **147**(8), 40–42.

Disasters: community preparedness

American Red Cross 1992. *Community disaster education guide*. Washington DC: American Red Cross.

Buckle, P. 1995. Community-based management of social disruption following disasters. *Australian Journal of Emergency Management* **10**, 31–8.

Burby, R. J. (ed.) 1998. *Cooperating with nature: confronting natural hazards with land-use planning for sustainable communities*. Washington DC: National Academy Press.

Kartez, J. D. & W. J. Kelley 1987. *Adaptive planning for community emergency management*. Olympia, Washington: Environmental Research Centre.

LaValla, P. & R. Stoffel 1983. *Blueprint for community emergency management: a text for managing emergency operations*. Olympia, Washington: Emergency Response Institute.

Lindell, M. K. & R. W. Perry 1992. *Behavioural foundations of community emergency planning*. Bristol, Pennsylvania: Taylor & Francis.

320

National Fire Protection Association Staff 1991. *Airport – community emergency planning*. Quincy, Massachusetts: US National Fire Protection Association.

World Health Organisation 1999. *Community emergency preparedness: a manual for managers and policy makers*. Geneva: World Health Organisation.

Disasters: information technology and computing

Balon, B. J. & H. W. Gardner 1988. Disaster planning for electronic records. *Records Management Quarterly* **22**(3), 20–25.

Carroll, J. M. 1983. *Simulation in emergency planning*. Simulation Series 11, Society for Computer Simulation, La Jolla, California.

—— (ed.) 1985. *Emergency planning*. Simulation Series 15, Society for Computer Simulation, La Jolla, California.

Chartrand, R. L. & K. C. Chartrand 1989. *Strategies and systems for disaster survival: the symposium on Information Technology and Emergency Management*. Rockville, Maryland: Research Alternatives.

Lyons, A. J. 1996. Contingency planning data processing. *Disaster Recovery Journal* **9**(1), 16–20.

Masri, A. & J. E. Moore II 1995. Integrated planning information systems: disaster planning analysis. *Journal of Urban Planning and Development* **121**(1), 19–39.

Pember, M. E. 1996. Information disaster planning: an integral component of corporate risk management. *Records Management Quarterly* **30**(2), 31–6.

Tobin, R. & R. Tobin 1997. *Emergency planning on the internet*. Rockville, Maryland: Government Institutes Inc.

Toigo, J. W. 1989. *Disaster recovery planning: managing risk and catastrophe in information systems*. Englewood Cliffs, New Jersey: Prentice-Hall.

Wallace, W. A. & F. De Balogh 1985. Decision support for disaster management. *Public Administration Review* **45**(special issue), 134–46.

Disasters: search and rescue and logistics

Cosgrove, J. 1996. Decision making in emergencies. *Disaster Prevention and Management* **5**(4), 28–35.

—— 1997. Estimating the capacity of warehouses. *Disasters* **21**(2), 155–65.

De Ville De Goyet, C. 1993. Post-disaster relief: the supply-management challenge. *Disasters* **17**(2), 169–76.

—— 1996. SUMA (Supply Management Project), a management tool for post-disaster relief supplies. *World Health Statistics Quarterly* **49**, 253–61.

Doyle, C. J. 1986. Helicopter transport in disaster care. *Journal of the World Association of Emergency and Disaster Medicine* **2**(1–4), 45–50.

Emergency Response Institute 1990. *Managing search operations*. Olympia, Washington: Emergency Response Institute.

Handmer, J. & J. Behrens 1992. Rescuers and the law: the legal rights and responsibilities of rescuers in Australia. *Disaster Management* **4**, 138–46.

Hebard, C. 1993. Use of search-and-rescue dogs. *Journal of the American Veterinary Medical Association* **203**(7), 999–1001.

Nylén, L. 1996. The role of the police in the total management of disaster. *Disaster Prevention and Management* **5**(5), 23–30.

Olson, R. S. & R. A. Olson 1987. Urban heavy rescue. *Earthquake Spectra* **3**(4), 645–58.

Payne, C. F. 1999. Contingency plan exercises. *Disaster Prevention and Management* **8**(2), 111–17.

Pierce, W. 1986. Search and rescue. *Journal of the World Association of Emergency and Disaster Medicine* **2**(1–4), 63–72.

Scanlon, J. T. 1994. The role of EOCs in emergency management: a comparison of Canadian and American experience. *International Journal of Mass Emergencies and Disasters* **12**(1), 51–75.

Waugh Jr, W. L. 1993. Co-ordination or control: organizational design and the emergency management function. *Disaster Prevention and Management* **2**(4), 17–31.

Emergency planning and management: general texts

Auf der Heide, E. 1989. *Disaster response: principles of preparation and co-ordination*. St Louis, Missouri: Mosby.

Carter, W. N. 1992. *Disaster management: a disaster manager's handbook*. Manila: Asian Development Bank.

Charles, M. T. & J. C. K. Kim (eds) 1988. *Crisis management: a casebook*. Springfield, Illinois: Charles C. Thomas.

Comfort, L. K. (ed.) 1988. *Managing disaster: strategies and policy perspectives*. Durham, North Carolina: Duke University Press.

Fink, S. 1986. *Crisis management: planning for the inevitable*. New York: American Management Association.

Foster, H. D. 1980. *Disaster planning: the preservation of life and property*. New York: Springer.

Garb, S. & E. Eng 1969. *The disaster handbook*. New York: Springer.

Gigliotti, R. & R. C. Jason 1990. *Emergency planning for maximum protection*. Boston: Butterworth-Heinemann.

Godschalk, D. R., T. Beatley, P. Berke, D. J. Brower, E. J. Kaiser 1999. *Natural hazard mitigation: recasting disaster policy and planning*. Washington DC: Island Press.

Griffiths, R. F. (ed.). 1982. *Dealing with risk: the planning, management and acceptability of technological risk*. New York: John Wiley.

Healy, R. J. 1969. *Emergency and disaster planning*. New York: John Wiley.

Hodgkinson, P. E. & M. Stewart 1991. *Coping with catastrophe: a handbook of disaster management*. New York: Routledge.

ICDO 1998. *Disaster management guide*. Geneva: International Civil Defence Organisation.

ICMA 1980. *Organizing for comprehensive emergency management*. Washington DC: International City Managers Association.

Laford, R. 1999. *Planning and practice: a guide for emergency services – planning and operations*. Tampa, Florida: Responder Publications.

Lagadec, P. 1993. *Preventing chaos in a crisis: strategies for prevention, control and damage limitation*. New York: McGraw-Hill.

Nudell, M. & N. Antokol 1988. *The handbook for effective emergency and crisis management*. Lexington, Massachusetts: D. C. Heath.

Perry, R. W. 1985. *Comprehensive emergency management*. Greenwich, Connecticut: JAI Press.

Sikich, G. W. 1996. *Emergency management planning handbook*. New York: McGraw-Hill.

Skeet, M. H. 1977. *Manual for disaster relief work*. New York: Churchill-Livingstone.

Spirgi, E. H. 1979. *Disaster management: comprehensive guidelines for disaster relief*. Bern: H. Huber.

Thygerson, A. L. 1979. *Disaster survival handbook*. Orem, Utah: Brigham Young University Press.

UNDRO, 1979–86. *Disaster prevention and mitigation: a compendium of current knowledge* (12 vols). Geneva: Office of the United Nations Disaster Relief Coordinator.

—— 1990. *Mitigating natural disasters: phenomena, effects and options*. Geneva: Office of the United Nations Disaster Relief Coordinator.

Waugh Jr, W. L. 2000. *Living with hazards, dealing with disasters: an introduction to emergency management*. Armonk, New York: M. E. Sharpe.

Waugh Jr, W. L. & R. J. Hy 1990. *Handbook of emergency management: programs and policies dealing with major hazards and disasters*. Westport, Connecticut: Greenwood Press.

Emergency planning and management: specialized texts

Bell, J. K. 2000. *Disaster survival planning: a practical guide for businesses* (2nd edn). Port Hueneme, California: Disaster Survival Planning Inc.

Buchanan, S. A. 1988. *Disaster planning: preparedness and recovery for libraries and archives.* UNISIST, UNESCO, Paris.

Building Owners and Managers Association International Staff 1994. *Emergency planning guidebook: a blueprint for preparing your building's response.* Washington DC: Building Owners and Managers Association International.

Drabek, T. E. 1990. *Emergency management: strategies for maintaining organisational integrity.* New York: Springer.

—— 1987. *The professional emergency manager: structures and strategies for success.* Monograph 44, Natural Hazards Research and Applications Information Centre, Boulder, Colorado.

Erickson, P. A. 1999. *Emergency response planning for corporate and municipal managers.* Orlando, Florida: Academic Press.

Faupel, C. E., D. Wenger, T. James 1985. *Disaster beliefs and emergency planning.* New York: Irvington.

Fortson, J. 1992. *Disaster planning and recovery: a how-to-do-it manual for librarians and archivists.* New York: Katz.

George, S. C. 1994. *Emergency planning and management in college libraries.* CLIP Note Series 17, Association of College and Research Libraries, New York.

Heath, S. E. 1999. *Animal management in disasters.* St Louis, Missouri: Mosby.

Littlejohn, R. F. 1983. *Crisis management: a team approach.* New York: American Management Association.

Martin, L. R. G. & G. Lafond (eds) 1988. *Risk assessment and management: emergency planning perspectives.* Waterloo, Ontario: University of Waterloo Press.

Morentz, J. W., H. C. Russell, J. A. Kelly 1982. *Practical mitigation: strategies for managing disaster prevention and reduction.* Rockland, Maryland: Research Alternatives, Inc.

Noji, E. K. (ed.) 1997. *The public health consequences of disasters.* New York: Oxford University Press.

Parker, D. & J. Handmer (eds) 1996. *Hazard management and emergency planning: perspectives on Britain.* London: James & James.

Parker, D. J. & E. C. Penning-Rowsell 1991. *Institutional design for effective hazard management.* Publication 184. Flood Hazard Research Centre, Middlesex Polytechnic, England.

Perry, R. W. & H. Hirose 1991. *Volcano management in the United States and Japan.* Greenwich, Connecticut: JAI Press.

Stringfield, W. H. 2000. *Emergency planning and management: ensuring your company's survival in the event of a disaster.* Rockville, Maryland: ABS Group (Government Institutes Division).

Toft, B. & S. Reynolds 1995. *Learning from disasters: a management approach.* Woburn, Massachusetts: Butterworth-Heinemann.

United Nations 1995. *Yokohama strategy and plan of action for a safer world: guidelines for natural-disaster prevention, preparedness and mitigation.* New York: United Nations Organisation.

Wahle, T. & G. C. Beatty 1994. *Emergency management guide for business and industry: a step-by-step approach to emergency planning, response and recovery for companies of all sizes.* New York: DIANE Publications.

Waugh Jr, W. L. 1990. *Terrorism and emergency management.* New York: Marcel Dekker.

Wenger, D. E., T. F. James, C. F. Faupel 1985. *Disaster beliefs and emergency planning.* New York: Irvington.

Emergency planning and management: articles

Benini, J. B. 1998. Getting organized pays off for disaster response. *Journal of Contingencies and Crisis Management* 6(1), 61–3.

Blanco, J., J. H. Lewko, D. Gillingham 1996. Fallible decisions in management: learning from errors. *Disaster Prevention and Management* **5**(2), 5–11.

Britton, N. R. 1992. Uncommon hazards and orthodox emergency management: toward a reconciliation. *International Journal of Mass Emergencies and Disasters* **10**(2), 329–48.

—— 1999. Whither the emergency manager? *International Journal of Mass Emergencies and Disasters* **17**(2), 223–35.

Buckle, P. 1990. Prospects for public sector disaster management in the 1990s: an indication of current issues, with particular reference to Victoria, Australia. *International Journal of Mass Emergencies and Disasters* **8**(3), 301–324.

Burby, R. J. & L. C. Dalton 1994. Plans can matter: the role of land use plans and state planning mandates in limiting the development of hazardous areas. *Public Administration Review* **54**(3), 229–38.

Burling, W. K. & A. E. Hyde 1997. Disaster preparedness planning: policy and leadership issues. *Disaster Prevention and Management* **6**(4), 234–44.

Comfort, L. K. 1985. Integrating organizational action in emergency management: strategies for change. *Public Administration Review* **45**(special issue), 155–64.

Cuny, F. C. 1993. Introduction to disaster management, lesson 2: concepts and terms in disaster management. *Prehospital and Disaster Medicine* **8**(1), 87–93.

Denis, H. 1995. Scientists and disaster management. *Disaster Prevention and Management* **4**(2), 14–19.

Doyle, J. C. 1996. Improving performance in emergency management. *Disaster Prevention and Management* **5**(3), 32–46.

Drabek, T. E. 1985. Managing the emergency response. *Public Administration Review* **45**(special issue), 85–92.

Evans, H. H. 1993. Emergency management at federal and state level in the United States of America: an outline. *Disaster Prevention and Management* **2**(1), 45–53.

Fischer III, H. W. 1996. What emergency management officials should know to enhance mitigation and effective disaster response. *Journal of Contingencies and Crisis Management* **4**(4), 208–217.

Gillespie, D. F. & C. L. Streeter 1987. Conceptualizing and measuring disaster preparedness. *International Journal of Mass Emergencies and Disasters* **5**(2), 155–76.

Godschalk, D. R. & D. J. Brower 1985. Mitigation strategies and integrated emergency management. *Public Administration Review* **45**(special issue), 64–71.

Horlick-Jones, T. 1994. Planning and coordinating urban emergency management: some reflections on New York City and London. *Disaster Management* **6**(3), 141–6.

IEEE 1998. Special issue on emergency management and engineering. *IEEE Transactions on Engineering Management* **45**(2).

Kartez, J. D. 1984. Crisis response planning: toward a contingent analysis. *American Planning Association Journal* **50**, 9–21.

Kelly, C. 1995. A framework for improving operational effectiveness and cost efficiency in emergency planning and response. *Disaster Prevention and Management* **4**(3), 25–31.

Kovel, J. & R. Kangari 1995. Planning for disaster-relief construction. *Journal of Professional Issues in Engineering Education and Practice* **121**(4), 207–215.

Kreps, G. A. 1983. The organisation of disaster response: core concepts and processes. *International Journal of Mass Emergencies and Disasters* **1**(4), 439–65.

—— 1990. The Federal Emergency Management System in the United States: past and present. *International Journal of Mass Emergencies and Disasters* **8**(3), 275–300.

Lange, S. 1998. Disaster planning: the challenge within. *Risk Management* **45**(5), 34–7.

Lindell, M. K. & R. W. Perry 1980. Evaluation criteria for emergency response plans. *Journal of Hazardous Materials* **3**, 349–61.

Lindsey, D. 1987. You can write a disaster plan on one sheet of paper. *Journal of the World Association of Emergency and Disaster Medicine* **3**(2), 147–56.

McLoughlin, D. 1985. A framework for integrated emergency management. *Public Administration Review* **45**(special issue), 165–72.

Moran, C., N. R. Britton, B. Correy 1992. Characterizing voluntary emergency responders. *International Journal of Mass Emergencies and Disasters* **10**(1), 207–216.

Neal, D. M. & B. D. Phillips 1995. Effective emergency management: reconsidering the bureaucratic approach. *Disasters* **19**(4), 327–37.

Noji, E. K. 1994. Progress in disaster management. *The Lancet* **343**(8908), 1239–40.

Perry, R. W. 1984. Evaluating emergency response plans. *Emergency Management Review* **1**(1), 20–23.

Perry, R. W. & J. Nigg 1985. Emergency management strategies for communicating hazard information. *Public Administration Review* **45**, 72–7.

Public Management 1989. The value of emergency preparedness (special issue). *Public Management* **71**(12), 2–27.

Quarantelli, E. L. 1982. Ten research-derived principles of disaster planning. *Disaster Management* **2**(1), 23–5.

—— 1988. Disaster crisis management: a summary of research findings. (Special Issue: Industrial Crisis Management: Learning from Organisational Failures) *Journal of Management Studies* **25**(4), 373–85.

—— 1988. Assessing disaster preparedness planning. *Regional Development Dialogue* **9**, 48–69.

—— 1992. The case for a generic rather than agent-specific agent approach to disasters. *Disaster Management* **2**, 191–6.

Rosenthal, U. & A. Kouzmin 1997. Crises and crisis management: toward comprehensive government decision making. *Journal of Public Administration Research and Theory* **7**(2), 277–304.

Siegel, G. B. 1985. Human resource development for emergency management. *Public Administration Review* **45**(special issue), 107–117.

Smith, D. 1990. Beyond contingency planning: towards a model of crisis management. *Industrial Crisis Quarterly* **4**, 263–75.

Sutphen, S. & W. L. Waugh Jr 1998. Organisational reform and technological innovation in emergency management. *International Journal of Mass Emergencies and Disasters* **16**(1), 9–12.

Tufekci, S. & W. A. Wallace 1998. The emerging area of emergency management and engineering. *IEEE Transactions on Engineering Management* **45**(2), 103–105.

Emergency planning and management: hazardous materials

Diane Publications 1996. *Hazardous materials emergency planning guide*. New York: DIANE Publications.

Gow, H. B. & R. W. Kay (eds) 1989. *Emergency planning for industrial hazards: proceedings of the European conference, Villa Ponti, Varese, Italy, 4–6 November 1987*. Amsterdam: Elsevier.

Himmelman, W A. & J. E. Zajic 1978. *Highly hazardous materials spills and emergency planning* (Hazardous and Toxic Substances Series 1). New York: Marcel Dekker.

LaPlante, J. M. & J. S. Kroll-Smith 1989. Coordinated emergency management: the challenge of the chronic technological disaster. *International Journal of Mass Emergencies and Disasters* **7**, 134–50.

OECD staff 1990. *Emergency planning in case of nuclear accident: proceedings of NEA–CEC workshop*. Paris: Organisation for Economic Co-operation and Development.

Rogers, G. O., J. H. Sorensen, J. F. Long Jr, D. Fisher 1989. Emergency planning for chemical agent releases. *The Environmental Professional* **11**, 396–408.

Sorensen, J. H. & B. M. Vogt 1988. Emergency planning for nuclear accidents: contentions and issues. *Journal of the Washington Academy of Sciences* **78**, 210–25.

Emergency planning and management: health/medicine

Friedman, E. 1994. Coping with calamity: how well does healthcare disaster planning work? *Journal of the American Medical Association* **272**(23), 1875–9.

PAHO 1982. *Environmental health management after natural disasters*. Washington DC: Pan American Health Organization.

Waeckerle, J. F. 1991. Disaster planning and response. *New England Journal of Medicine* **324**(12), 815–22.

Emergency planning and management: incident-command system (ICS)

D'Acchioli, R. 1986. Disaster management: an introduction to the incident-command system. *Journal of the World Association of Emergency and Disaster Medicine* **2**(1–4), 112–31.

Irwin, R. L. 1989. The incident-command system (ICS). In *Disaster responses: principles of preparation and coordination*, E. Auf Der Heide (ed.), 133–63. St Louis, Missouri: Mosby.

Yates, J. 1999. Improving the management of emergencies: enhancing the ICS. *Australian Journal of Emergency Management* **14**(2), 18–24.

Emergency planning and management: journals

Disaster Management (quarterly), FMJ International Publications Ltd., Redhill, England.

Disaster Prevention and Management (bi-monthly), University Press, Bradford, England.

Disaster Recovery Journal (quarterly), Disaster Recovery Journal, St Louis, Missouri.

Disasters: the International Journal of Disaster Studies and Management (quarterly), Blackwell, Oxford.

Earthquake Spectra (quarterly), Earthquake Engineering Research Institute, Oakland, California.

Electronic Journal of Emergency Management (on line), University of Richmond, Richmond, Virginia.

Emergency (quarterly), Brodie Publishing, Liverpool.

Emergency Preparedness Digest (quarterly), Emergency Preparedness Canada, Ottawa, Ontario.

Hazard Technology (bi-monthly), IES International, Rockville, Maryland.

Industrial and Environmental Crisis Quarterly (quarterly), Bucknell University, Lewisburg, Pennsylvania.

International Civil Defence Journal (quarterly), International Civil Defence Organisation, Geneva, Switzerland.

International Journal of Mass Emergencies and Disasters (quarterly), Research Committee on Disasters, c/o University of North Texas, Denton, Texas.

Journal of the American Society of Professional Emergency Planners (annual), ASPEP, Falls Church, Virginia.

Journal of Contingencies and Crisis Management (quarterly), Blackwell, Oxford.

Natural Hazards (bi-monthly), Kluwer Academic Publishers, Dordrecht, Netherlands.

Natural Hazards Observer (bi-monthly) and *Natural Hazards Informer*. Natural Hazards Research and Applications Information Centre, University of Colorado, Boulder.

Natural Hazards Review (quarterly), American Society of Civil Engineers, New York.

Response! The Journal of Search, Rescue and Emergency Response (quarterly), National Association of Search and Rescue Headquarters, Fairfax, Virginia.

Risk Management (quarterly). Scarman Centre for the Study of Public Order, University of Leicester, England.

SARScene: the Canadian Search and Rescue Newsletter (quarterly), National Search and Rescue Secretariat, Ottawa, Ontario.

Stop Disasters (bi-monthly), International Institute Stop Disasters, Naples.

Emergency planning and management: local government

Britton, N. R. & J. Lindsay 1995. Integrated city planning and emergency preparedness: some of the reasons why. *International Journal of Mass Emergencies and Disasters* **13**(1), 67–92.

Drabek, T. E. & G. J. Hoetmer (eds) 1991. *Emergency management: principles and practice for local government*. Washington DC: International City Management Association.

Herman, R. E. 1982. *Disaster planning for local government*. New York: Universe Books.

Kartez, J. D. & M. K. Lindell 1987. Planning for uncertainty: the case of local disaster planning. *Journal of the American Planning Association* **53**(4), 487–98.

Labadie, J. R. 1984. Problems in local emergency management. *Environmental Management* **8**(6), 489–94.

Mountain, R. H. 1993. A system for local authority management in a major emergency. *Disaster Prevention and Management* **2**(1), 145–55.

Mushkatel, A. H. & L. Wechsler 1985. Emergency management and the intergovernmental system. *Public Administration Review* **45**(special issue), 49–56.

Petak, W. J. 1985. Emergency management: a challenge for public administration. *Public Administration Review* **45**(special issue), 3–6.

Emergency planning and management: seismic hazards

Arnold, C. 1984. Planning against earthquakes in the United States and Japan. *Earthquake Spectra* **1**(1), 75–88.

Berke, P. R. & T. Beatley 1992. *Planning for earthquakes: risk, politics and policy*. Baltimore: Johns Hopkins University Press.

Bolton, P., S. Heikkala, M. M. Greene, P. May 1986. *Land-use planning for earthquake hazard mitigation: a handbook for planners*. Special Publication 14, Institute of Behavioural Science, University of Colorado, Boulder.

Coburn, A. & R. Spence 1992. *Earthquake protection* [fig. 8.2, p.258]. Chichester, England: John Wiley.

Eguchi, R. T. 1997. Real-time loss estimation as an emergency response decision support system: the early post-earthquake damage assessment tool (EPEDAT). *Earthquake Spectra* **13**(4), 815–32.

Emergency planning and management: sociology and psychology applied

Denis, H. 1997. Technology, structure and culture in disaster management: coping with uncertainty. *International Journal of Mass Emergencies and Disasters* **15**(2), 293–308.

Flin, R. 1996. *Sitting in the hot seat: leaders and teams for critical incident management*. New York: John Wiley.

Flin, R. & G. Slaven 1996. Personality and emergency command ability. *Disaster Prevention and Management* **5**(1), 40–46.

Janis, I. L. & L. Mann 1977. Emergency decision-making. *Journal of Human Stress* **3**, 35–45.

Mileti, D. S. & J. H. Sorensen 1987. Determinants of organizational effectiveness in responding to low-probability catastrophic events. *Columbia Journal of World Business* **22**, 13–21.

Paton, D. & R. Flin 1999. Disaster stress: an emergency management perspective. *Disaster Prevention and Management* **8**(4), 261–67.

Emergency planning and management: standards

NFPA 1600. *Standard on Disaster/Emergency Management and Business Continuity Programs* (2000). National Fire Protection Agency, Quincy, Massachusetts: www.nfpa.org.

Sphere Project, *Humanitarian Charter and Minimum Standards in Disaster Response* (1998). The SPHERE Project, Geneva: www.sphereproject.org.

UK Home Office *Standards for Civil Protection in England and Wales* (1999). Home Office Communication Directorate, London. www.homeoffice.gov.uk/epd/

Emergency planning and management: tourism

Barton, L. 1994. Crisis management: preparing for and managing disasters. *Cornell Hotel and*

Restaurant Administration Quarterly **35**(2), 59–65.

Drabek, T. E. 1994. *Disaster planning and the tourist industry*. Monograph 57, Natural Hazards Center, Boulder, Colorado.

—— 1995. Disaster responses within the tourist industry. *International Journal of Mass Emergencies and Disasters* **13**(1), 7–23.

Murphy, P. E. & R. Bayley 1989. Tourism and disaster planning. *Geographical Review* **79**(1), 36–46.

Emergency planning and management: training and education

Alexander, D. E. 2000. Scenario methodology for teaching principles of emergency management. *Disaster Prevention and Management* **9**(2), 89–97.

Baldwin, R. 1994. Training for the management of major emergencies. *Disaster Prevention and Management* **3**(1), 16–23.

Fischer, H. W. III & K. McCullough 1993. The role of education in disaster mitigation adjustment. *Disaster Management* **5**, 123–9.

Paton, D. 1996. Training disaster workers: promoting wellbeing and operational effectiveness. *Disaster Prevention and Management* **5**(5), 11–18.

Emergency planning and management: utilities

American Academy of Environmental Engineers 1995. *Natural disaster experiences: how to prepare environmental facilities for the worst*. Annapolis, Maryland: American Academy of Environmental Engineers.

AWWA 1994. *Emergency planning for water utility management*. Manual of Water Supply Practices Series M19. American Water Works Association, Washington DC .

Levitt, A. M. 1997. *Disaster planning and recovery: a guide for facility professionals*. New York: John Wiley.

Mullen, S. 1994. *Emergency planning guide for utilities*. New York: PennWell Books.

Water Pollution Control Federation Staff 1989. *Emergency Planning for Municipal Wastewater Facilities*. MOP Series SM–8. Water Pollution Control Federation, Washington DC.

Emergency planning and management: websites

The World Wide Web is a rich, if heterogeneous, source of information on emergency planning and management. Many hundreds of relevant sites exist. For the sake of brevity, the present list is limited to a few of the principal ones, with the proviso that sites tend to be ephemeral or may change their URLs. As the content of sites changes constantly and sometimes radically, only minimal information is given here; readers are encouraged to investigate for themselves and to use both general search engines and those available at specific sites in this list. In that context, a classified source of links to hundreds of sites concerned with emergencies and disasters has been set up by NASA and NOAA of the US government. It is called "A Disaster Finder" and is available at ltpwww.gsfc.nasa.gov/ndrd/disaster/links/.

American Red Cross: a useful site for emergency preparedness manuals.
  www.redcross.org
Australian Emergency Management Institute
  www.vifp.monash.edu.au/~davidt/aemi.html
Canadian Centre for Emergency Preparedness
  www.ccep.ca
Disaster Resource Guide
  www.disaster-resource.com
Earthquake Engineering Research Institute: publisher of *Earthquake Spectra*, whose abstracts are available on line at this site.
  www.eeri.org
Emergency Relief Web
  www.reliefweb.int

Emergency Management Australia: some important EMA publications on emergency management can be downloaded from this site.
www.ema.gov.au
Emergency Response and Research Institute
www.emergency.com
Emergency Resource Directory
www.clarknet.com/erd/
FEMA *Emergency Management Institute*: offering higher education course guides and model
course structures in many aspects of emergency preparedness and management.
www.fema.gov/emi/
Incident Command System forms and description: forms and manual are available in Adobe Acro-
bat format from the second of these two sites.
mindlink.net/sarinfo/sarinfo.htm
and
www.dot.gov/dotinfo/uscg/hq/g-m/nmc/response/forms/Default.htm
International Center for Disaster-Mitigation Engineering (INCEDE)
incede.iis.u-tokyo.ac.jp/default.html
International Federation of Red Cross and Red Crescent Societies: offers the on-line version of *World
Disasters Report* and worldwide disaster statistics.
www.ifrc.ch
Pan American Health Organization: provides reports and manuals that are freely available for
downloading.
www.paho.org
Simon Fraser University, Emergency Preparedness Information Exchange (EPIX): a large, multiple-
purpose site that is rich in information on disasters and emergencies.
hoshi.cic.sfu.ca/epix/
State University of New York at Buffalo, Multidisciplinary Center for Earthquake Engineering: offering publications and the Quakeline Internet-based bibliography.
mceer.buffalo.edu
US *Federal Centers for Disease Control and Prevention*: a site rich in information on the health and
medical effects of disasters.
www.cdc.gov
University of Wisconsin Disaster Management Center: on-line emergency training courses can be
accessed from this site.
epdwww.engr.wisc.edu/dmc/welcome.html
University of California at Berkeley, Earthquake Engineering Research Center
nisee.ce.berkeley.edu
University of Colorado at Boulder, Natural Hazards Center: a multiple-purpose site with publications, bibliographies and information on hazards and disasters.
www.colorado.edu/hazards
US *Geological Survey – hazards*: with gateways to the other USGS sites on hazards and disasters
(floods, earthquakes, landslides, etc.)
www.usgs.gov/themes/hazard.html
US *Federal Emergency Management Agency*: the largest and most richly endowed site on hazards,
emergencies and disasters.
www.fema.gov
Volunteers in Technical Assistance: an important source of information on current disasters and
relief appeals.
www.vita.org
World Association of Disaster and Emergency Medicine
hypnos.m.ehime-u.ac.jp/GHDNet/WADEM/index.html

Index

Page numbers in *italics* refer to figures and in **bold** to tables.